T0305616

BIOMEDICAL APPLICATIONS OF NANOTECHNOLOGY

THE WILEY BICENTENNIAL—KNOWLEDGE FOR GENERATIONS

$\mathcal{E}$ach generation has its unique needs and aspirations. When Charles Wiley first opened his small printing shop in lower Manhattan in 1807, it was a generation of boundless potential searching for an identity. And we were there, helping to define a new American literary tradition. Over half a century later, in the midst of the Second Industrial Revolution, it was a generation focused on building the future. Once again, we were there, supplying the critical scientific, technical, and engineering knowledge that helped frame the world. Throughout the 20th Century, and into the new millennium, nations began to reach out beyond their own borders and a new international community was born. Wiley was there, expanding its operations around the world to enable a global exchange of ideas, opinions, and know-how.

For 200 years, Wiley has been an integral part of each generation's journey, enabling the flow of information and understanding necessary to meet their needs and fulfill their aspirations. Today, bold new technologies are changing the way we live and learn. Wiley will be there, providing you the must-have knowledge you need to imagine new worlds, new possibilities, and new opportunities.

Generations come and go, but you can always count on Wiley to provide you the knowledge you need, when and where you need it!

WILLIAM J. PESCE
PRESIDENT AND CHIEF EXECUTIVE OFFICER

PETER BOOTH WILEY
CHAIRMAN OF THE BOARD

BIOMEDICAL APPLICATIONS OF NANOTECHNOLOGY

EDITED BY

Vinod Labhasetwar
Department of Biomedical Engineering
Lerner Research Institute
Cleveland Clinic, Cleveland, Ohio

Diandra L. Leslie-Pelecky
Department of Physics and Astronomy
University of Nebraska—Lincoln

WILEY-INTERSCIENCE
A JOHN WILEY & SON, INC., PUBLICATION

For general information on our other products and services or for technical support, please contact our
Customer Care Department within the United States at (800) 762-2974, outside the United States at
(317) 572-3993 or fax (317) 572-4002.

Wiley Bicentennial Logo: Richard J. Pacifico

Wiley also publishes its books in a variety of electronic formats. Some content that appears in print may not
be available in electronic formats. For more information about Wiley products, visit our Web site at
www.wiley.com.

Library of Congress Cataloging-in-Publication Data

Biomedical applications of nanotechnology / [edited by] Vinod Labhasetwar,
Diandra L. Leslie-Pelecky.
    p.; cm.
    Includes bibliographical references.
    ISBN 978-0-471-72242-7 (cloth)
    1. Nanotechnology. 2. Biomedical engineering. I. Labhasetwar, Vinod. II.
Leslie-Pelecky, Diandra L.
    [DNLM: 1. Nanotechnology. 2. Biomedical Engineering–methods. QT 36.5
B6152 2007]
    R857.N34B5566 2007
    610.28–dc22
                                                                            2006103522
Printed in the United States of America

10 9 8 7 6 5 4 3 2 1

CONTENTS

PREFACE

Nanotechnology is poised to make potentially revolutionary innovations in areas of biomedical science such as diagnostics, drug therapy, and imaging. In the future, nanotechnology using different biomarkers will be able to diagnose patients in much earlier stages of disease. Microchip-based diagnostic tests using biomarkers conjugated to nanoparticles or quantum dots can detect abnormalities at molecular levels that potentially can lead to disease progression. Nanotechnology can overcome anatomical and physiological barriers to deliver drugs more effectively to the target sites to reduce nonspecific effects. Many drugs, especially modern therapeutics, cannot be successful unless mechanisms for their effective delivery are developed. Nanotechnology can be a powerful tool to address delivery-related issues such as poor solubility or stability in biological environments. Imaging plays an important role in detection of pathologies such as tumors or vascular pathologies. Magnetic nanoparticles are under extensive investigation to enhance and improve the magnetic resonance imaging (MRI) capability for early detection of diseases.

Researchers in this area realize that the field of nanotechnology has matured over the last two decades of extensive research. We have developed the ability to design new systems, smart bioresponsive polymers that respond to changes in the bioenvironment stimulated by disease conditions, and we have a better understanding of their action mechanisms, interactions with cells and tissue, body distribution, and clearance. Also, we know how to assemble biomolecules into different nanostructures. We appreciate the pros and cons of each system and are making every effort to refine them to further enhance their therapeutic potential. The progress in the field of nanotechnology is evident from the range of nanotechnologies under various stages of clinical development from diagnostic to drug delivery applications. The field has certainly galvanized interdisciplinary research by bringing together polymer science, biology, pharmaceutical sciences, medicine, and physical science. Collaborative efforts address issues from various angles, and they may develop more effective solutions.

As we continue exploring nanotechnology for biomedical applications, it is essential for us to ensure that the nanotechnologies developed are safe. Nanotoxicity is an emerging field of research that will become an integral part of nanotechnology research; however, the burden for ensuring the safety of these technologies resides with all of us. We are pleased to cover some of the above important aspects of nanotechnology in this book.

Cleveland, Ohio   VINOD LABHASETWAR
Lincoln, Nebraska   DIANDRA L. LESLIE-PELECKY
March 2007

CONTRIBUTORS

Samuel D. Bader, Materials Science Division and Center for Nanoscale Materials, Argonne National Laboratory, Argonne, Illinois

Liaohai Chen, Biosciences Division, Argonne National Laboratory, Argonne, Illinois

Seok-Hwan Chung, Materials Science Division and Center for Nanoscale Materials, Argonne National Laboratory, Argonne, Illinois

Edward J. Felton, Department of Physics and Astronomy, Johns Hopkins University, Baltimore, Maryland

Axel Hoffmann, Materials Science Division and Center for Nanoscale Materials, Argonne National Laboratory, Argonne, Illinois

Sarah E. Johnson, Department of Pharmaceutical Sciences, College of Pharmacy, University of Nebraska Medical Center, Omaha, Nebraska

Raghuraman Kannan, Director, Nanoparticle Product Core Facility, Department of Radiology, University of Missouri, Columbia, Missouri

Rangaramanujam M. Kannan, Department of Chemical Engineering and Material Science and Biomedical Engineering, Wayne State University, Detroit, Michigan

Sujatha Kannan, Critical Care Medicine, Department of Pediatrics, Children's Hospital of Michigan, Wayne State University, Detroit, Michigan

Kattesh V. Katti, Nanoparticle Product Core Facility, Department of Radiology, University of Missouri, Columbia, Missouri

Mahima Kaushik, Department of Pharmaceutical Sciences, College of Pharmacy, University of Nebraska Medical Center, Omaha, Nebraska

Irine Khutsishvili, Department of Pharmaceutical Sciences, College of Pharmacy, University of Nebraska Medical Center, Omaha, Nebraska

Robert H. Kraus, Jr., Biophysics and Quantum Physics Group, Los Alamos National Laboratory, Los Alamos, New Mexico

Vinod Labhasetwar, Department of Biomedical Engineering, Lerner Research Institute, Cleveland Clinic, Cleveland, Ohio

Diandra L. Leslie-Pelecky, Department of Physics and Astronomy, Nebraska Center for Materials Nanoscience and Nanoscience, University of Nebraska—Lincoln, Lincoln, Nebraska

Souvik Maiti, Department of Pharmaceutical Sciences, College of Pharmacy, University of Nebraska Medical Center, Omaha, Nebraska

Lee Makowski, Biosciences Division, Argonne National Laboratory, Argonne, Illinois

Luis A. Marky, Department of Pharmaceutical Sciences, College of Pharmacy, University of Nebraska Medical Center, Omaha, Nebraska

Olga Mykhaylyl, Institute of Experimental Oncology, Technische Universität München

Chris Olsen, Department of Pharmaceutical Sciences, College of Pharmacy, University of Nebraska Medical Center, Omaha, Nebraska

Omathanu P. Perumal, Omathanu P. Perumal College of Pharmacy, South Dakota State University, Brookings, South Dakota

Christian Plank, Institute of Experimental Oncology, Technische Universität München, Munich, Germany

Maram K. Reddy, Department of Biomedical Engineering, Lerner Research Institute, Cleveland Clinic, Cleveland, Ohio

Daniel Reich, Department of Physics and Astronomy, Johns Hopkins University Baltimore, Maryland

Ulrike Schillinger, Institute of Experimental Oncology, Technische Universität München, Munich, Germany

Ronald Shikiya, Department of Pharmaceutical Science, College of Pharmacy, University of Nebraska Medical Center, Omaha, Nebraska

Dialekti Vlaskou, Institute of Experimental Oncology, Technische Universität München, Munich, Germany

Bradford Wright, Biophysics and Quantum Physics Group, Los Alamos National Laboratory, Los Alamos, New Mexico

Murali Mohan Yallapu, Department of Biomedical Engineering, Lerner Research Institute, Cleveland Clinic, Cleveland, Ohio

1

BIOLOGICAL APPLICATIONS OF MULTIFUNCTIONAL MAGNETIC NANOWIRES

Edward J. Felton and Daniel H. Reich

1.1 INTRODUCTION

Nanoscale magnetic particles are playing an increasingly important role as tools in biotechnology and medicine, as well as for studying biological systems. With appropriate surface functionalization, they enable the selective application of magnetic forces to a wide range of cells, subcellular structures, and biomolecules, and have been applied to or are being developed for areas including magnetic separation, magnetic biosensing and bioassays, drug delivery and therapeutics, and probes of the mechanical and rheological properties of cells [1–10]. Despite these successes, however, the structure of the magnetic particles in common use limits the range of potential applications. Most biomagnetic particles available today are spherical, with either (a) a "core-shell" structure of concentric magnetic and nonmagnetic layers or (b) magnetic nanoparticles randomly embedded in a nonmagnetic matrix [2, 11]. These geometries constrain the range of magnetic properties that can be engineered into these particles, as well as their chemical interactions with their surroundings, because such particles typically carry only a single surface functionality. A new and more versatile approach is to use asymmetric, multisegment magnetic nanoparticles, such as the metal nanowires shown in Figure 1.1.

Biomedical Applications of Nanotechnology.   Edited by Vinod Labhasetwar and Diandra L. Leslie-Pelecky
Copyright © 2007 John Wiley & Sons, Inc.

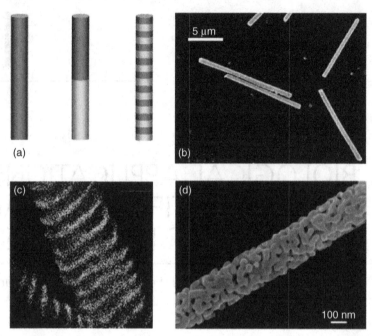

Figure 1.1. (a) Schematic illustration of magnetic nanowires, showing single-segment, two-segment, and two-component multisegment nanowires. (b) SEM image of 15 μm Ni nanowires (from Ref. 24, reproduced with permission of The Royal Society of Chemistry). (c) EELS image of Ni–Cu multisegment nanowires (reprinted with permission from Ref. 15, Copyright 2003, American Institute of Physics). (d) Nanoporous Au–Ag nanowire with Ag etched away. (Reprinted with permission from Ref. 16. Copyright 2003 American Chemical Society.)

The multisegment architecture of these particles, along with the ability to vary the aspect ratio and juxtaposition of dissimilar segments, allows the nanowires to be given a wide range of magnetic, optical, and other physical properties. In addition, differences in the surface chemistry between segments can be exploited to selectively bind different ligands to those segments, enabling the development of magnetic nanoparticle carriers with spatially resolved biochemical functionality that can be programmed to carry out multiple tasks in an intracellular environment.

This chapter provides an overview of recent results of a research program, centered at Johns Hopkins University, that is aimed at development of multifunctional magnetic nanowires for biotechnology applications. Section 1.2 provides a brief introduction to the fabrication process, and this is followed in Section 1.3 by an overview of the physical properties of the nanowires that are important in a biotechnological context. Sections 1.4–1.6 describe our development of the needed "tool-kit" for biological applications: manipulation of the nanowires in suspension, chemical functionalization, and self-assembly techniques. Section 1.7 discusses prospects for magnetic biosensing using nanowires, and Sections 1.8 and 1.9 discuss the major biological applications of the nanowires explored to date: novel approaches to magnetic separations, new tools for cell positioning and patterning, and new carrier particles for drug and gene delivery.

1.2 NANOWIRE FABRICATION

Nanowires are fabricated by electrochemical deposition in nanoporous templates. Originally developed for fundamental studies of the electrical and magnetic properties of modulated nanostructures [12], this method offers control of both nanowire size and composition and thus allows the nanowires' magnetic and chemical properties to be tailored for specific biological applications. To make the nanowires, a copper or gold conductive film is sputtered on one side of the template to create the working electrode of a three-electrode electrodeposition cell. Metal is then deposited from solution into the template's pores to form the wires. The nanowires' diameter is determined by the template pore size and can range from 10 nm to approximately 1 μm. The wires' length is controlled by monitoring the total charge transferred and is only limited by the thickness of the template. After the nanowire growth is complete, the working electrode film is etched away and the template is dissolved, releasing the nanowires into suspension.

Ferromagnetic nickel nanowires were commonly used in the work reported here. Grown in commercially available 50 μm-thick alumina templates, they have a radius of 175 ± 20 nm and lengths ranging from 5 to 35 μm. An SEM image of 15 μm-long nickel nanowires is seen in Figure 1.1b. The high pore density of the alumina templates $(3 \times 10^8 \, \text{cm}^{-2}$ [13]) enables fabrication of large numbers of nanowires. In addition to single-component nanowires such as these, nanowires comprised of multiple segments can be made by changing the deposition solution during growth. This technique has been used with alumina templates to create two-segment Ni–Au nanowires [14]. Alternatively, multisegment nanowires of certain materials can be grown from a single solution by varying the deposition potential. One example is the alternating ferromagnetic and nonmagnetic layers of the Ni–Cu nanowire shown in Figure 1.1c [15]. Nanowires incorporating two metals can also be synthesized as alloys. In one example, this technique has been used to produce high-surface area nanoporous Au wires by selectively etching away the Ag from Au–Ag alloy nanowires, as shown in Figure 1.1d [16].

1.3 PHYSICAL PROPERTIES

The elongated architecture of the nanowires and the flexibility of the fabrication method permit the introduction of various magnetic and other physical properties. The magnetic properties can be tuned and controlled through the size, shape, and composition of magnetic segments within the wires. For example, due to their high magnetic shape anisotropy, single-segment magnetic nanowires form nearly single-domain states with large remanent magnetizations for a wide range of nanowire lengths. This is illustrated in Figure 1.2, which shows magnetic hysteresis curves for 175 nm-radius nickel nanowires of different lengths [17]. The shape of the hysteresis curves is

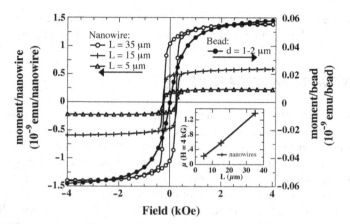

Figure 1.2. Room temperature magnetization versus field curves for 1 to 2 μm beads, 5, 15, and 35 μm nanowires. Inset: Saturation moment versus nanowire size. (Reprinted with permission from Ref. 17. © 2004 IEEE.)

nearly independent of nanowire length, with coercive field $H_C \sim 250$ Oe and remanent magnetization $M_R \sim 0.8 M_S$, where M_S is the saturation magnetization. These large, stable, and well-aligned moments make such nanowires useful for low-field manipulations of cells and biomolecules, as discussed in Section 1.8. As seen in the inset, M_S scales linearly with the wire length, and at high fields the nanowires have moment per unit length $\mu/L = 3.9 \times 10^{-11}$ emu/μm. For comparison, Figure 1.2 also shows the magnetic moment of commercially available 1.5 μm-diameter magnetic beads. Note that while the volume of the longest nanowires shown here is only 1.5 times that of the beads, their high-field moment is 20 times that of the beads. Thus the nanowires can provide significantly larger forces per particle in magnetic separations and other high-field applications.

There are, of course, biomagnetic applications in which large magnetic moments in low field are not desirable. These include situations in which it is important to control interactions among particles to reduce agglomeration in suspension. The remanent magnetization of multisegment nanowires such as those shown in Figure 1.1c can be tuned by controlling the shape of the magnetic segments [15, 18, 19]. If the magnetic segments within a multisegment nanowire have an aspect ratio greater than unity, shape anisotropy favors the adoption of a high-remanence state with the segments' moments parallel to the wire axis, even if they are short compared to the length of the nanowire, as shown in Figure 1.3a. In contrast, if the magnetic segments are disk-shaped (aspect ratio < 1), the shape anisotropy of the individual segments favors alignment of their moments perpendicular to the nanowire axis. Dipolar interactions between the segments then favor antiparallel alignment of the moments of neighboring segments, leading to a low-moment state in zero field, as shown in Figure 1.3b.

In addition to defining the magnetic properties, the segment composition can be exploited for other purposes. For example, the high-surface-area nanoporous gold segments previously mentioned (Figure 1.1d) can be used for efficient chemical functionalization, or for biosensing applications. Optical properties of the nanowires can also

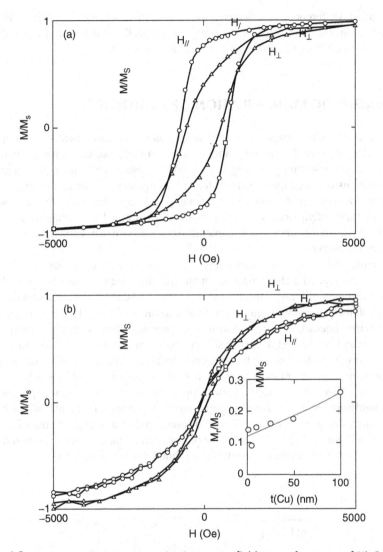

Figure 1.3. Room temperature magnetization versus field curves for arrays of Ni–Cu multi-layer nanowires in the template. (a) Ni–Cu nanowires with rod-shaped Ni segments (aspect ratio 2.5) and easy axis parallel to the nanowire axis. (b) Ni–Cu nanowires with disk-shaped Ni segments (aspect ratio 0.1) and easy axis perpendicular to the nanowire axis. The inset shows the remanence for Ni–Cu nanowires with disk-shaped Ni segments as a function of Cu layer thickness. (Reprinted with permission from Ref. 15. Copyright 2003, American Institute of Physics.)

be controlled. Differences in reflectivity in Au–Ag multisegment nanowires are being exploited for "nano-barcoding" of molecules and subcellular structures [20], and oxide segments with intrinsic fluorescence can also be introduced.

1.4 MAGNETIC MANIPULATION OF NANOWIRES

The large and tunable magnetic moments of nanowires allow precise manipulation of molecules and bound cells, with applications ranging from cell separations to two-dimensional cell positioning for diagnostics and biosensing, and to the potential creation of three-dimensional cellular constructs for tissue engineering. The approaches we have developed for these applications take advantage of nanowire–nanowire interactions, as well as their interactions with lithographically patterned micromagnet arrays and external fields. To illustrate these capabilities, we first discuss manipulation of the nanowires themselves.

In liquid suspensions, the nanowires readily orient with their magnetic moments parallel to an applied field. Single-segment and multisegment nanowires with long magnetic segments align with the wire axis parallel to the field, and multisegment wires with disk-shaped segments align perpendicular to the field [15, 21]. When magnetized, the nanowires interact through dipole–dipole magnetic forces. Self-assembly of the nanowires can be achieved, either in suspension or by allowing the wires to settle on flat substrates. This process can be controlled by an external field. Without an applied field, the nanowires are randomly oriented in the suspension, and they will assemble into random collections due to the anisotropy of the dipolar interaction. Application of a small field suppresses this random aggregation by prealigning the nanowires parallel to each other. The nanowires then form end-to-end chains as they settle out of solution, as shown in Figure 1.4 [22]. The addition of descending nanowires to chains settled on the substrate can yield chains that extend over hundreds of micrometers.

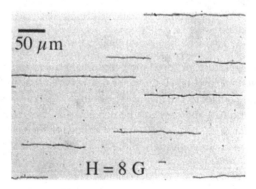

Figure 1.4. Optical micrograph of Ni nanowire chain formation after precipitation from a water suspension in an 8-G external magnetic field. (Reprinted with permission from Ref. 22. Copyright 2002, American Institute of Physics.)

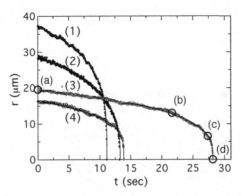

Figure 1.5. Separation versus time for four chain-formation events in a 4-Oe external field. Events (1) and (2) were in water, and events (3) and (4) were in ethylene glycol. (Reprinted from *J Magn Mater*, 249, C. L. Chien et al., Electrodeposited magnetic nanowires: Arrays, field-induced assembly, and surface functionalization, 146–155. Copyright 2002, with permission from Elsevier.)

The motion of both bare nanowires and nanowires bound to cells in suspension is governed by low Reynolds number hydrodynamics, and a nanowire's velocity is given by $\mathbf{v} = \mathbf{F}/D$, where $\mathbf{F}$ is the net force due to external fields, neighboring nanowires, and gravity, and D is the appropriate viscous drag coefficient. Integrating this equation of motion allows precise prediction and modeling of the nanowires' dynamics [21, 23]. For example, Figure 1.5 shows an analysis of a video microscopy study of nanowire chaining dynamics. For all the events shown in Figure 1.5 the wires or chains are nearly collinear. In this case, the force between two wires or chains of lengths L_1 and L_2 is

$$f(r) = -Q_m^2 \left(\frac{1}{r^2} - \frac{1}{(r + L_1)^2} - \frac{1}{(r + L_2)^2} + \frac{1}{(r + L_1 + L_2)^2} \right),$$

where r is the end-to-end separation. The nanowires are described very accurately in this and in all subsequent modeling discussed below as extended dipoles with magnetic charges $\pm Q_m = \pm M\pi a^2$ separated by L, where M is the wire's magnetization. The solid curves are fits to $r(t)$ based on the (somewhat involved) analytic form determined from the one-dimensional equation of motion $dr/dt = \tilde{D}f(r)$, where $\tilde{D} = D_1 D_2/(D_1 + D_2)$ is the reduced drag coefficient. Full details are given in Ref. 21. These results demonstrate that quantitative predictions of the nanowire–nanowire interactions and dynamics can be made.

Another important manipulation tool involves using the strong local fields generated by micrometer-size magnetic features patterned by microlithography on substrates to capture and position nanowires and cells [22, 24, 25]. This "magnetic trapping" process works because the nanowires are drawn into regions of strong local field gradients produced by the patterned micromagnets, such as those at the ends of the Ni ellipses shown in Figure 1.6. The snapshots show video frames from a trapping event, and

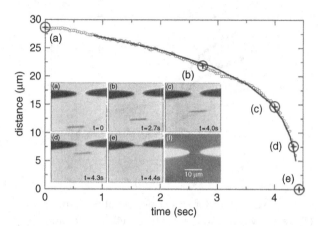

Figure 1.6. Distance from the center of a 10 μm Ni nanowire to the center of the gap between two elliptical micromagnets versus time. A 10 G external magnetic field is oriented parallel to the long axis of the micromagnets. Points (a)–(e) correspond to the inset videomicroscopy images. Inset image (f) is a reflected light image taken after the solvent dried. (Reprinted with permission from Ref. 24. Copyright 2003, American Institute of Physics.)

the trace shows the distance $z(t)$ of the wire from the trap versus time. Analysis of the force produced on the wire by the micromagnets again yields a simple model that can be integrated to obtain $z(t)$ (solid curve). A SEM image of a nanowire trapped by micromagnets is presented in Figure 1.7.

1.5 CHEMICAL FUNCTIONALIZATION

The ability to chemically functionalize nanowires enhances their utility in biological applications. Selectively binding ligands to the surface of nanowires allows additional control of interactions between nanowires, between nanowires and surfaces, and between cells and nanowires, as well as control of the wires' optical characteristics.

We have built on prior knowledge of surface chemistry on planar metallic films [26–28] to selectively functionalize both single- and multicomponent nanowires. Functionalization of nickel utilizes binding between carboxylic acids and metal oxides, in

Figure 1.7. SEM image of a nanowire magnetically trapped between two micromanets. (Reprinted with permission from Ref. 22. Copyright 2002, American Institute of Physics.)

this case the native oxide layer present on the nanowires' surface [29], while gold functionalization makes use of the well-known selective binding of thiols to gold [30]. It is therefore possible to attach various molecules possessing a compatible binding ligand to a particular metallic surface. This has been demonstrated with single-segment nickel nanowires that have been functionalized with hematoporphyrin IX dihydrochloride (HemIX), a fluorescent molecule with two carboxylic acid groups [14, 21, 23], as well as 11-aminoundecanoic acid and subsequently a fluorescent dye (Alexa Fluor 488 or fluorescein-5-isothiocyanate (FITC)) [14]. Single-segment gold nanowires have been functionalized with thiols including the thioacetate-terminated thiol P-SAc [24] and 1,9-nonanedithiol with the fluorescent dye Alexa Fluor 546 [14].

Multisegment nanowires are attractive because the differences in surface chemistry between different segments makes possible spatially resolved chemical functionalization with multiple molecules on the same nanowire, with different ligands directed to different segments. Our work with two-segment Ni–Au nanowires serves as an example of this spatially resolved functionalization. In one scheme, after exposure to HemIX, Ni–Au nanowires showed strong fluorescence from the Ni segments, and the Au segment exhibited weak fluorescence due to nonspecific HemIX adsorption. However, after simultaneous functionalization with HemIX and Au-specific nonylmercaptan, only the Ni segments showed fluorescence, indicating that the nonylmercaptan had attached to the Au segment to prevent nonspecific binding of HemIX [14, 24]. Bauer and co-workers also reacted Ni–Au nanowires with 11-aminoundecanoic acid and nonylmercaptan, and then subsequently with Alexa Fluor 488, which binds only to the 11-aminoundecanoic acid. Selective functionalization caused only the Ni segment to fluoresce, as shown in Figures 1.8a and 1.8b. Conversely, reacting the Ni–Au wires with 1,9-nonanedithiol and palmitic acid (for specific binding to Ni), and then with Alexa Fluor 546, which binds only to the 1,9-nonanedithiol, resulted in fluorescence of only the Au segment. Lastly, exposing Ni–Au nanowires to both 11-aminoundecanoic acid and 1,9-nonanedithiol, and then adding the fluorescent markers Alexa Fluor 488 and 546, resulted in fluorescence of both segments.

Selective surface functionalization of nanowires has also been accomplished with biomolecules. In one study, single-segment nickel and gold nanowires were functionalized with palmitic acid and an ethylene glycol-terminated alkanethiol, respectively, to render the nickel hydrophobic and the gold hydrophilic. Two-segment Ni–Au nanowires were exposed to both reagents. The nanowires were then exposed to Alexa Fluor 594 goat anti-mouse IgG protein, an antibody with an attached fluorescent tag. It is known that proteins are able to attach noncovalently to hydrophobic surfaces, but are prevented from such binding to hydrophilic surfaces. As seen in Figures 1.8c and 1.8d, only the nickel surfaces showed fluorescence, confirming that they had been selectively functionalized with the protein [31].

Other experiments involving functionalization with biomolecules have used two-segment Ni–Au nanowires as synthetic gene-delivery systems [32]. DNA plasmids encoding fluorescent proteins were bound to the nickel segment through a carboxylic acid intermediary, while the cell-targeting protein transferrin was bound to the gold segment through a thiolate linkage. This application of biomolecule-functionalized nanowires to gene delivery is detailed in Section 1.9.

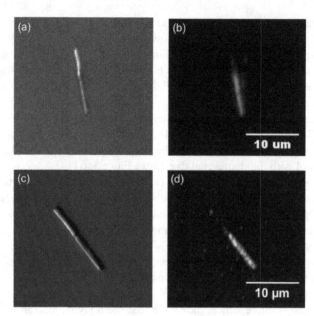

Figure 1.8. (a, b) Two-segment Ni-Au nanowire, functionalized with 11-aminoundecanoic acid and nonylmercaptan, and reacted with Alexa Fluor 488. (Reprinted with permission from Ref. 14. Copyright 2003 American Chemical Society.) (a) Reflected light image. (b) Fluorescent image showing fluorescence from Ni segment only. (c, d) Two-segment Ni–Au nanowire functionalized with palmitic acid and an ethylene glycol-terminated alkanethiol followed by exposure to a fluorescent protein. (c) Reflected light image, (d) Fluorescence image of Ni segment. (Reprinted with permission from Ref. 31. Copyright 2003 American Chemical Society.)

1.6 RECEPTOR-MEDIATED SELF-ASSEMBLY OF NANOWIRES

Chemical functionalization has also been used as a means to position nanowires using receptor-mediated binding to tether nanowires to specific regions of a substrate. This technique again has many potential biological applications, ranging from cell positioning to biosensing. Salem et al. [33] have demonstrated this technique using two-segment Ni–Au nanowires, with 8 μm Ni segments and 1 μm Au segments. These nanowires were functionalized by exposure to a solution of palmitic acid and thiol-terminated biotin. The biotin bound preferentially to the gold segment, and the palmitic acid coated the nickel segment to prevent nonspecific binding of the biotin to the remainder of the nanowire. Stripes of avidin were patterned via microfluidics on a silver film that was first coated with thiol-terminated biotin. Upon introduction of the biotinylated nanowires, the gold ends of the nanowires were anchored to the patterned regions of the substrate by the strong linkage between avidin and biotin, resulting in directed assembly of nanowires in the striped regions, as shown in Figure 1.9. These robust and flexible linkages allowed the nanowires to be pivoted about their binding points to align with an external magnetic field.

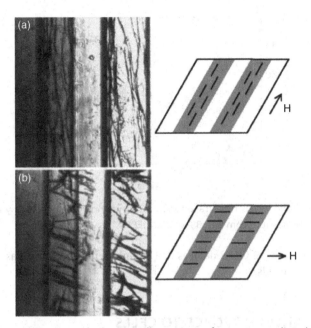

Figure 1.9. Optical images and schematic illustrations of receptor-mediated nanowire self-assembly. Chemically functionalized Ni–Au nanowires are selectively anchored by their gold ends to patterned stripes of avidin on the substrate. A magnetic field is applied parallel (A) and perpendicular (B) to the stripes, resulting in pivoting about the nanowire binding points. (Reprinted with permission from Ref. 33.)

1.7 MAGNETIC SENSING OF NANOWIRES

Detection and identification of biomolecules is becoming a crucial component of many biotechnological applications, and magnetic biosensing is a rapidly developing and evolving field [9,10,34–38]. The typical approach to magnetic biosensing uses integrated arrays of magnetic field sensors, such as GMR devices or magnetic tunnel junctions [35,36], whose surfaces are functionalized with receptor ligands for analytes of interest. If an analyte is labeled with a magnetic nanoparticle, its presence can then be detected by the action of the nanoparticle's magnetic field on the sensor when the analyte binds to the surface of the sensor. This thus provides an alternative approach to commonly used immunofluorescence-based detection techniques, such as ELISA.

Nanowires have shown potential for use in biosensing applications. Many of the magnetic biosensing schemes currently in development use superparamagnetic beads as the tagging particles, which requires the use of an external magnetic field to magnetize the beads. However, due to their large remanent moment, magnetic detection of nanowires can be performed in the absence of a large external magnetic field. We have demonstrated the feasibility of detection of ferromagnetic nanowires, as shown in Figure 1.10. Using GMR sensors as detectors [39], we find that both the presence and orientation of single wires are readily detectable, which may make possible a number of

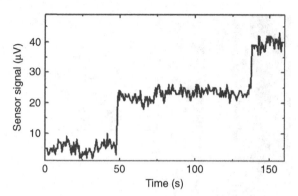

Figure 1.10. Voltage versus time trace for two 5 μm nanowires detected by a GMR sensor. (Reprinted with permission from Ref. 39 © 2004 IEEE.)

biological applications of GMR devices that complement the coverage assays currently implemented with beads.

1.8 APPLICATION OF FORCE TO CELLS

The tunable magnetic and chemical properties of nanowires make them an excellent vehicle for applying forces to cells. Superparamagnetic beads have been in use for some time as means of force application in biological systems [1–8], but are generally available only in a spherical geometry and with a single surface chemistry. Nanowires, however, offer several advantages through the tunability of their magnetic properties. Ferromagnetic nickel nanowires, for example, feature a large remanent magnetization; therefore, they can apply large forces in small magnetic fields, as well as a saturation magnetization that allows for large forces in increased magnetic fields. Furthermore, the nanowires' dimensions can be adjusted to span relevant biological length scales. As we will show, this latter property offers additional versatility in controlling cell–nanowire interactions.

The binding and internalization of nanowires by cells was investigated by immunofluorescent staining of NIH 3T3 mouse fibroblast cells with attached nanowires for the focal adhesion protein paxillin [40]. The results indicate the presence of focal adhesions containing paxillin along the length of the nanowires on short timescales, as seen in Figures 1.11A and 1.11B. The focal adhesions disappear within several hours, suggesting that the nanowires have been internalized inside the cell membrane. This was confirmed by coating nanowires with mouse IgG protein and then incubating them with cells for different durations. After short incubation times, exposure to Alexa Fluor 488 conjugated goat anti-mouse IgG fluorescently labeled nanowires that were attached to the cell membrane but not internalized (Figures 1.11C and 1.11D), while after longer incubation times nanowires remained unlabeled, indicating that they had been internalized and thus protected from the stain. This indicates that the nanowires that were internalized into the cell through integrin-mediated phagocytosis.

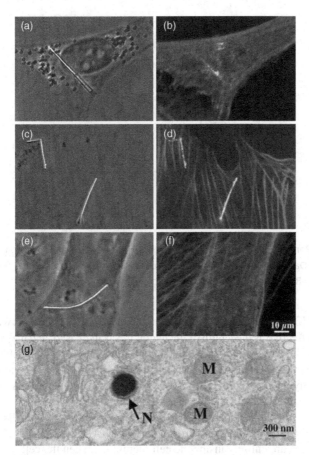

Figure 1.11. Binding of cell to nanowire. Top row: (a) Phase contrast image of a cell incubated with a 35 μm nanowire for 30 min and (b) fluorescence image of the same cell indicating paxillin focal adhesions. Second row: (c) Phase contrast image of a cell after a 30 min incubation with mouse IgG-coated nanowires and (d) fluorescence image of the same cell showing immunofluorescent staining of mouse IgG on the nanowire, indicating that the nanowire is external to the cell. Third row: (e) Phase contrast image of a cell after a 24 hr incubation with mouse IgG-coated nanowires and (f) fluorescense image of the same cell. The mouse IgG on the nanowire is unstained, indicating that the nanowire is internalized. (g) TEM image of a cell incubated with a nanowire for 24 hr. N, nanowire; M, mitochondria. (Reprinted with permission from Ref. 40. Copyright 2005 American Chemical Society.)

Attaching magnetic nanowires to cells has enabled their use as a tool for performing magnetic cell separations. In this process, magnetic particles are bound to cells in a mixture, which is then put in suspension in a magnetic field gradient. The field gradient creates a force that collects only cells with attached magnetic particles. Due to their higher magnetic moment, nickel nanowires have been shown to outperform superparamagnetic beads when separating cells with attached magnetic particles from

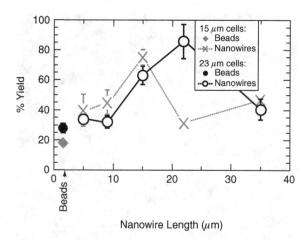

Figure 1.12. Percent yield versus nanowire length for separations using 15 and 23 μm-diameter cells. Data for superparamagnetic beads provided for reference. (Reprinted with permission from Ref. 40. Copyright 2005 American Chemical Society.)

those without [13,40]. Further studies have demonstrated that separations with nanowires become more effective as the length of the nanowires is increased, and they are optimized when the nanowire length matches the diameter of the cell [17]. This has been confirmed with 3T3 cells whose average size was increased by exposure to the cell division inhibitor mitomycin-C. Figure 1.12 shows that the maximum in the separation yield versus nanowire length tracks the increase in cell diameter.

The manner in which nanowires of different lengths interact with cells was explored to account for this finding. For nanowires with lengths less than the cell diameter, internalized nanowires are entirely inside the cell membrane when the cell is in suspension, as is the case in Figures 1.13A and 1.13B. However, nanowires with lengths greater than the cell diameter cannot be enclosed by the suspended cells, as seen in Figures 1.13C–E. It is likely that mechanical forces on the nanowire ends protruding from the cell membrane of suspended cells cause these nanowires to detach from the cells in some cases, accounting for the reduction in cell separation yield encountered when the nanowire length exceeds the suspended cell's diameter [40].

This dependence on nanowire length in nanowire–cell interactions may provide a new way to select among cells in a heterogeneous population. For example, cell separations have been performed with cell mixtures, in which half of the cells have had their diameter increased through exposure to mitomycin-C. Nanowires with lengths matched to the cells of smaller diameter separated cells of both diameters in equal proportions, while nanowires matched to the cells of larger diameter separated a higher proportion of larger cells [40]. This result suggests that it is possible to use magnetic nanowires to separate heterogeneous cell mixtures based on differences in the physical size of the cells.

Magnetic nanowires attached to cells can also be used to assemble multicellular constructs and microarrays of cells. We have used nanowires to direct the self-assembly

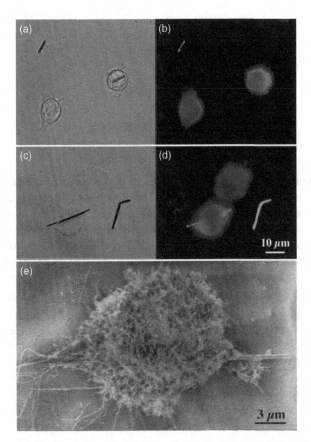

Figure 1.13. Optical images of suspended 3T3 cells. Top row: Suspended cell bound to a 9 μm mouse IgG coated nanowire. (a) Transmitted light and (b) composite fluorescense image of the same cell showing actin filaments and staining of the mouse IgG on an isolated nanowire (upper left) but not of the bound nanowire that is in the cell. Second row: Suspended cell bound to a 22 μm mouse IgG-coated nanowire. (c) Transmitted light and (d) composite fluorescent image of the same cell showing actin filaments and staining of the mouse IgG on both the portion of the nanowire that is no lkonger internal to the cell and on the isolated nanowire (right). (e) SEM image of a cell bound to a 22 μm nanowire. (Reprinted with permission from Ref. 40. Copyright 2005 American Chemical Society.)

of one-dimensional chains of cells by placing suspended cells with attached nanowires in a uniform magnetic field [25]. The chaining process is similar to the one previously described in Section 1.4, the difference being that the viscous drag due to the attached cells is significantly larger than that of just the nanowires, resulting in reduced motion during the chaining. Figure 1.14 illustrates the cell chaining process with a schematic, and shows images of chained 3T3 cells.

Magnetic manipulation has also been demonstrated as a technique for organizing cells into two-dimensional microarrays [25]. These experiments utilized elliptically

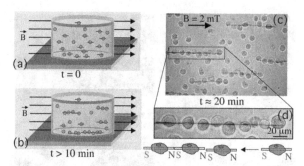

Figure 1.14. Magnetic cell chaining. (a) Schematic of nanowires bound to suspended cells and aligned in a magnetic field B. (b) Schematic of chain formation process due to magnetic dipole–dipole interactions between prealigned nanowires. (c) Cell chains formed on the bottom of a culture dish with $B = 2$ mT. (d) Close-up of a single-cell chain detailing wire–wire alignment. Interactions of North and South poles of adjacent nanowires are indicated schematically below. (From Ref. 25. Reprinted with permission of The Royal Society of Chemistry.)

shaped permalloy micromagnets that were patterned on substrates. A uniform external magnetic field applied parallel to the long axis of the micromagnets magnetized them, and it also co-aligned nanowires with attached cells in suspension in the same direction. The local magnetic field of the micromagnets attracted nearby nanowires to their poles, where their field was most intense, while repelling them from the area above them. An example of cells trapped on the ends of micromagnets in this way is seen in Figure 1.15, along with a graphical representation of the nanowire–micromagnet interaction energy that indicates the repulsive region above the micromagnet and the deep, attractive "wells" at each end.

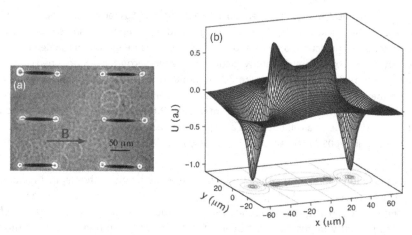

Figure 1.15. (a) Trapping of single cells by ellipsiodal micromagnets. Aligning field $B = 2$ mT. (b) Calculated wire–ellipse interaction energy U_1 at a wire height $z = 3$ μm. Ellipse footprint is shown on the floor of the figure. (From Ref. 25. Reprinted with permission of The Royal Society of Chemistry.)

Patterning hundreds or thousands of micromagnets into rectangular arrays enabled ordered cell trapping in two dimensions, and variation of the spacing between magnets allowed additional control over the type of cell patterns created. As described above, each micromagnet created an attractive region on its ends and a repulsive region above. For isolated micromagnets, as in Figures 1.16a and 1.16d, this resulted in individual points of attraction, whose interaction energy is seen in Figure 1.16g. Micromagnets with their long axes close together exhibited a row of such points that merge together to become

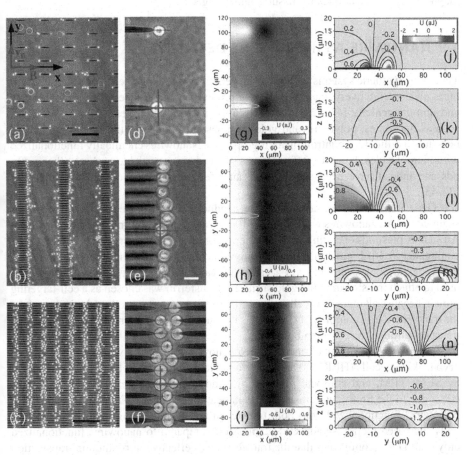

Figure 1.16. (a–c) Overview images of cell trapping on magnetic arrays. The direction of the external field $B = 10$ mT and the fluid flow $Q_f = 1.7$ μL sec^{-1} are shown in (a). The array lattice parameters are (a) $a = 125$ μm, $b = 100$ μm; (b) $a = 260$ μm, $b = 17$ μm; (c) $a = 32$ μm, $b = 17$ μm. Scale bars in (a–c) = 200 μm. (d–f) Close-up images of panels (a–c). Scale bars in (d–f) = 20 μm. (g–i) Calculated magnetic energy for a cell with a wire at a height $z = 8$ μm above the regions shown in (d–f). The wire is attracted to dark regions and repelled from white regions. Selected micromagnets are outlined. (j–o) Calculated magnetic energy of wire and cell in vertical planes above the lines in (d–f). The micromagnets appear as thick black lines at the bottom of (j), (l), and (n). (From Ref. 25. Reprinted with permission of The Royal Society of Chemistry.)

an attractive "trough," trapping cells into lines as seen in Figures 1.16b, 1.16e, and 1.16h. Patterning the micromagnets close together in both dimensions effectively merged together the troughs of two adjacent columns of micromagnets to create alternating bands of attractive and repulsive regions. The resultant patterned bands of cells are seen in Figures 1.16c, 1.16f, and 1.16i. The micromagnets can thus be used to create a variety of attractive and repulsive regions on a substrate that lead to two-dimensional organization of cells. Such organization has numerous possible applications, including biosensing, diagnostics, and techniques for tissue engineering.

1.9 NANOWIRE-ASSISTED GENE DELIVERY

The use of synthetic systems for delivering genetic material into cells has several advantages over viral delivery, including ease of production and reduced risk of cytotoxicity and immunogenicity [41, 42]. Previously introduced nonviral gene delivery methods, such as liposomes, polymers, and gold particles, are limited in transfection effectiveness due to difficulties controlling their properties. Nanowires chemically functionalized with biomolecules, however, offer degrees of freedom unavailable to these other methods and indeed have been shown to be an effective tool for nonviral gene delivery [32, 43].

Gene transfection experiments were conducted by Salem et al. [33] using two-segment Ni–Au nanowires with diameter 100 nm and length 200 nm (100 nm Ni, 100 nm Au). DNA plasmids encoding firefly luciferase or green fluorescent protein (GFP) were conjugated to 3-[(2-aminoethyl)dithio] propionic acid (AEDP) that was selectively attached to the nickel surfaces via a carboxylic acid terminus. A thiolate linkage was used to attach transferrin to the gold surfaces. Transferrin was chosen as a cell-targeting protein because all metabolic cells take in iron through receptor-mediated endocytosis of the transferrin–iron complex [44], and thus it would increase the probability of nanowire transport through the cell membrane.

In *in vitro* experiments, functionalized nanowires were incubated with human embryonic kidney cells (HEK293) and were found to be internalized by the cells, as shown by SEM, TEM, and confocal microscopy. Additionally, the cells exhibited fluorescence due to the firefly luciferase or GFP, indicating genetic transfection (Figure 1.17). Nanowires functionalized with both transferrin and DNA plasmid displayed substantially increased luciferase and GFP expression compared to nanowires functionalized only with DNA, confirming that the transferrin was effective in promoting transfection through receptor-mediated endocytosis.

To test the potential utility of multifunctional nanowires for genetic immunization applications, an *in vivo* study was done to measure the immune response in mice to cutaneous delivery via gene gun bombardment of nanowires functionalized with the model antigen ovalbumin and a DNA plasmid that further stimulates the immune response [43]. Both a strong antibody response in the bloodstream and a strong CD8+ T-cell response in the spleen were observed. A comparison of the antibody response for the nanowires with that obtained using 1.6 μm gold particles as the carriers is shown in Figure 1.18. These gold particles are the state of the art in inorganic carriers

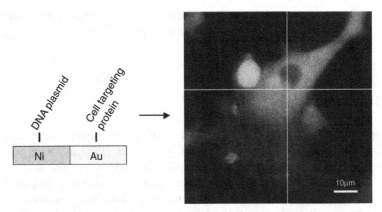

Figure 1.17. Drug delivery using multisegment nanowires, showing HEK cell expressing GFP after transfection by two-segment nanowires carrying GFP plasmid, and the cell-targeting agent transferring. (Reprinted by permission from Macmillan Publishers Ltd, *Nature Materials* [32], copyright 2003.)

for gene gun bombardment, and delivery methods using them have been optimized over a number of years. It is thus quite encouraging that the nanowires generate comparable antibody responses in these initial experiments. In the CD8+ T-cell studies, DNA bound to the nanowires alone generated very low or no response. However, nanowires with both DNA and the ovalbumin produced an eightfold increase in the CD8+ response compared to nanowires carrying only the ovalbumin. These results suggest

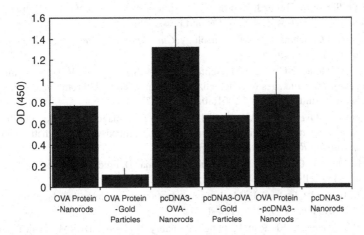

Figure 1.18. Ovalbumin-specific antibody responses in C57BL/6 mice immunized with functionalized Ni–Au nanowires gold particles: Nanowire formulations included ovalbumin protein only (OVA Protein), DNA plasmid encoding ovalbumin only (pcDNA3-OVA), ovalbumin protein and control plasmid (not encoding ovalbumin) (OVA Protein–pcDNA3), and control plasmid only (pcDNA3). Gold particle formulations included ovalbumin protein (OVA Protein) and DNA plasmid encoding ovalbumin protein (pcDNA3-OVA). (Reprinted with permission from Ref. 43.)

that with further optimization, the ability of the nanowires to deliver multiple immunostimulants to the same cell in controlled doses may prove highly effective for genetic immunization.

1.10 SUMMARY

This chapter has presented an overview of recent work on the development and applications of multifunctional magnetic nanowires for biotechnology, including areas such as cell separations, cell positioning and manipulation, and intracellular drug and gene delivery. A wide range of other applications are envisioned or under development, in areas such as tissue engineering, biomagnetic control of cellular function, and subcellular force transduction. It is also important to note that the concept of asymmetric magnetic nanoparticles with spatially resolved chemical functionality need not be limited to electrodeposited nanowires; many other architectures are possible, and indeed some are already under exploration [45]. Thus it seems likely that there will be continued innovation and development of new applications of multifunctional magnetic nanoparticles in biotechnology in the foreseeable future.

REFERENCES

1. Safarik I, Safarikova M. Use of magnetic techniques for the isolation of cells. *J Chromatogr B* 1999;722:33–53.
2. Häfeli U, Schütt W, Teller J, Zborowski M, editors. *Scientific and Clinical Applications of Magnetic Microspheres*. New York: Plenum Press; 1997.
3. MacKintosh FC, Schmidt CF. Microrheology. *Curr Opin Coll Interface Sci* 1999;4:300–307; and references cited therein.
4. Bausch AR, Hellerer U, Essler M, Aepfelbacher M, Sackmann E. Rapid stiffening of integrin receptor-actin linkages in endothelial cells stimulated with thrombin: A magnetic bead microrheology study. *Biophys J* 2001;80:2649–2657.
5. Matthews BD, Overby DR, Alenghat FJ, Karavitis J, Numaguchi Y, Allen PG, Ingber DE. Mechanical properties of individual focal adhesions probed with a magnetic microneedle. *Biochem Biophys Res Commun* 2004;313:758–764.
6. Fabry B, Maksym GN, Butler JP, Glogauer M, Navajas D, Fredberg JJ. Scaling the microrheology of living cells. *Phys Rev Lett* 2001;87:148102.
7. Wang N, Butler JP, Ingber DE. Mechanotransduction across the cell surface and through the cytoskeleton. *Science* 1993;260:1124–1127.
8. Meyer CJ, Alenghat FJ, Rim P, Fong JH, Fabry B, Ingber DE. Mechanical control of cyclic AMP signalling and gene transcription through integrins. *Nat Cell Biol* 2000;2:666–668.
9. Baselt DR, Lee GL, Natesan M, Metzger SW, Sheehan PE, Colton RJ. A biosensor based on magnetoresistance technology. *Biosens Bioelectron* 1998;13:731–739.
10. Li GX, Wang SX, Sun SH. Model and experiment of detecting multiple magnetic nanoparticles as biomolecular labels by spin valve sensors. *IEEE Trans Magn* 2004;40:3000–3002.

11. Pellegrino T, Kudera S, Liedl T, Javier AM, Manna L, Parak WJ. On the development of colloidal nanoparticles towards multifunctional structures and their possible use for biological applications. *Small* 2005;1:48–63; and references therein.

12. Fert A, Piraux L. Magnetic nanowires. *J Magn Magn Mater* 1999;200:338–358; and references therein.

13. Hultgren A, Tanase M, Chen CS, Meyer GJ, Reich DH. Cell manipulation using magnetic nanowires. *J Appl Phys* 2003;93:7554–7556.

14. Bauer LA, Reich DH, Meyer GJ. Selective functionalization of two-component magnetic nanowires. *Langmuir* 2003;19:7043–7048.

15. Chen M, Sun L, Bonevitch JE, Reich DH, Chien CL, Searson PC. Tuning the response of magnetic suspensions. *Appl Phys Lett* 2003;82:3310–3312.

16. Ji C, Searson PC. Synthesis and characterization of nanoporous gold nanowires. *J Phys Chem B* 2003;107:4494–4499.

17. Hultgren A, Tanase M, Chen CS, Reich DH. High-yield cell separations using magnetic nanowires. *IEEE Trans Magn* 2004;40:2988–2990.

18. Chen M, Searson PC, Chien CL. Micromagnetic behavior of electrodeposited Ni/Cu nanowires. *J Appl Phys* 2003;93:8253–8255.

19. Chen M, Hao Y, Chien CL, Searson PC. Tuning the properties of magnetic nanowires. *IBM J Res Dev* 2005;49:79–102.

20. Nicewarner-Pena SR, Freeman RG, Reiss BD, He L, Pena DJ, Walton ID, Cromer R, Keating CD, Natan MJ. Submicrometer metallic barcodes. *Science* 2001;294:137–141.

21. Tanase M, Bauer LA, Hultgren A, Silevitch DM, Sun L, Reich DH, Searson PC, Meyer GJ. Magnetic alignment of fluorescent nanowires. *Nano Lett* 2001;1:155–158.

22. Tanase M, Silevitch DM, Hultgren A, Bauer LA, Searson PC, Meyer GJ, Reich DH. Magnetic trapping and self-assembly of multicomponent nanowires. *J Appl Phys* 2002;91:8549–8551.

23. Chien CL, Sun L, Tanase M, Bauer LA, Hulgren A, Silevitch DM, Meyer GJ, Searson PC, Reich DH. Electrodeposited magnetic nanowires: Arrays, field-induced assembly, and surface functionalization. *J Magn Magn Mater* 2002;249:146–155.

24. Reich DH, Bauer LA, Tanase M, Hultgren A, Chen CS, Meyer GJ. Biological applications of multifunctional magnetic nanowires (invited). *J Appl Phys* 2003;93:7275.

25. Tanase M, Felton EJ, Gray DS, Hultgren A, Chen CS, Reich DH. Assembly of multicellular constructs and microarrays of cells using magnetic nanowires. *Lab Chip* 2005;5:598–605.

26. Roberts C, Chen CS, Mrksich M, Martichonok V, Ingber DE, Whitesides GM. Using mixed self-assembled monolayers presenting RGD and (EG)3OH groups to characterize long-term attachment of bovine capillary endothelial cells to surfaces. *J Am Chem Soc* 1998;120:6548–6555.

27. Baker M, Jennings G, Laibinis P. Underpotentially deposited copper promotes self-assembly of alkanephosphonate monolayers on gold substrates. *Langmuir* 2000;16:3288–3293.

28. Folkers J, Gorman C, Laibinis P, Buchholz S, Whitesides G. Self-assembled monolayers of long-chain hydroxamic acids on the native oxides of metals. *Langmuir* 1995;11:813–824.

29. Martin CR. Nanomaterials: A membrane-based synthetic approach. *Science* 1994;266:1961–1966.

30. Penner RM, Martin CR. Preparation and electrochemical characterization of ultramicroelectrode ensembles. *Anal Chem* 1987;59:2625–2630.

31. Birenbaum NS, Lai BT, Chen CS, Reich DH, Meyer GJ. Selective noncovalent adsorption of protein to bifunctionalized metallic nanowire surfaces. *Langmuir* 2003;19:9580–9582.
32. Salem AK, Searson PC, Leong KW. Multifunctional nanorods for gene delivery. *Nat Mater* 2003;2:668–671.
33. Salem AK, Chao J, Leong KW, Searson PC. Receptor mediated self-assembly of multi-component magnetic nanowires. *Adv Mater* 2004;16:268–271.
34. Reiss G, Brueckl H, Huetten A, Schotter J, Brzeska M, Panhorst M, Sudfield D, Becker A, Kamp PB, Puehler A, Wojczykowski K, Jutzi P. Magnetoresistive sensors and magnetic nanoparticles for biotechnology. *J Mater Res* 2005;20:3294–3302.
35. Shen WF, Liu XY, Mazumdar D, Xiao G. *In situ* detection of single micron-sized magnetic beads using magnetic tunnel junction sensors. *Appl Phys Lett* 2005;86:253901.
36. Tondra M, Popple A, Jander A, Millen RL, Pekas N, Porter MD. Microfabricated tools for manipulation and analysis of magnetic microcarriers. *J Mang Mater* 2005;293:725–730.
37. Wang SX, Bae SY, Li GX, Sun SH, White RL, Kemp JT, Webb CD. Towards a magnetic microarray for sensitive diagnositics. *J Mang Magn Mater* 2005;293:731–736.
38. Graham DL, Ferreira HA, Feliciano N, Freitas PP, Clarke LA, Amaral MD. Magnetic field-assisted DNA hybridisation and simultaneous detection using micron-sized spin-valve sensors and magnetic nanoparticles. *Sens Actuators B Chem* 2005;107:936–944.
39. Anguelouch A, Reich DH, Chien CL, Tondra M. Detection of magnetic nanowires using GMR sensors. *IEEE Trans Mag* 2004;40:2997–2999.
40. Hultgren A, Tanase M, Felton EJ, Bhadriraju K, Salem AK, Chen CS, Reich DH. Optimization of yield in magnetic cell separations using magnetic nanowires of different lengths. *Biotechol Prog* 2005;21:509–515.
41. Carter PJ, Samulski RJ. Adeno-associated viral vectors as gene delivery vehicles (review). *Int J Mol Med* 2000;6:17–27.
42. Pouton CW, Seymour LW. Key issues in non-viral gene delivery. *Adv Drug Deliv Rev* 2001;46:187–203.
43. Salem AK, Hung CF, Kim TW, Wu TC, Searson PC, Leong KW. Multi-component nanorods for vaccination applications. *Nanotechnology* 2005;16(4):484–487.
44. Wagner E, Curiel D, Cotten M. Delivery of drugs, proteins and genes into cells using transferrin as a ligand for receptor-mediated endocytosis. *Adv Drug Deliv Rev* 1994;14:113–135.
45. Yu H, Chen M, Rice PM, Wang SX, White RL, Sun SH. Dumbbell-like bifunctional Au–Fe3O4 nanoparticles. *Nano Lett* 2005;5m379–382.

2

NUCLEIC ACID DELIVERY AND LOCALIZING DELIVERY WITH MAGNETIC NANOPARTICLES

Christian Plank, Ulrike Schillinger, Dialekti Vlaskou,
and Olga Mykhaylyk

2.1 INTRODUCTION

This book focuses on biomedical applications of nanotechnology. The shuttles developed for nucleic acid delivery, the so-called vectors, are nanoparticles themselves and are either genetically modified viruses or virus-like synthetic constructs. Efficient delivery is one of the major bottlenecks in nucleic acid based therapy. Localized delivery is an important objective for pharmaceuticals in general, including viral and nonviral shuttles for nucleic acid delivery. We and others have found that magnetic nanoparticles can be advantageously combined with gene vectors, which thus become amenable to guidance by magnetic fields. Before discussing magnetically targeted nucleic acid delivery, we will give an overview of nucleic acid delivery and localized delivery in general with a focus on synthetic vectors. The synthesis of magnetic nanoparticles suitable for gene delivery will be discussed briefly, followed by an overview of magnetic targeting. Finally, we will describe recent developments in the field and present some of our own latest research.

It is our intent to highlight the importance of localization of delivery and the eminent role that evolving physical techniques play in this context.

Biomedical Applications of Nanotechnology.   Edited by Vinod Labhasetwar and Diandra L. Leslie-Pelecky
Copyright © 2007 John Wiley & Sons, Inc.

2.2 NUCLEIC ACID DELIVERY—WHAT FOR?

Nucleic acids carry the building plans of living systems. Nucleic acid sequences are translated into structures and functions of cellular molecules, which, together with the biochemical reactions in which they participate, constitute the material aspect of life. Located at the head of this cellular information flow, nucleic acids occupy a distinguished position among biological molecules. As the original information carriers, they participate, in an indirect manner, in any cellular process. Beyond their role as carriers and transmitters of information, nucleic acids also participate in a direct manner in cellular reactions. They have been known for a long time as structural and functional elements of multienzyme complexes such as ribosomes. In splicing reactions, for example, nucleic acids themselves carry some of the active ingredients of their own processing. In recent years it has become evident that nucleic acids participate directly in a multitude of cellular processes and thus contribute, maybe to an extent equal to that of proteins, to the coordinated and regulated network of cellular chemical reactions. RNA species (micro RNAs, short hairpin RNAs, small interfering RNAs) in particular have been recognized as natural regulators of cellular processes [1,2].

Given the distinguished role of nucleic acids in living systems, it is justified to conclude that any cellular process may be influenced to some particular purpose by the introduction of nucleic acids into cells from outside. During the past almost 40 years, very efficient tools for nucleic acid delivery into cells have been developed. Nowadays, the introduction of nucleic acids into cells is a well-established and widely used research tool. The other major field of nucleic acid application is nucleic acid therapy. The objectives of nucleic acid delivery for therapeutic purposes are (1) complementation and overexpression of genes, (2) on/off regulation of genes, and (3) repair of genes. These objectives were formulated in 1966, the year that the deciphering of the genetic code was concluded [3]. Despite almost 40 years of research, formidable challenges remain in the field of gene therapy. Magnetic drug targeting applied with nucleic acids or gene vectors can overcome some of the remaining obstacles.

2.3 NUCLEIC ACID DELIVERY—HOW?

It was believed for a long time that cells do not incorporate nucleic acids voluntarily, at least not in a manner that would result in the expression of an engulfed gene. Only relatively recently have we learned that, in certain cases, "naked" nucleic acids are efficiently taken up into cells in functional form [4]. Before, there was agreement that shuttles for nucleic acid delivery would be required, because these polyelectrolyte macromolecules are unable to cross cellular membranes by passive mechanisms such as diffusion. Nature itself has provided the ideal solution for this delivery problem in the form of viruses, which are nanoparticles composed of biomolecules. These obligatorily parasitic entities need to cross cellular membranes and ultimately need to shuttle their genetic information into cell nuclei in order to propagate. Consequently, genetically engineered viruses were among the earliest shuttles used for nucleic acid delivery and in many respects are still the most efficient. In the meantime, a multitude of viral vectors

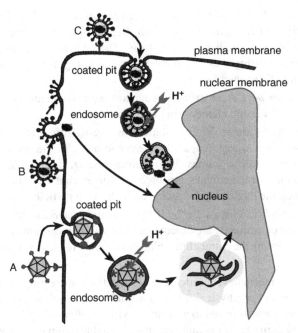

Figure 2.1. Viral entry into cells, schematic overview. Viruses bind to cell surfaces via receptor–ligand interactions. Many virus species are taken up into cells by endocytosis, like adenovirus (A) or membrane-coated viruses such as influenza virus (C). Other membrane-coated viruses such as retroviruses (B) directly fuse their membranes with the plasma membrane. Endocytosis proceeds via segregation of membrane-surrounded vesicles (endosomes) from the plasma membrane. A proton pump in the endosomal membrane mediates the acidification of the endosomal lumen. This pH change triggers conformational rearrangements of viral proteins, which then by interaction with the endosomal membrane can lead to the disruption of endosomes (like in the case of adenoviruses) or to the fusion of endosomal and viral membranes (like in the case of influenza virus). These membrane disruption/fusion events are essential parts of viral entry into cells, which ultimately leads to uptake/transport of viral genetic information into the cell nucleus. Details can be found in textbooks of virology. See insert of color representation of this figure.

have been described, such as recombinant retro-, lenti-, adeno-, adeno-associated, or vaccinia viruses, each of them having their specific advantages and shortcomings. For nonviral vector engineers, it has been highly instructive to take a closer look at the major features of the naturally evolved solution to the nucleic acid delivery problem. The major functions of viral infectivity are as follows (see also Figure 2.1):

1. *Viral Genomes Are Packaged.* Nucleic acids are compacted, such that the sizes of these macromolecules are compatible with the requirements of natural transport mechanisms. Packaging also protects the genome from degradation.

2. *Receptor–Ligand Interactions.* Viruses bind specifically to cell surface molecules, thereby gaining specificity in terms of host tropisms.

3. *Exploitation of Natural Cellular Uptake Mechanisms Such as Endocytosis and Mechanisms of Escaping Intracellular Degradation.* Many virus species enter cells via receptor-mediated endocytosis. The endosomal acidification process is exploited to trigger escape mechanisms, resulting in the release of the viral capsid from these internal vesicles.

4. *Nuclear Transport.* Active transport across the nuclear membrane is exploited to localize viral genetic elements in the cell nucleus.

5. *Genome Organization.* Viral genomes are organized in a manner as to exploit the information storage capacity of nucleic acids in the most efficient ways (overlapping reading frames, bidirectional coding, differential splicing, etc.). Furthermore, viral genomes are organized to exploit host functions, thereby minimizing the payload to be packaged in the viral particle.

6. *Biocompatibility.* Viruses are built from natural materials. Although immunogenic, their constituents are biocompatible enough to warrant sufficient stability in the host during the extracellular phase of delivery to secure target cell infection.

Nonviral vector engineers have mimicked these functions in creating synthetic modules that can be chemically or physically (self-) assembled to result in synthetic virus-like nanoparticles (often also referred to as artificial viruses; Figure 2.2).

Nucleic acid compaction is quite easily achieved by simple mixing with cationic molecules in solution. Early examples are the DEAE dextrane and calcium phosphate precipitation methods [5,6]. Examining chromosome structure and function, it has been realized early on that DNA compaction is brought about by cationic sequences in histones and that such compaction can be achieved with synthetic oligo- and polycations [7–14]. Wu and Wu [15, 16] and later on Wagner et al. [17] reported about nonviral vector nanoparticles prepared from polylysine and plasmid DNA capable of transfecting cells *in vitro* and *in vivo*. DNA packaging for delivery purposes has also been attempted by incorporation in the aqueous lumen of liposomes [18]. The liposomal approach experienced a breakthrough when cationic lipids were first introduced as DNA binding and compacting agents [19,20]. Today, the literature discriminates between lipoplexes, when referring to cationic lipid–DNA complexes, and polyplexes, when referring to polycation–DNA complexes [21]. Important results of early nonviral vector research are that it is a natural property of nucleic acids as polyelectrolytes to "condense" to nanostructures upon mixing with polyelectrolytes of opposite charge (for further reading and review see [22–24]) and that the resulting complexes are able to transfect cells. As we know now, the potency in transfecting cells is not strictly dependent on DNA compaction. One important function of the cationic modules for DNA binding is that they mediate in a nonspecific manner the binding of vector particles to cell surfaces.

Receptor–Ligand Interactions. Wu and Wu first introduced nonviral receptor-mediated gene delivery [15, 16, 25]. By coupling asialoorosomucoid, a natural ligand of the asialoglycoprotein receptor on liver cells, to the DNA compacting moiety polylysine, they generated vectors with increased target cell specificity that are taken

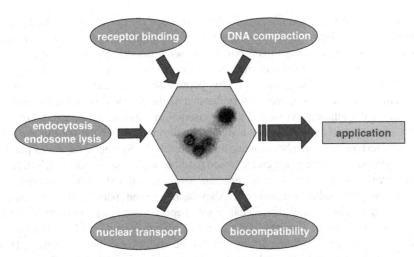

Figure 2.2. Nonviral vectors for nucleic acid delivery (sometimes called artificial viruses) are prepared by self-assembly of synthetic modules that mimic essential viral functions that allow them to infect cells. The self-assembly process is mostly based on noncovalent interactions of the individual modules such as electrostatic and hydrophobic interactions. The most important interaction is the one between the nucleic acid and a polycation or a cationic lipid, which can lead to the formation of a charged nanoparticle that is able to transfect cells. The functionalities of receptor binding, membrane destabilization (such as endosome lysis), nuclear targeting, and biocompatibility can be covalently coupled to a DNA binding/compacting moiety or can be incorporated into the complex as individual molecules by noncovalent interactions. The center of the figure shows toroidal nanoparticles that are typically formed upon mixing of plasmid DNA and polycations. See insert of color representation of this figure.

up into cells by receptor-mediated endocytosis. Following a similar concept, Wagner et al. established *transferrinfection*, based on bioconjugates of transferrin and polycations that enter cells by transferrin receptor mediated endocytosis [17, 26, 27]. In the meantime a multitude of suitable receptor ligands attached to nucleic acid binding moieties have been described. These include synthetic carbohydrates, synthetic peptides, recombinant proteins, immunoglobulins (antibodies), and other molecules such as folate. For recent reviews see Refs. 28–30.

Endocytosis and Endosomal Escape. Both the unspecific binding of vector particles to cell surfaces by electrostatic interaction and the specific receptor-ligand type binding lead to uptake into cells by a natural transport process called *endocytosis*. Membrane-surrounded vesicles, the so-called endosomes, containing the cell bound material are segregated from the plasma membrane and internalized. This pathway subjects the internalized material to the cellular breakdown machinery within endosomes and lysosomes unless specific measures are taken to trigger endosomal escape. For synthetic virus-like delivery systems, the required module has been provided in biological and chemical ways. Wagner and co-workers have first described and later refined the use of pH-specific membrane-disrupting peptides for endosomal escape [31,32]. Synthetic peptides with sequence analogy to the N-terminal sequence of the influenza virus hemagglutinin subunit 2 (HA-2) were chemically or noncovalently coupled to polylysine or other DNA compacting modules [31]. This peptide sequence, capable of adopting an amphipathic α-helix as its active conformation at acidic pH, is responsible for inducing the fusion of the influenza virus and the endosomal membranes in the natural context, triggered by the acidic pH within endosomes. It is important to note that this sequence does not interfere with membrane integrity at neutral pH. Synthetic influenza peptides trigger endosomal release when endocytosed together with gene vectors [33,34]. In this manner, greatly improved transfection efficiencies compared with vectors not containing an endosomolytic module are achieved. Also the bee venom peptide melittin has been used for endosomal release [35]. Because this peptide displays membrane disrupting activity also at neutral pH, suitable gene vector formulations and coupling strategies are required to minimize membrane disturbance by the vector as a whole at neutral pH and maximize it at acidic pH [36]. A breakthrough in terms of transfection efficiency was achieved when chemically inactivated adenovirus particles were coupled to polylysine-DNA complexes [37–39]. The genome of the virus was inactivated by psoralen treatment which leaves the virus capsid and its endosome disruptive function intact [40]. Coupled to an otherwise nonviral vector, this function mediates highly efficiently the release of the vector from endosomes.

Synthetic polymers on polyacrylic acid derivative basis with pH-specific membrane disruptive properties have been described [41,42] and are useful in promoting drug and nucleic acid delivery across endosomal membranes [43–46].

Boussif et al. [47] achieved endosomal escape based on the chemical structure of the DNA compacting cationic moiety. Polyethylenimine (PEI), a cationic chemical produced on industrial scale, binds and compacts DNA and, by virtue of its secondary and tertiary amines, has buffering capacity at physiological pH. In consequence, if a PEI–DNA particle is internalized into cells by endocytosis, it will buffer the acidification process within endosomes. This means that the endosomal proton pump needs to pump way more protons into the endosome until the natural endosomal pH of about 5.5 to 6.5 is reached. The so-called "proton sponge hypothesis" postulates enhanced gene delivery due to the buffering capacity of polymers with structural features like PEI through enhanced endosomal chloride accumulation and consequent osmotic swelling/lysis. Sonawane et al. [48] have provided experimental evidence supporting this hypothesis. They directly measured the previously postulated chloride accumulation and swelling of endosomes with elegant fluorescence techniques in living cells.

Earlier than PEI, polyamidoamine dendrimers were described as useful agents mediating gene delivery [49, 50]. Similar mechanisms probably account for the activity of these polymers as is the case for PEI. A variety of other cationic polymers with protonatable amino groups have been described for nucleic acid delivery [51–60]. Some of them display reduced toxicity compared to PEI. Interesting alternatives to PEI also include poly(2-(dimethylamino)ethyl methacrylate) [61] and biodegradable poly(2-(dimethylaminoethylamino phosphazene) [62]. In terms of gene transfer efficiency, no single polymer outperforms the others to a degree such that it can be considered the polymer or lead structure of choice.

The endosomal escape of nucleic acids formulated as lipoplexes is thought to be mediated by lipid exchange reactions between the endosomal membrane and the lipoplex; that is, anionic lipids from the endosomal membrane compete with the nucleic acid for binding to the cationic lipid moieties and thereby release the nucleic acid from the complex. Through this process, the endosomal membrane is destabilized [63–65]. It is generally accepted that endocytosis is the major cellular uptake mechanism for lipoplexes. However, depending on the biophysical properties of lipoplexes, direct fusion with the cytoplasmic membrane can occur as well [66,67]. Recent work by Safinya's group has led to an improved understanding of structure–function relationships in lipoplex-mediated nucleic acid delivery [67,68]. The charge densities of lipid–DNA complexes are essential factors governing transfection efficiencies, at least if the lipids in the DNA complex are in lamellar configuration.

Nuclear Transport. It is still not well understood how and in which form nonviral vectors gain access to the nucleus. In any case, it is clear that the nuclear membrane represents a major barrier and bottleneck to gene delivery. In many cases, the breakdown of the nuclear membrane during cell division is a prerequisite for access to the nucleus. Nevertheless, the coupling of nuclear localization peptides directly to nucleic acids or the incorporation of such peptides into vector formulations has generated improvements to the delivery process [69,70]. Background and recent progress in targeting to the cell nucleus are discussed in more detail elsewhere [71].

Genome Organization. No major efforts have been invested in directly mimicking viral genome organization. Nevertheless, researchers have used viral genomic elements in order to enhance the persistence of transfected gene expression. Viral promoters such as the CMV promoter are widely used to drive the expression of a transfected gene. Plasmids have been constructed that contain elements of Epstein–Barr virus in order to achieve extrachromosomal plasmid replication in eukaryotic cells (reviewed in Refs. [72] and [73]). Elements from adeno-associated virus responsible for the site-specific genomic integration of the virus have been used to generate a hybrid AAV-adenovirus vector carrying a double-reporter gene integration cassette flanked by AAV ITRs and a tightly regulated, drug-inducible Rep expression cassette [74]. Similar constructs can be delivered with nonviral technology. Site-specific genomic integration has also been achieved with the øC31 integrase system. This is a recombinase found in a *Streptomyces* phage that mediates stable chromosomal integration of genes into host genomes without any additional co-factors [75]. The genomic integration is unidirectional and sequence-specific [76]. The øC31 integrase mediates the integration of *att*B attachment sites of the transgenic DNA into *att*P attachment sites in the host genome, which occur as pseudo-*att*P attachment sites in mammalian genomes [76].

Biocompatibility. Viruses are recognized as foreign by their host organism. Nevertheless, their constituents are biologic materials and viruses are biocompatible enough to secure their replication in the host even though they may kill the host in doing so. In a biomaterial scientist's view, viruses are nanoparticles that are stable enough (biocompatible) during the delivery phase, yet their constituents are assembled in a manner labile enough to allow disassembly and biological processing once they have reached their target. It is not surprising that synthetic constructs for nucleic acid delivery are also recognized as foreign by the host organism. This recognition takes place on a systemic level during the extracellular delivery phase but also at the target cell level. First-generation nonviral vectors undergo strong interactions with blood components and are strong activators of the complement system [77]. These vector particles are mostly cleared from the systemic circulation by the reticuloendothelial system. At the target cell level, the nucleic acid components of nonviral vectors can be recognized as foreign. One example is the interaction of unmethylated CpG sequences with toll-like receptor 9 (TLR9) in intracellular compartments initiating a signaling cascade leading to the production of proinflammatory cytokines [78]. Another example is the induction of innate immune pathways by long double-stranded RNA leading to a generalized repression of protein synthesis [79].

Although no entirely satisfying solutions concerning the biocompatibility limitations to nonviral vectors are available, partial solutions have been provided. Inactivating interactions of vector particles with blood components can be reduced or even eliminated by appropriate surface modifications. These include the attachment of PEG chains either covalently [80, 81] or noncovalently [82] or surface modifications by poly(acrylic acid) derivatives [41, 42] that are useful in promoting drug and nucleic acid delivery across endosomal membranes [43–46] or N-(2-hydroxypropyl)methacrylamide [83]. The resulting vector nanoparticles are sterically stabilized, meaning that their interactions with each other and third components are minimized by limiting the accessibility of their surfaces. Such surface modifications reduce the acute toxicities of vector particles that can be lethal (in animal experiments [84]).

In summary, considerable progress has been made toward the construction of artificial virus-like systems for nucleic acid delivery. Nonviral transfection has become an important tool in biological research and promises great potential in nucleic acid therapies. Reagents for vector construction which approach the efficiency of viral vectors are commercially available to anyone. Since the concepts of gene therapy were first formulated almost 40 years ago, this field has experienced scientific breakthroughs, enthusiastic expectations, and serious setbacks. The validity and feasibility of the concepts have been demonstrated in thousands of animal experiments and in human clinical studies. Given the tremendous potential of nucleic acid-based therapies, the obvious question is why such therapies have not developed into widely practiced, state-of-the-art treatments, at least in specialized hospitals, all over the world. The answer is that most current tools for the genetic modification of cells are still neither efficient enough or safe enough, nor are they affordable enough, simple to practice, or well understood. In consequence, similar limitations hold true for envisaged therapeutic strategies involving such tools. Nucleic acid delivery for therapeutic purposes is a highly complex challenge where multiple parameters can have a major impact on the therapeutic outcome. One such parameter is the ability to localize nucleic delivery.

2.4 WHY IS LOCALIZATION OF DRUG AND NUCLEIC ACID DELIVERY IMPORTANT?

The maximum drug dose a patient can be given is that which he/she can ultimately tolerate, not the one that may be required to cure his/her disease. An instructive example is chemotherapy of cancer. Cytostatics have well-defined potentials to kill cells in culture: A given dose will eradicate a given percentage of a cell population under consideration. In the patient, however, complex biodistribution patterns, drug metabolism, drug resistance, and the pharmacokinetics of a drug can limit its bioavailability at a target site. The patient is systemically "flooded" with a drug in order to achieve its threshold of action at the site of disease. Drugs are designed to act preferentially on selected biological processes in target cells, but absolute specificity in terms of target cell and target process is virtually impossible to achieve. Therefore, in the case of systemic administration of a drug, the threshold dose for target site action is often close to the threshold dose for undesired action at nontarget sites. In other words, the target-specific full dose–response range of a drug cannot be exploited to the level of saturation of the biologic process at which the drug is designed to act (Figure 2.3) [85]. Put yet another way, therapeutic windows of drugs are often narrow and undesired side effects are frequent. Therefore, localization (targeting) of drug delivery is an important objective and mainly serves three related purposes: first, to exceed the local threshold of drug action at the target site while remaining below this threshold at nontarget sites; second, to avoid side effects in this manner; and third, to enlarge the therapeutic window (i.e., to exploit the full dose–response range of a drug locally).

A closer look at nucleic acid delivery highlights the importance of vector targeting. The probability of vector success (functional delivery of a nucleic acid to the desired subcellular localization) is the product of the probabilities of overcoming the individual barriers to delivery. These barriers may weigh differently on the final probability, depending on vector type, but if the probability of vector–target cell contact is low to start with, the efficacy of the overall delivery process will be low as well, independent of vector type. Nonviral plasmid delivery with lipoplexes has been reported to be a mass action process [86], a statement that certainly also applies to other vector types (and drugs in general) if the frequency (or probability) of vector-target cell contact is a limiting barrier. For polyethylenimine(PEI)–DNA vectors it has been estimated that of about 700,000 plasmid copies applied per cell in a standard transfection, roughly 50,000 copies per cell will be present in the cell after 7 hr of incubation [87]. In another publication, it was estimated that one out of 100 microinjected cytoplasmic pDNA copies in a PEI–DNA formulation reaches the nucleus [88]. These two estimates together would predict that at least 1400 plasmid copies in PEI formulation per cell would be required in order to have one copy reach the nucleus. Because it cannot be assumed that each cell-associated copy is located in the cytoplasm, a more realistic estimate would predict rather that 10,000 or more copies in PEI formulation per cell would be required for this purpose. These estimates apply for one particular vector type in cell culture, where rapid vector inactivation, degradation, or clearance before it has a chance of target cell contact do not represent the major limiting barrier (although nonviral in vitro transfections are often

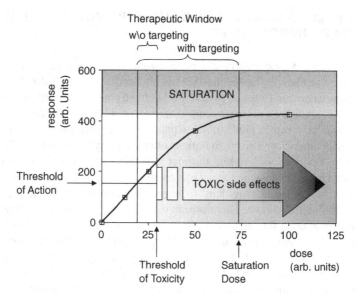

Figure 2.3. Toxic side effects often restrict the possibility of exploiting the full dose–response range of a drug up to (local) saturation levels. One objective of targeting is to achieve target site saturation levels while pushing the non-target-site toxicity threshold to higher doses. In this manner the therapeutic window widens enough to achieve a maximum local effect. Shown is a hypothetic dose–response relationship with arbitrary toxicity and saturation levels just to illustrate the potential of drug targeting. Reproduced from Ref. 85.

carried out in serum-free medium to reduce vector inactivation). It is obvious that *in vivo*, where stability during the extracellular delivery phase represents a limiting factor, the required nucleic acid copy number per cell will be much higher than in cell culture. This applies to viral vectors as well.

In summary, the threshold of action for nucleic acid delivery in terms of required copynumber per target cell can be quite high. Thresholds of action are related to (a) the dynamics of delivery processes, (b) residence times in individual compartments along the delivery pathway, (c) the physiological characteristics of such compartments, and (d) the boundaries between them. Methods for accumulating or holding an applied vector dose at a target site can be expected to improve the overall efficacy of nucleic acid drugs.

2.5 METHODS OF LOCALIZATION AND OF LOCAL CONTROL

Many authors discriminate between active and passive targeting. The latter refers to the preferred accumulation of a drug formulation or a gene vector in a particular tissue based on the biophysical properties of the formulation. Traditionally, active targeting is characterized as involving some form of molecular recognition that allows a formulation to specifically interact with target cells. This definition would mostly be limited to the biological methods of drug localization listed in Table 2.1. In a more comprehensive

Table 2.1. Biological Methods of Localization

Biological	Selected References[a]	Physical	Selected References[a]
		Localization of Delivery	
Receptor–ligand interaction	28, 30, 89	Passive targeting via biophysical properties of vector	90
Localization sequences	73, 91–94	Physical force used for vector accumulation:	95
Site-specific genomic integration	72, 74, 96	Gravitational force	
		Precipitate formation	5, 6, 97, 98
		Centrifugation	99–102
		Magnetic fields	85, 103
		Hydrodynamic force (vector flow towards target cells, direct injection into target tissue)	104–109
		Aerosolization	110, 111
		Ballistic methods	95 and references therein; 112
		Carrier-mediated (implants)	113, 114
		Injectable implants	115, 116
		Solid implants	114, 117–129
		Electric fields	95, 130, 131
Local Control of Delivery and Expression			
Tissue-specific and inducible promoters ("transcriptional targeting")	132–134	Tissue-specific and inducible promoters ("transcriptional targeting" by electromagnetic radiation)	135–139
		Controlled release depots	115, 140
		Controlled release by electromagnetic radiation (heat)	140
		Ultrasound	141–144

[a]Preferably, review papers and not the primary literature are cited here.

definition, modalities of active targeting comprise not only providing a formulation with a molecular recognition element but also any active procedure exerted on a formulation that will lead to localized drug action. This would also include techniques for local *control* of delivery and nucleic acid expression, although such techniques do not qualify as methods of delivery in a strict sense. At least for nucleic acid delivery, it is useful to discriminate between biological and physical methods of localization. Both comprise various subtypes that often can be combined in a flexible manner including the combination of biological and physical subtypes. Most of the physical localization and drug activation methods listed in Table 2.1 would qualify as active targeting and would primarily serve accumulating a formulation in a target tissue and establishing target cell contact.

2.6 ESTABLISHING TARGET CELL CONTACT

Equally important as overcoming cellular barriers to delivery is establishing vector–target cell contact in the first place. All downstream events are dependent on the frequency of this first step. Passive targeting based on the biophysical properties of vectors can be sufficient to achieve preferred transfection of certain tissues [90]. This is observed, for example, upon intravenous administration of PEI polyplexes [145] or of lipoplexes [146], resulting in high transfection levels in the lungs in mice. Interestingly, the pattern of transfection levels in the various organs does not match the actual biodistribution of the administered vectors. The major fraction of the applied vector dose is rapidly cleared by the reticuloendothelial system [147], highlighting the importance of unspecific interactions *in vivo*. Providing vectors with targeting ligands can greatly improve transfection efficiencies and specificities if at the same time nonspecific interactions can be reduced. This has been demonstrated *in vitro* and *in vivo*, particularly in tumor targeting upon intravenous adminstration. Vectors were shielded from nonspecific interactions by PEGylation while epidermal growth factor (EGF) or transferrin provided targeting specificity [81, 148, 149]. Another example is a particular class of lipid-based nanoparticles with bound nucleic acids, provided with an $\alpha_v\beta_3$-targeting ligand that mediated efficient and therapeutically relevant gene delivery to tumor endothelium [150].

Despite the encouraging success with targeted nucleic acid delivery in animal models, it is worth reconsidering the basic physics of the extracellular delivery phase from an administration site to the target cell surface. Cell culture serves as an instructive model from which conclusions for *in vivo* applications can be drawn. Luo and Saltzman [98] have pointed out that DNA transfection efficiency is limited by a simple physical barrier: low DNA concentration at the cell surface. Generalizing this observation, one can state that for a drug added to cell culture supernatants, drug–cell contact is driven by diffusion, no matter whether or not the drug carries a targeting ligand. As a first approximation, diffusion toward the target equals diffusion away from the target in the absence of binding or uptake events. The probability of cell–drug contact increases with drug concentration, incubation time, and temperature (which cannot be chosen arbitrarily). This explains why standard transfection protocols suggest more than one hour of transfection time. In the presence of binding and uptake, the internalized drug amount will be proportional to some order of the drug concentration in the vicinity of the cell surface over a concentration range up to the saturation of the uptake process. The obvious prediction is that below the saturation limit, any measure that increases the drug concentration at the target cell surface at a given drug dose will increase the response to the drug. Luo and Saltzman have verified this prediction for gene delivery and have substantiated it with theoretical analysis by associating vectors with dense silica particles that sedimented vectors on the cell surfaces. Generalizing their observations, one can state that physical force acting on vectors directed in a manner to overcome motion away from the target enhances the delivery process. Suitable physical forces and delivery methods are listed in Table 2.1. The most convenient force for *in vitro* experiments is gravitation as exploited by Luo and Saltzman and almost three decades earlier by Graham and van der Eb [6] in establishing the calcium phosphate precipitation method. For PEI–DNA vectors, it has been found that large DNA complexes transfect more efficiently than smaller ones [97]. In fact, gravitation is exploited unwittingly by most researchers performing *in vitro* transfections with commercially available reagents. Most cationic lipids and polycations form precipitates with nucleic acids in salt-containing solution. Not surprisingly, also centrifugal force enhances nucleic acid delivery by accelerating vectors toward the cells to be transfected [99–102].

The options for drug administration *in vivo* are oral and parenteral. For obvious reasons, gravitation and centrifugation are not suitable for targeting in this case. Oral administration of gene vectors localizes delivery to the gastrointestinal tract and promises great potential for genetic vaccination. Bacterial vectors [151–155], viral vectors [156, 157], chitosan–DNA complexes [158–160], and microencapsulated nucleic acids or viruses [161–167] are used for this purpose. For details, the reader is referred to the cited literature. A complete review volume has been dedicated recently to microencapsulated DNA formulations for vaccination purposes [168].

In parenteral administration, the choice is between local (orthotopic) and systemic routes. Success with biological vector targeting exploiting receptor–ligand-type interactions upon systemic administration has been discussed briefly before. However, the above considerations for cell culture, where diffusion has been defined as a limiting barrier, suggest that the probability of vector–target-cell contact upon systemic administration will be even orders of magnitude lower than *in vitro*. Depending on the target tissue, the accessibility of target cells can be limited, diffusion may be restricted, and hydrodynamic forces (e.g. blood flow) can carry vectors away from the target site. In this respect it is particularly encouraging that even without further provisions for retention at the target, site-specific transfection is possible. Nevertheless (with restrictions), the prediction holds that any measure that increases the vector concentration at the target cell surface at a given administered dose will increase the response (e.g., level of transfected gene expression). The restrictions are that the applied measure must not interfere with vector integrity, uptake, and intracellular processing. Our own work with magnetic-field-guided delivery confirms this prediction.

2.7 VECTOR LOCALIZATION BY MAGNETIC FORCE (MAGNETOFECTION)

We define magnetofection as nucleic acid delivery guided and mediated by magnetic force acting on associates of magnetic particles and nucleic acids (Figure 2.4). This comprises "naked" nucleic acids as well as "packaged" nucleic acids where the packaging can be in the form of a synthetic nucleic acid vector but also in the form of a virus.

We developed magnetofection [169] after learning about the concept of magnetic drug targeting. In contrast to applying gravitational or centrifugal force, this concept is

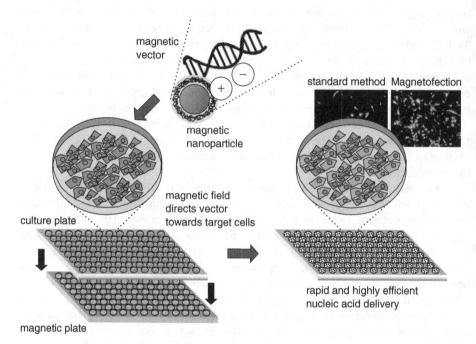

Figure 2.4. Principle of magnetofection in cell culture. Polyelectrolyte-coated magnetic nanoparticles are mixed with naked nucleic acids or synthetic or viral nucleic acid vectors in salt-containing buffer. The particles associate with nucleic acids and vectors by electrostatic interaction and/or salt-induced colloid aggregation. The mixtures are added to cells in culture. The cell culture plate is positioned on a magnetic plate during 5–30 min of incubation. The magnetic field(s) rapidly sediment vectors on the cells to be transfected/transduced. The result is rapid kinetics and high-efficiency nucleic acid delivery. Shown is a cell culture plate and a magnetic plate in 96-well format. The magnetic plate consists of 96 individual neodymium–iron–boron magnets (IBS Magnets, Berlin, Germany) inserted in drill holes in an acrylic glass or PVC plate in strictly alternating polarization. The plate was designed for application with 96-well cell culture plates but is also applicable for 24-, 12- and 6-well layouts, petri dishes of various diameters, and culture flasks of various sizes. Detailed protocols can be found at www.ozbiosciences.com. Reproduced from Ref. 209.

applicable *in vivo* for increasing the concentration of a drug formulation at the target cell surface. Drugs are associated with magnetically responsive materials in the nano- to micrometer-size range and in that manner can be "navigated" by magnetic force. By the mid-1960s, researchers had already attempted the first steps to produce magnetically localized thrombi in intracranial aneurisms in both animals and humans [170–173] using carbonyl iron. Pioneering work by Widder et al. [174] has inspired research into magnetically accumulating drugs mostly in tumors upon administration into the circulation. The magnetic carrier materials are mostly iron oxides of various compositions which can be of natural or synthetic origin [85,175–177]. Magnetic albumin microspheres with entrapped doxorubicin were magnetically accumulated in a Yoshida sarcoma in a rat model. A 100-fold-higher dose of free doxorubicin was required to achieve the same drug level in the tumor as with the magnetically targeted drug [178]. The treatment was therapeutically effective in that it resulted in total tumor remission in a high percentage of experimental animals. In contrast, animals treated with free doxorubicin, placebo microspheres, or nonlocalized doxorubicin microspheres exhibited a significant increase in tumor size with metastases and subsequent death in 90–100% of the animals [179,180]. Other researchers obtained similar results [181–187]. After extensive preclinical examinations, Lübbe et al. applied magnetic drug targeting in cancer patients [188–191]. Retardation of tumor growth and even local remissions were observed [191]. A different type of magnetic particles (MTCs, magnetic targeted carriers) [192–195] are used in another clinical study with magnetically targeted doxorubicin where reportedly 32 patients have been enrolled [196]. However, in the meantime a phase II/III clinical trial involving this technology has been discontinued as the clinical endpoints could not be met with statistical significance (http://freshnews.com/news/biotech-biomedical/article_17775.html).

This highlights the difficulties encountered when proceeding from animal to clinical studies, and hopefully a thorough failure analysis will be published at some point. Nevertheless, at least in animal models it has been clearly demonstrated that (i) magnetic drug targeting is feasible even if the drug administration site is remote from the target site under magnetic field influence [189,197], (ii) the magnetic particles can extravasate under the influence of the magnetic field [183,193,198], and (iii) the magnetic carriers are well-tolerated.

Magnetic targeting of nucleic acid pharmaceuticals is in an early preclinical phase. It was required to associate nucleic acids or vectors with magnetic particles in a manner that is compatible with cellular uptake and the desired intracellular processing. Surprisingly, this has been a relatively simple task. We have used magnetic iron oxide nanoparticles for this purpose which are coated with cationic or anionic polyelectrolytes [85,103,169].

Core/shell-type magnetic nanomaterials for magnetically targeted nucleic acid delivery (which we call magnetofection) can be easily synthesized using precipitation of the iron hydroxides from aqueous iron salt solution and transformation into magnetite in an oxygen-free atmosphere with immediate spontaneous adsorption of the appropriate stabilizer/modifier. An example protocol is as follows: Typically, 200 ml of a 0.2-μm filtered aqueous solution of 50 mmol ferric chloride hexahydrate and 25 mmol ferrous chloride tetrahydrate is cooled to 2–4°C in a 2000-ml four-neck flask. Argon is bubbled continuously through the solution. While being rapidly stirred, 25 ml of 28–30% ammonium hydroxide solution and a solution of coating compound (for example, severals grams of 12-kD dextran in 100 ml of deionized water kept under argon) are quickly added. A primary precipitate forms which is not magnetic yet. The material is heated to 90°C over a 15-min interval and stirred at this temperature for the next 30 min. Magnetic material forms. The mixture is then cooled and incubated at 25°C for 2 h with continuous stirring. No more inert gas bubbling is needed. The product is sonicated for 5 min using 35 mW and a resonance frequency of about 20 kHz, and it is dialyzed extensively against water using Spectrapor 12- to 14-kD MW cutoff membrane to remove excess

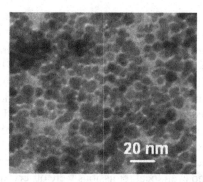

Figure 2.5. Electron micrograph of a representative example of magnetic iron oxide nanoparticles synthesized according to the general protocol outlined in the text.

unbound stabilizer. A large variety of particles can be generated using this basic protocol with different stabilizers such as derivatized dextrans or starch or polycations such as polyethylenimine or polylysine, citrate, polyaspartic acid, and other compounds. Figure 2.5 shows an electron micrograph of a nanomaterial synthesized in this manner. Apart from electron microscopy, the materials are characterized by X-ray diffraction, dynamic light scattering, zeta potential determination, determination of composition (iron content, weight fraction of coating compound), determination of magnetization curves, determination of nucleic acid or gene vector binding isotherms, and finally characterization of their suitability in magnetofection.

The natural tendency of charged colloidal particles to aggregate in salt-containing solution is usually considered to be an annoying characteristic because it limits the stability of colloidal suspensions under physiolgical conditions. The same problem prevails for nonviral nucleic acid vectors which are charged nanoparticles as well. However, we used the otherwise undesired salt-induced aggregation to associate vectors with magnetic nanoparticles. Simple mixing of the vector components (polycation and/or lipid, nucleic acid or viruses) with polyelectrolyte-coated magnetic nanoparticles in salt-containing solution (such as cell culture media or physiological buffers) is sufficient to obtain the desired magnetic vectors. In our own work we have used predominantly polyethylenimine-coated iron oxide nanoparticles, but we have also shown that other polycationic and polyanionic surface coatings are suitable for magnetofection [85]. Most recently, Haim et al. [199] have used negatively charged magnetic nanoparticles coated with derivatized starch to associate these with lentivirus preparations in a noncovalent manner. This is achieved, as the authors argue, by colloidal clustering facilitated by positively charged ions in solution [200]. Other researchers have used colloidally stable streptavidin-coated magnetic particles and biotinylated vectors for the same purpose [201–205]. In cell culture, these magnetic vectors can be sedimented on the cells to be transfected by magnetic fields within a few minutes. For this purpose, commercially available neodymium–iron–boron magnets are attached to the downside of cell culture dishes. Also, magnetic plates which we have designed for this purpose are commercially available (see also Figure 2.4). The magnetic attraction has the

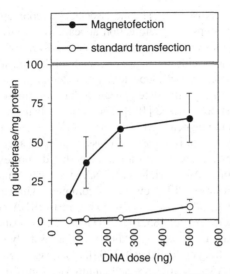

Figure 2.6. Standard transfection and magnetofection of B16F10 mouse melanoma cells. The figure shows typical dose–response relationships observed when comparing magnetofection and standard transfection. In this case, the cells were seeded in a 96-well plate at a density of 6.000 cells per well the day prior to transfection. For the standard transfection, plasmid DNA coding for luciferase was mixed with DOTAP–cholesterol liposomes (1:0.9 mol/mol) to result in a charge ratio of 1.25 (positive charges of DOTAP over negative charges of DNA). For magnetofection, DOTAP–cholesterol liposomes were mixed with DEAE dextran-coated magnetic iron oxide nanoparticles (obtained from chemicell, Berlin, Germany) followed by mixing with plasmid DNA. The w/w ratio of magnetic particles to DNA was 2, and the charge ratio of DOTAP–cholesterol to DNA was 1.25. After addition of the DNA complexes to the cells, the culture plate was positioned on a magnetic plate for 20 min (see Figure 2.4). Luciferase expression was determined 24 hr after transfection. The figure shows that saturation levels of transfection are achieved with Magnetofection while the standard reagent at the same dosage remains considerably below this level. For achieving the same effect with the standard reagent, high doses are required which would lead to toxictiy (compare to Figure 2.3).

consequence that the full vector dose rapidly gets in contact with the target cells (the diffusion limitation is overcome). As predicted, this greatly improves the dose–response profiles of most examined gene vectors (an example is shown in Figure 2.6). Incubation times thus can be limited to minutes instead of hours. We have shown that at least in the case of antisense oligonucleotide delivery, the rapid transfection kinetics helps reduce transfection-associated toxicity to the cells [206]. Another consequence of magnetically guided nucleic acid delivery is that it can be confined to cells under the influence of the magnetic field within one cell culture dish. We have discussed the details and benefits of the magnetofection method in several reviews and research papers [85, 102, 103, 169, 206–209]. Briefly summarized: The linkage between magnetic particles and vectors can be established in a reversible manner. Therefore, cells can obviously dissociate the components, and the association is compatible with the required intracellular processing steps. Magnetofection appears to be universally applicable to viral and nonviral vectors and, among the latter, to the delivery of large (plasmid DNA) and small synthetic nucleic acids (antisense oligonucleotides and siRNA [85,206,209]). The uptake into cells proceeds via endocytotic processes similar to those for the parent vectors, and the applied magnetic force appears to have no further effect than localizing vectors at the target cell surface [102]. The only mechanistic differences between standard transfection and magnetofection observed so far were with adenoviral vectors and with siRNA delivery [85, 169, 209]. The association of adenovirus with cationic magnetic particles allows the vector to infect cells that do not express the coxsackie and adenovirus receptor (CAR). Synthetic siRNA molecules cannot be delivered in a functional manner with linear PEI. However, if combined with cationic magnetic particles and magnetofected, otherwise inactive linear PEI–siRNA complexes efficiently knock down target gene expression [85,209]. The mechanistic basis for this has not been elucidated so far. We have used magnetofection very successfully for the transfection of primary cells including lung epithelial cells [208], blood vessel endothelial cells [207], keratinocytes, chondrocytes, osteoblasts, and amniocytes (unpublished results) as well as with whole tissue specimens of airways [208] and with blood vessels ([169,206] and unpublished results). In the meantime, magnetofection reagents are commercially available from OZ Biosciences (Marseille, France; www.ozbiosciences.com) and chemicell (Berlin, Germany; www.chemicell.com). Accordingly, more publications involving the method can be expected in the near future.

An important question is whether magnetic nucleic acid targeting is feasible *in vivo* and whether magnetofection is useful beyond research applications in nucleic acid based therapies. We have provided proof of principle in demonstrating magnetically localized transfections in segments of blood vessels and in the gastrointestinal tract [169]. We have also demonstrated therapeutic potential in an ongoing veterinary clinical study of immuno gene therapy of feline fibrosarcoma [209]. This is one of the most common feline tumors with a relapse rate of 75% within 6 months upon surgical resection, which is the standard therapy (for more details on feline fibrosarcoma see Ref. 210). We inject a plasmid construct with the human GM-CSF gene under the control of the CMV promoter associated with magnetic particles directly into the tumor twice with a one-week interval starting two weeks prior surgical resection of the tumor. During the application, a neodymium–iron–boron magnet is placed on the tumor adjacent to the injection site in order to retain the injected dose within the tumor tissue. So in this case, magnetic field guidance is not used to direct the vector to the target tissue upon remote administration but rather to keep a locally applied dose in the target tissue. The interim result of this study is that tumor-free survival of the cats is raised from only 23% at the one-year time point in the case of standard therapy (surgery only) to 52% with presurgical magnetofection of the human GM-CSF gene (20 patients treated).

One can conclude that magnetically guided nucleic acid delivery has potential *in vivo*. At the same time, limitations are clearly evident, some of which may be overcome by appropriate formulations and novel magnetic field technologies. Magnetic nanoparticles in a magnetic field move in a preferred direction of space only if they experience a magnetic field gradient. The magnetic force acting on a particle is proportional to the magnetic flux density, to the volume (and thus the third power of the radius) of the particle, and to the field gradient. During *in vivo* applications, hydrodynamic forces counteract magnetic retention. An example is the viscous drag force according to Stoke's law in the bloodstream which is proportional to the first power of the particle radius. Detailed theoretical considerations substantiated with experimental evidence have been published [211–214]. The study by Nagel [214] shows that magnetic particles with diameters in the lower nanometer range (around 50 nm) are not suitable for magnetic drug targeting. In agreement with theoretical predictions, only a minor percentage of magnetic particles could be trapped using rare earth permanent magnets even at low flow rates of up to 4 mm/sec, which is prevalent in small capillaries. Increasing particle diameters helps, but upper limits are set by the anatomy of blood vessels (capillary diameter of about 5 μm). Magnetic drug targeting appears impossible at flow rates around 20 cm/sec, like in the human aorta. Another limitation is that magnetic flux density and field gradients decrease rapidly with increasing distance from a magnetic pole shoe. Gradients cannot be generated arbitrarily in space. Hence, for the moment, magnetic drug targeting is limited to superficial or surgically accessible areas of an organism. Nevertheless, even with the given constraints, numerous applications of magnetic targeting can be envisaged. Blood flow rates can be reduced locally and temporarily, the vasculature of major organs is accessible to catheters, strong electromagnets with tailored field gradients are constructed, and suitable formulations containing magnetic particles are developed. The study by Nagel suggests that magnetic deposition of magnetic particles against hydrodynamic force is a cooperative process. Particles, once deposited, generate additional local field gradients in an external field that facilitate the deposition of further particles. Babincova et al. [215] have suggested the positioning of ferromagnetic materials close to a target site. In a strong external homogeneous field like the one present in magnetic resonance imaging equipment, such material will generate strong local gradients that may be exploited for magnetic drug targeting.

It can be concluded that important developments in the field of magnetic drug targeting can be expected from a collaboration of experts in magnetic field physics, electrical engineering, magnetic particle synthesis, drug formulation, and medical sciences. Such an interdisciplinary network has been installed in Germany in the federal government funded program Nanobiotechnology (www.nanobio.de). Engineers from Siemens Corporate Technology currently develop electromagnets that are tailor-made for magnetic drug targeting under clinical settings [216]. These magnets are designed to provide high and scalable magnetic force in an application volume around and inside a tumor and are designed to have low mass, small size, and be easy to handle. A cross section of a prototype for preclinical experiments is shown in Figure 2.7. This particular magnet produces a field gradient ranging from 100 T/m at the pole tip to 10 T/m in a prechosen cylindrical application volume of 20 mm in diameter and height. This magnet is currently under evaluation for magnetic drug targeting in animal tumor models. For the research field it is highly important that standardized hardware (magnets, magnetic particles) is available. Until now, most researchers had to rely on permanent magnets

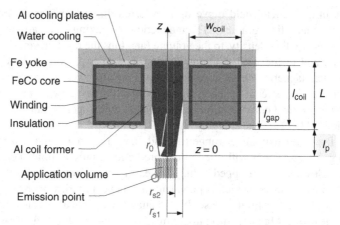

Figure 2.7. Design of an open-pot-type magnet developed by Siemens Corporate technology with parameters varied during optimization. The specification was that the magnet ought to produce a normalized magnetic force at $z = 20$ mm and $r = 10$ mm of up to 10 T/m on magnetic nanoparticles used in our magnetic drug targeting experiments. These particles are multidomain particles or clusters of single-domain particles with a typical susceptibility χ of 0.2 with typical particle sizes of 100–150 nm. Saturation magnetization of these particles is reached above $\approx$0.2–0.4 T. The magnetic field near the pole tip easily exceeds this value. Hence the saturation of the magnetic particle clusters within the application volume can be assumed, with the particles consequently having a constant magnetic moment. Therefore, the magnet had to be optimized with respect to the force per magnetic moment only. Details will be published elsewhere [216]. Parameters that were optimized: w_{coil}, width of the coil (radial); l_{coil}, length of the coil (axial); f, aspect ratio of coil: w_{coil}/l_{coil}; r_0, radius curvature at pole tip; r_{s1}, radius of iron–cobalt core piece; r_{s2}, radius of core at pole tip; l_{gap}, length of gap between core and coil former; l_p, length of core outside the coil; L, total coil length including aluminum side plates; NI, ampere-turns (number of turns times current); N, number of turns for Al tape winding; F/m, normalized magnetic force at $z = 20$ mm and $r = 10$ mm; m, mass of the magnet. (These data were kindly provided by Dr. Wolfgang Schmidt, Siemens AG, Corporate Technology, Erlangen, Germany).

with poorly defined characteristics and geometries based on availability rather than on specific requirements for the given application.

Further improvements in magnetic drug targeting can arise from optimized formulations. The magnetic force acting on a magnetic drug carrier is proportional to the third power of its radius, while the counteracting viscous drag force is proportional to only the first power. Hence, a magnetic carrier for drug targeting needs to be of large diameter (or, in other words must contain a maximum amount of magnetic material) while maintaining enough flexibility to migrate through blood capillaries like blood cells. Liposomes containing an optimized amount of magnetic nanoparticles in their aqueous lumen appear to be suitable for this purpose [217, 218]. It is our goal to generate an alternative magnetic carrier which, in addition to magnetic force, is susceptible to an independent second physical force which is also relevant in drug targeting and medical imaging. We have developed magnetic microbubbles.

2.8 USE OF ULTRASOUND TO TRIGGER LOCALIZED DELIVERY, MICROBUBBLES, AND MAGNETIC MICROBUBBLES

Microbubbles are gas-filled microspheres that were developed as contrast agents for medical ultrasound imaging purposes. Micrometer-sized gas bubbles that resonate at a diagnostic frequency are ideal reflectors for ultrasound [219]. For application in the systemic circulation, these bubbles should be smaller than 5–7 μm in order not to obstruct blood capillaries. In the simplest case, microbubbles are nothing more than air bubbles [220,221] or gas emulsions [222] in an aqueous phase. Such bubbles can be stabilized if the air–liquid interface is provided with a shell. The shell can consist of renografin, indocyanin green, carbohydrates (such as dextrose), proteins, denatured proteins, surfactants, lipids, or synthetic polymers (such as polylactides) [223,224]. Various compositions are discussed in comprehensive reviews [141,143]. First-generation microbubbles, which were air-filled, suffered from limited stability. Upon intravenous infusion, the air dissolves rapidly in the blood, hence the bubbles are lost for imaging or drug delivery. The physical background for these phenomena has been discussed by Schutt et al. [219]. Using gases with low Ostwald coefficients greatly improves bubble stability. Perfluorocarbons have turned out to be ideal gases for microbubble preparation due to low aqueous solubility and sufficient volatility. More recent compositions are so-called nanoemulsions consisting of a bubble shell filled with a liquid perfluorocarbon. These compositions can be designed such that the fluorocarbon will undergo a phase transition from liquid to gaseous states at a range of different temperatures [143]. Preparation procedures for microbubbles include simple shaking, emulsion procedures, application of shear forces, or sonication. Protocols can be found in the scientific and patent literature.

Microbubbles have been used as drug carriers. Both low-molecular-weight drugs and high-molecular-weight drugs such as nucleic acids can be associated with microbubbles. The association can be realized in various ways. The drug can be bound covalently or noncovalently to the surface of a bubble shell, the drug can be integral part of the shell itself, or it can reside in the interior of the bubble. In all cases, drug molecules can optionally interact via chemical bonds or physically (noncovalently) with each other and/or other components of the shell. For association with microbubbles, a drug may also be provided in the form of a pro-drug [225,226]. The various modes of drug association have been discussed in comprehensive reviews [141,143,227,228].

Drug-loaded microbubbles hold potential as "magic bullet" agents to deliver drugs to precise locations in the body; these precise locations are determined by where the ultrasound energy is focused [143]. The physical basis is that gas-filled microbubbles can be induced to "pop" using ultrasound of appropriate frequency and energy. Ultrasound probes operating in the low megahertz range have been found optimal for this purpose [229]. The interaction of microbubbles and ultrasound leads to cavitation, bubble burst, and consequent drug release. In addition, cavitation can lead to microvessel rupture leading to increased permeability of the endothelial barrier [230]. This effect has been used to deliver nanoparticles and red blood cells to the interstitium of rat skeletal muscle [231]. Cavitation nuclei formed by microbubbles have also been used to permeabilize the blood–brain barrier [232]. It can be envisaged that simple co-delivery of a drug with microbubbles and local ultrasound irradiation is sufficient to achieve locally enhanced delivery [143]; that is, the drug to be delivered may not need to be associated with microbubbles.

Microbubbles have been used successfully in nucleic acid delivery [141, 143, 144]. Interestingly, ultrasound alone has been shown to enhance gene delivery to cell lines [233, 234], skeletal muscle [235], and tumors [236]. Associating nucleic acids with microbubbles, applying such compositions *in vitro* and *in vivo* under exposure of the target tissue to ultrasound is a highly effective method for triggering localized delivery of nucleic acids and drugs in general in a variety of tissues [142, 226, 237–246]. In addition, microbubbles can be targeted exploiting receptor–ligand-type interactions (including antigen–antibody interactions) [141, 143].

Based on this state of the art, we reasoned that microbubbles may be ideal carriers to incorporate a high quantity of magnetic nanoparticles as well as active agents such as drugs or nucleic acids. The flexibility of this carrier should be sufficient to "squeeze" through blood capillaries. Magnetic accumulation of the carrier against hydrodynamic forces ought to be facilitated compared with "free," physically uncoupled magnetic nanoparticles. This should occur because in magnetic microbubbles one carrier object would comprise a plurality of magnetic particles which would be immobilized in the bubble shell in close vicinity and thus Brownian relaxation would be greatly reduced. Active agents, once magnetically accumulated at a target site, would be released locally by the application of ultrasound which might also improve tissue penetration by the active agent. Last but not least, magnetic microbubbles might be highly useful contrast agents both for ultrasound and magnetic resonance imaging. The only remaining question was how to prepare magnetic microbubbles. The patent literature provides a multitude of microbubble compositions and preparation procedures. We chose to focus on albumin and on lipid microbubbles because for both preparations their suitability as nucleic acid carrier in gene delivery had been demonstrated previously. We found that for the preparation of magnetic microbubbles the choice of a suitable magnetic nanoparticles coating is essential. It is obvious that this surface coating must be such that a strong interaction with other bubble constituents, particularly with the main constituent, is enabled. We found that the preparation of albumin microbubbles using sonication such as described in the literature is simple in principle. However, it was difficult to achieve stable and well-defined magnetic albumin microbubbles. The simplest solution was to mix a commercially available microbubble preparation such as Optison with magnetic nanoparticles. In our study, the best quality of magnetic microbubbles was obtained with lipids and an agitation procedure. The detailed compositions and preparation procedures will be published in scientific journals. In brief, the major component of our magnetic microbubbles is soybean oil; a drop of it is provided in a glass vial. Also, a mixture of nucleic acid, a cationic lipid transfection reagent and magnetic nanoparticles is added in a buffer containing glycerol and propylene glycol. The gaseous space above the liquid surface is filled with perfluoropropane. Subsequently the glass vial is sealed and vigorously shaken at a defined frequency for a specified period of time. Using detergent-coated magnetic nanoparticles, we achieved quantitative magnetic nanoparticle incorporation and, due to the cationic lipid comprised in the preparation, also quantitative nucleic acid incorporation. Plasmid DNA survives the shaking procedure undamaged as determined by gel electrophoresis. Figures 2.8 and 2.9 show examples of magnetic microbubbles with incorporated plasmid DNA. We confirmed that indeed the magnetic accumulation of magnetic nanoparticles against hydrodynamic forces is greatly superior in microbubble formulation than in the form of free magnetic nanoparticles. Such as has been described for classical microbubbles, also magnetic microbubbles can be destroyed by ultrasound of 1-MHz frequency (Figure 2.10). These preparations are suitable to transfect cells in culture (Figure 2.11), which serves as a model for later *in vivo* applications. Not surprisingly, similarly as in standard magnetofection, the highest impact on transfection efficiency is contributed by the presence of a magnetic field that warrants microbubble–target cell contact. In the absence of a magnetic field, the microbubbles float and thus hardly get in contact with adherent cells in culture. The application of ultrasound and magnetic field in combination yields the highest efficiency in transfections in cell culture. Currently we are examining magnetically targeted and ultrasound triggered delivery of nucleic acids *in vivo*. In one study, magnetic microbubbles are injected into the circulation, and magnetic accumulation and ultrasound-triggered deposition of nucleic acids are observed by intravital microscopy in a mouse dorsal skin chamber model. The other model, exemplary for topical administration, is a ventral skin flap model. Magnetic microbubbles carrying the gene of an angiogenic factor are injected subcutaneously. Later on, a skin flap is excised and repositioned. Without disclosing details here (they will be published in scientific journals), the encouraging result is that only the combination of magnetic field and ultrasound yields a local deposition of nucleic acids in the circulation model or an improved skinflap survival in the orthotopic model, respectively. One of our goals is to target endothelial cells in the vasculature of tumors with the help of magnetic microbubbles and in this manner deposit anti-cancer agents there, be it classical pharmaceuticals or nucleic acids. The coming months will show whether this can be achieved.

2.9 CONCLUDING REMARKS

Almost 40 years ago initial concepts of gene therapy have been conceived [3]. The validity of the concept and therapeutic efficacy has been demonstrated in humans using viral vectors [247]. However, this success has been overshadowed by severe adverse events [248]. This and another tragic setback to nucleic acid therapy concepts, the death of a treated patient [249], highlight how little we understand the complex biology being the basis of the therapeutic concept we pursue. Despite more than 30 years of continued efforts, the final breakthrough in nucleic acid-based therapies has not been achieved yet. However, these decades of research have greatly contributed to an improved understanding of the involved biology and to an appreciation of the complex challenges presented by nucleic acid delivery. Nucleic acid delivery has become an important research tool in the biomedical sciences. Concerning the important major step still to be

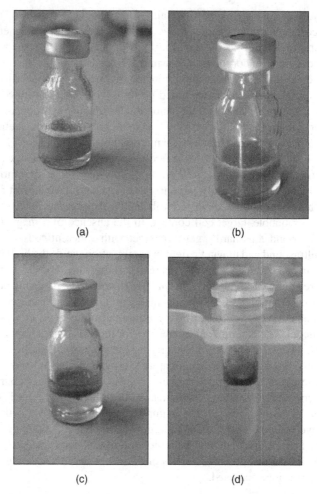

(a)   (b)

(c)   (d)

Figure 2.8. Stages of magnetic microbubble preparation. The microbubble components are combined in a Wheaton glass vial. The vial is sealed and the headspace is filled with perfluoropropane. Subsequently, the vial is shaken for 1 min at 2.500 rpm using a Mini BeadBeater (Biospec Products, Inc., Bartlesville, Oklahoma). (a) Directly after shaking of the components in the Wheaton vial using the Mini BeadBeater; (b) 1 min after shaking; (c) 5 min after shaking; (d) after washing with buffer. The sequence parts a–d shows that within 5 min after preparation, the magnetic microbubbles float on the surface of the preparation, as is typical for perfluoropropane-filled microbubbles. Parts c and d also demonstrate the efficient incorporation of magnetic particles into the bubbles by virtue of the brown color floating on the surface. If the magnetic particles were not incorporated in the bubble preparation, the whole preparation would be turbid.

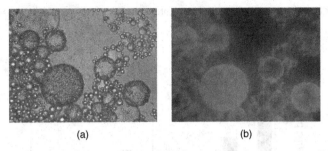

(a)   (b)

Figure 2.9. Plasmid-loaded magnetic microbubbles in the light microscope. Bright field (a) and fluorescence microscopy (b) lipid magnetic microbubbles (magnification > 400×) where the DNA component of the bubbles was fluorescence-labeled with YOYO-1 iodide. The figure shows that both magnetic particles and DNA are efficiently incorporated into the complex. (a) The granular appearance of the bubbles shows the incorporation of magnetic particles. This was further evidenced by a rapid movement of the particles toward a neodymium–iron–boron permanent magnet positioned in an area close to what is the lower left corner of the photograph. Accordingly, the photograph shows magnetic microbubbles concentrated in the lower left corner. (b) The fluorescence image demonstrates that all microbubbles seen in the phase contrast image contain DNA (YOYO-1 only gives significant fluoresencence when intercalated in the nucleic acid). The bubbles in this figure appear quite heterogeneous. This is due to bubble fusion under magnetic field influence as occurred here. The original bubble preparation is relatively homogeneous with bubble diameters of 2–4 μm (determined using a Coulter Counter).

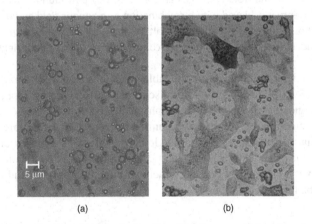

(a)   (b)

Figure 2.10. Magnetic microbubble destruction by ultrasound. An aliquot of the bubble preparation was diluted with phosphate buffered saline in a 6-well cell culture plate and observed under a light microscope. The magnetic microbubbles were magnetically sedimented to the bottom of the well using a disk-shaped neodymium–iron–boron permanent magnet (NeoDelta NE 155, IBS Magnet, Berlin, Germany). (a) Photograph taken directly after magnetic sedimentation before sonication and (b) after sonication at 2 W/cm^2, 20% duty cycle, for 60 sec at room temperature with a 6-mm ultrasound probe of a Sonitron 2000D ultrasound device operating at 1 MHz (Rich Mar Inc., Inola, Oklahoma). The figure demonstrates that magnetic lipid microbubbles can be destroyed efficiently using ultrasound. The granular structures in part b are magnetic particles.

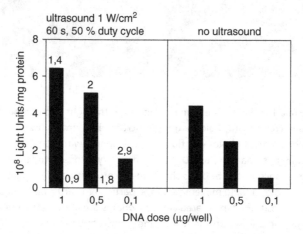

Figure 2.11. Transfection with magnetic microbubbles in cell culture. NIH 3T3 cells were transfected with magnetic microbubbles carrying a luciferase reporter plasmid in the presence or absence of magnetic field influence and ultrasound. Transfection with ultrasound application at 1 W/cm^2 for 60 sec at 50% duty cycle (left) and without ultrasound application (right) in the presence and absence of a magnetic field. The numbers above the bars (left part of the figure) indicate the -fold enhancements achieved by ultrasound application. Highest transfection efficiency is achieved with a combination of magnetic field and ultrasound with appropriately chosen ultrasound parameters, although in cell culture experiments the major enhancing impact on transfection efficiency is contributed by the magnetic field.

taken toward efficient and widely applicable nucleic acid-based therapies, we now know that it has to be taken by an interdisciplinary effort. Medical, pharmaceutical, chemical, biological, and importantly also physical aspects need to be considered, and the respective scientific efforts need to be united in order to generate safe and efficient nucleic acid pharmaceuticals. The ability to localize delivery is an important step in this direction, in terms of both efficacy and safety. It is likely that a combination of physical control of delivery, of localization, and of activation with the corresponding biological concepts will be the way to success.

REFERENCES

1. Mello CC, Conte D, Jr. Revealing the world of RNA interference. *Nature* 2004;431(7006):338–342.

2. He L, Hannon GJ. MicroRNAs: Small RNAs with a big role in gene regulation. *Nat Rev Genet* 2004;5(7):522–531.

3. Tatum EL. Molecular biology, nucleic acids, and the future of medicine. *Perspect Biol Med* 1966;10(1):19–32.

4. Wolff JA, Malone RW, Williams P, Chong W, Acsadi G, Jani A, Felgner PL. Direct gene transfer into mouse muscle *in vivo*. *Science* 1990;247(4949):1465–1468.

5. Tovell DR, Colter JS. Observations on the assay of infectious viral ribonucleic acid: Effects of DMSO and DEAE-dextran. *Virology* 1967;32(1):84–92.

6. Graham FL, Van Der Eb AJ. Transformation of rat cells by DNA of human adenovirus 5. *Virology* 1973;54(2):536–539.

7. Sober HA, Schlossman SF, Yaron A, Latt SRG. Protein–nucleic acid interaction. I. Nuclease–resistant polylysine-ribonculeic acid complexes. *Biochemistry* 1966; 5(11):3608–3616.

8. Latt SA, Sober HA. Protein–nucleic acid interactions. 3. Cation effect on binding strength and specificity. *Biochemistry* 1967;6(10):3307–3314.

9. Latt SA, Sober HA. Protein–nucleic acid interactions. II. Oligopeptide-polyribonucleotide binding studies. *Biochemistry* 1967;6(10):3293–3306.

10. Latt SA, Sober HA. Protein–nucleic acid interactions. III. Cation effect on binding strength and specificity. *Biochemistry* 1967;6(10):3307–3314.

11. Porschke D. The binding of Arg- and Lys-peptides to single stranded polyribonucleotides and its effect on the polymer conformation. *Biophys Chem* 1979;10(1):1–16.

12. Haynes M, Garrett RA, Gratzer WB. Structure of nucleic acid-polybase complexes. *Biochemistry* 1970;9:4410–4416.

13. Gosule LC, Schellman JA. Compact form of DNA induced by spermidine. *Nature* 1976;259(5541):333–335.

14. Gosule LC, Schellman JA. DNA condensation with polyamines I. Spectroscopic studies. *J Mol Biol* 1978;121(3):311–326.

15. Wu GY, Wu CH. Receptor-mediated *in vitro* gene transformation by a soluble DNA carrier system. *J Biol Chem* 1987;262:4429–4432.

16. Wu GY, Wu CH. Receptor-mediated gene delivery and expression *in vivo*. *J Biol Chem* 1988;263:14621–14624.

17. Wagner E, Zenke M, Cotten M, Beug H, Birnstiel ML. Transferrin–polycation conjugates as carriers for DNA uptake into cells. *Proc Natl Acad Sci USA* 1990;87(9):3410–3414.

18. Nicolau C, Cudd A. Liposomes as carriers of DNA. *Crit Rev Ther Drug* 1989;6(3):239–271.

19. Felgner PL, Gadek TR, Holm M, Roman R, Chan HW, Wenz M, Northrop JP, Ringold GM, Danielsen M. Lipofection: A highly efficient, lipid-mediated DNA-transfection procedure. *Proc Natl Acad Sci USA* 1987;84(21):7413–7417.

20. Behr JP, Demeneix BA, Loeffler JP, Perez-Mutul J. Efficient gene transfer into mammalian primary endocrine cells with lipopolyamine-coated DNA. *Proc Natl Acad Sci USA* 1989;86:6982–6986.

21. Felgner PL, Barenholz Y, Behr JP, Cheng SH, Cullis P, Huang L, Jessee JA, Seymour L, Szoka F, Thierry AR, Wagner E, Wu G. Nomenclature for synthetic gene delivery systems. *Hum Gene Ther* 1997;8(5):511–512.

22. Tang MX, Szoka FC. The influence of polymer structure on the interactions of cationic polymers with DNA and morphology of the resulting complexes. *Gene Ther* 1997;4(8):823–832.

23. Tang MX, Szoka FC, Jr. Characterization of polycation complexes with DNA. In Kabanov AV, Felgner PL, Seymour L, editors. *Self-Assembling Complexes for Gene Delivery: From Laboratory to Clinical Trial.* New York: John Wiley & Sons; 1998, pp. 169–196.

24. Tang MX, Li W, Szoka FC, Jr. Toroid formation in charge neutralized flexible or semi-flexible biopolymers: Potential pathway for assembly of DNA carriers. *J Gene Med* 2005;7(3):334–342.

25. Wu GY, Wu CH. Evidence for targeted gene delivery to HepG2 hepatoma cells *in vitro*. *Biochemistry* 1988;27:887–892.

26. Zenke M, Steinlein P, Wagner E, Cotten M, Beug H, Birnstiel ML. Receptor-mediated endocytosis of transferrin polycation conjugates—an efficient way to introduce DNA into hematopoietic cells. *Proc Natl Acad Sci USA* 1990;87(10):3655–3659.

27. Cotten M, Langlerouault F, Kirlappos H, Wagner E, Mechtler K, Zenke M, Beug H, Birnstiel ML. Transferrin polycation-mediated introduction of DNA into human leukemic cells—stimulation by agents that affect the survival of transfected dna or modulate transferrin receptor levels. *Proc Natl Acad Sci USA* 1990;87(11):4033–4037.

28. Gust TC, Zenke M. Receptor-mediated gene delivery. *Sci World J* 2002;2:224–229.

29. Pardridge WM. Brain drug targeting and gene technologies. *Jpn J Pharmacol* 2001; 87(2):97–103.

30. Varga CM, Wickham TJ, Lauffenburger DA. Receptor-mediated targeting of gene delivery vectors: insights from molecular mechanisms for improved vehicle design. *Biotechnol Bioeng* 2000;70(6):593–605.

31. Wagner E, Plank C, Zatloukal K, Cotten M, Birnstiel ML. Influenza virus hemagglutinin HA-2 N-terminal fusogenic peptides augment gene transfer by transferrin-polylysine-DNA complexes: Toward a synthetic virus-like gene-transfer vehicle. *Proc Natl Acad Sci USA* 1992;89(17):7934–7938.

32. Plank C, Oberhauser B, Mechtler K, Koch C, Wagner E. The influence of endosome-disruptive peptides on gene transfer using synthetic virus-like gene transfer systems. *J Biol Chem* 1994;269(17):12918–12924.

33. Plank C, Zauner W, Wagner E. Application of membrane-active peptides for drug and gene delivery across cellular membranes. *Adv Drug Del Rev* 1998;34(1):21–35.

34. Prchla E, Plank C, Wagner E, Blaas D, Fuchs R. Virus-mediated release of endosomal content *in vitro*: Different behavior of adenovirus and rhinovirus serotype 2. *J Cell Biol* 1995;131(1):111–123.

35. Ogris M, Carlisle RC, Bettinger T, Seymour LW. Melittin enables efficient vesicular escape and enhanced nuclear access of nonviral gene delivery vectors. *J Biol Chem* 2001; 276(50):47550–47555.

36. Boeckle S, Wagner E, Ogris M. C- versus N-terminally linked melittin–polyethylenimine conjugates: The site of linkage strongly influences activity of DNA polyplexes. *J Gene Med* 2005;7(10):1335–1347.

37. Curiel DT, Agarwal S, Wagner E, Cotten M. Adenovirus enhancement of transferrin polylysine-mediated gene delivery. *Proc Natl Acad Sci USA* 1991;88(19):8850–8854.

38. Wagner E, Zatloukal K, Cotten M, Kirlappos H, Mechtler K, Curiel DT, Birnstiel ML. Coupling of adenovirus to transferrin polylysine DNA complexes greatly enhances receptor-mediated gene delivery and expression of transfected genes. *Proc Natl Acad Sci USA* 1992;89(13):6099–6103.

39. Curiel DT, Wagner E, Cotten M, Birnstiel ML, Agarwal S, Li CM, Loechel S, Hu PC. High-efficiency gene transfer mediated by adenovirus coupled to DNA-polylysine complexes. *Hum Gene Ther* 1992;3(2):147–154.

40. Cotten M, Saltik M, Kursa M, Wagner E, Maass G, Birnstiel ML. Psoralen treatment of adenovirus particles eliminates virus replication and transcription while maintaining the endosomolytic activity of the virus capsid. *Virology* 1994;205(1):254–261.

41. Murthy N, Robichaud JR, Tirrell DA, Stayton PS, Hoffman AS. The design and synthesis of polymers for eukaryotic membrane disruption. *J Control Release* 1999;61(1–2):137–143.

42. Kusonwiriyawong C, van de Wetering P, Hubbell JA, Merkle HP, Walter E. Evaluation of pH-dependent membrane-disruptive properties of poly(acrylic acid) derived polymers. *Eur J Pharm Biopharm* 2003;56(2):237–246.

43. Stayton PS, Hoffman AS, Murthy N, Lackey C, Cheung C, Tan P, Klumb LA, Chilkoti A, Wilbur FS, Press OW. Molecular engineering of proteins and polymers for targeting and intracellular delivery of therapeutics. *J Control Release* 2000;65(1–2):203–220.

44. Jones RA, Cheung CY, Black FE, Zia JK, Stayton PS, Hoffman AS, Wilson MR. Poly(2-alkylacrylic acid) polymers deliver molecules to the cytosol by pH-sensitive disruption of endosomal vesicles. *Biochem J* 2003;372(Pt 1):65–75.

45. Kyriakides TR, Cheung CY, Murthy N, Bornstein P, Stayton PS, Hoffman AS. pH-sensitive polymers that enhance intracellular drug delivery *in vivo*. *J Control Release* 2002;78(1–3):295–303.

46. Cheung CY, Murthy N, Stayton PS, Hoffman AS. A pH-sensitive polymer that enhances cationic lipid-mediated gene transfer. *Bioconjug Chem* 2001;12(6):906–910.

47. Boussif O, Lezoualc'h F, Zanta MA, Mergny MD, Scherman D, Demeneix B, Behr JP. A versatile vector for gene and oligonucleotide transfer into cells in culture and *in vivo*: polyethylenimine. *Proc Natl Acad Sci USA* 1995;92(16):7297–7301.

48. Sonawane ND, Szoka FC, Jr., Verkman AS. Chloride accumulation and swelling in endosomes enhances DNA transfer by polyamine-DNA polyplexes. *J Biol Chem* 2003; 278(45):44826–44831.

49. Haensler J, Szoka FC. Polyamidoamine cascade polymers mediate efficient transfection of cells in culture. *Bioconjug Chem* 1993;4(5):372–379.

50. Tang MX, Redemann CT, Szoka FC. In vitro gene delivery by degraded polyamidoamine dendrimers. *Bioconjug Chem* 1996;7(6):703–714.

51. Boussif O, Delair T, Brua C, Veron L, Pavirani A, Kolbe HV. Synthesis of polyallylamine derivatives and their use as gene transfer vectors *in vitro*. *Bioconjug Chem* 1999;10(5):877–883.

52. Van Craynest N, Santaella C, Boussif O, Vierling P. Polycationic telomers and cotelomers for gene transfer: Synthesis and evaluation of their *in vitro* transfection efficiency. *Bioconjug Chem* 2002;13(1):59–75.

53. Midoux P, Monsigny M. Efficient gene transfer by histidylated polylysine pDNA complexes. *Bioconjug Chem* 1999;10(3):406–411.

54. Pack DW, Putnam D, Langer R. Design of imidazole-containing endosomolytic biopolymers for gene delivery. *Biotechnol Bioeng* 2000;67(2):217–223.

55. Putnam D, Gentry CA, Pack DW, Langer R. Polymer-based gene delivery with low cytotoxicity by a unique balance of side-chain termini. *Proc Natl Acad Sci USA* 2001;98(3):1200–1205.

56. Lim YB, Kim SM, Suh H, Park JS. Biodegradable, endosome disruptive, and cationic network-type polymer as a highly efficient and nontoxic gene delivery carrier. *Bioconjug Chem* 2002;13(5):952–957.

57. Lim DW, Yeom YI, Park TG. Poly(DMAEMA-NVP)-b-PEG-galactose as gene delivery vector for hepatocytes. *Bioconjug Chem* 2000;11(5):688–695.

58. Kichler A, Sabourault N, Decor R, Leborgne C, Schmutz M, Valleix A, Danos O, Wagner A, Mioskowski C. Preparation and evaluation of a new class of gene transfer reagents: poly(alkylaminosiloxanes). *J Control Release* 2003;93(3):403–414.

59. Coeytaux E, Coulaud D, Le Cam E, Danos O, Kichler A. The cationic amphipathic alpha-helix of HIV-1 viral protein R (Vpr) binds to nucleic acids, permeabilizes membranes, and efficiently transfects cells. *J Biol Chem* 2003;278(20):18110–18116.

60. Kichler A, Leborgne C, Marz J, Danos O, Bechinger B. Histidine-rich amphipathic peptide antibiotics promote efficient delivery of DNA into mammalian cells. *Proc Natl Acad Sci USA* 2003;100(4):1564–1568.

61. van de Wetering P, Cherng JY, Talsma H, Crommelin DJ, Hennink WE. 2-(Dimethylamino)ethyl methacrylate based (co)polymers as gene transfer agents. *J Control Release* 1998;53(1–3):145–153.

62. Luten J, van Steenis JH, van Someren R, Kemmink J, Schuurmans-Nieuwenbroek NM, Koning GA, Crommelin DJ, van Nostrum CF, Hennink WE. Water-soluble biodegradable cationic polyphosphazenes for gene delivery. *J Control Release* 2003;89(3):483–497.

63. Xu Y, Szoka FC, Jr. Mechanism of DNA release from cationic liposome/DNA complexes used in cell transfection. *Biochemistry* 1996;35(18):5616–5623.

64. Zelphati O, Szoka FC, Jr. Mechanism of oligonucleotide release from cationic liposomes. *Proc Natl Acad Sci USA* 1996;93(21):11493–11498.

65. Zelphati O, Szoka FC, Jr. Intracellular distribution and mechanism of delivery of oligonucleotides mediated by cationic lipids. *Pharm Res* 1996;13(9):1367–1372.

66. Pedroso de Lima MC, Simoes S, Pires P, Faneca H, Duzgunes N. Cationic lipid–DNA complexes in gene delivery: From biophysics to biological applications. *Adv Drug Deliv Rev* 2001;47(2–3):277–294.

67. Lin AJ, Slack NL, Ahmad A, George CX, Samuel CE, Safinya CR. Three-dimensional imaging of lipid gene-carriers: Membrane charge density controls universal transfection behavior in lamellar cationic liposome–DNA complexes. *Biophys J* 2003;84(5):3307–3316.

68. Ewert K, Slack NL, Ahmad A, Evans HM, Lin AJ, Samuel CE, Safinya CR. Cationic lipid–DNA complexes for gene therapy: Understanding the relationship between complex structure and gene delivery pathways at the molecular level. *Curr Med Chem* 2004;11(2):133–149.

69. Ritter W, Plank C, Lausier J, Rudolph C, Zink D, Reinhardt D, Rosenecker J. A novel transfecting peptide comprising a tetrameric nuclear localization sequence. *J Mol Med* 2003;81(11):708–717.

70. Rudolph C, Plank C, Lausier J, Schillinger U, Muller RH, Rosenecker J. Oligomers of the arginine-rich Motif of the HIV-1 TAT protein are capable of transferring plasmid DNA into cells. *J Biol Chem* 2003;278(13):11411–11418.

71. Plank C, Scherer F, Rudolph C. Localized nucleic acid delivery: A discussion of selected methods. In Schleef M, editor. *DNA Pharmaceuticals*. Weinheim: Wiley-VCH Verlag GmbH & Co. KGaA; 2005, pp. 55–116.

72. Stoll SM, Calos MP. Extrachromosomal plasmid vectors for gene therapy. *Curr Opin Mol Ther* 2002;4(4):299–305.

73. Conese M, Auriche C, Ascenzioni F. Gene therapy progress and prospects: Episomally maintained self-replicating systems. *Gene Ther* 2004;11(24):1735–1741.

74. Recchia A, Perani L, Sartori D, Olgiati C, Mavilio F. Site-specific integration of functional transgenes into the human genome by adeno/AAV hybrid vectors. *Mol Ther* 2004;10(4):660–670.

75. Groth AC, Olivares EC, Thyagarajan B, Calos MP. A phage integrase directs efficient site-specific integration in human cells. *Proc Natl Acad Sci USA* 2000;97(11):5995–6000.

76. Thyagarajan B, Olivares EC, Hollis RP, Ginsburg DS, Calos MP. Site-specific genomic integration in mammalian cells mediated by phage phi C31 integrase. *Mol Cell Biol* 2001;21(12):3926–3934.

77. Plank C, Mechtler K, Szoka FC, Wagner E. Activation of the complement system by synthetic DNA complexes: A potential barrier for intravenous gene delivery. *Hum Gene Ther* 1996;7(12):1437–1446.

78. Krieg AM. CpG motifs in bacterial DNA and their immune effects. *Annu Rev Immunol* 2002;20:709–760.

79. Williams BR. Role of the double-stranded RNA-activated protein kinase (PKR) in cell regulation. *Biochem Soc Trans* 1997;25(2):509–513.

80. Kursa M, Walker GF, Roessler V, Ogris M, Roedl W, Kircheis R, Wagner E. Novel shielded transferrin–polyethylene glycol–polyethylenimine/DNA complexes for systemic tumor-targeted gene transfer. *Bioconjug Chem* 2003;14(1):222–231.

81. Ogris M, Walker G, Blessing T, Kircheis R, Wolschek M, Wagner E. Tumor-targeted gene therapy: Strategies for the preparation of ligand–polyethylene glycol-polyethylenimine/DNA complexes. *J Control Release* 2003;91(1–2):173–181.

82. Finsinger D, Remy JS, Erbacher P, Koch C, Plank C. Protective copolymers for nonviral gene vectors: Synthesis, vector characterization and application in gene delivery. *Gene Ther* 2000;7(14):1183–1192.

83. Oupicky D, Ogris M, Howard KA, Dash PR, Ulbrich K, Seymour LW. Importance of lateral and steric stabilization of polyelectrolyte gene delivery vectors for extended systemic circulation. *Mol Ther* 2002;5(4):463–472.

84. Chollet P, Favrot MC, Hurbin A, Coll JL. Side-effects of a systemic injection of linear polyethylenimine–DNA complexes. *J Gene Med* 2002;4(1):84–91.

85. Plank C, Anton M, Rudolph C, Rosenecker J, Krotz F. Enhancing and targeting nucleic acid delivery by magnetic force. *Expert Opin Biol Ther* 2003;3(5):745–758.

86. Zabner J, Fasbender AJ, Moninger T, Poellinger KA, Welsh MJ. Cellular and molecular barriers to gene transfer by a cationic lipid. *J Biol Chem* 1995;270(32):18997–19007.

87. Kichler A, Leborgne C, Coeytaux E, Danos O. Polyethylenimine-mediated gene delivery: A mechanistic study. *J Gene Med* 2001;3(2):135–144.

88. Pollard H, Remy JS, Loussouarn G, Demolombe S, Behr JP, Escande D. Polyethylenimine but not cationic lipids promotes transgene delivery to the nucleus in mammalian cells. *J Biol Chem* 1998;273(13):7507–7511.

89. Wagner E, Kircheis R, Walker GF. Targeted nucleic acid delivery into tumors: New avenues for cancer therapy. *Biomed Pharmacother* 2004;58(3):152–161.

90. Takakura Y, Nishikawa M, Yamashita F, Hashida M. Influence of physicochemical properties on pharmacokinetics of non-viral vectors for gene delivery. *J Drug Target* 2002;10(2):99–104.

91. Boulikas T. Nuclear localization signal peptides for the import of plasmid DNA in gene therapy (review). *Int J Oncol* 1997;10(2):301–309.

92. Cartier R, Reszka R. Utilization of synthetic peptides containing nuclear localization signals for nonviral gene transfer systems. *Gene Ther* 2002;9(3):157–167.

93. Escriou V, Carriere M, Scherman D, Wils P. NLS bioconjugates for targeting therapeutic genes to the nucleus. *Adv Drug Deliv Rev* 2003;55(2):295–306.

94. Hebert E. Improvement of exogenous DNA nuclear importation by nuclear localization signal-bearing vectors: A promising way for non-viral gene therapy? *Biol Cell* 2003;95(2):59–68.

95. Wells DJ. Gene therapy progress and prospects: electroporation and other physical methods. *Gene Ther* 2004;11(18):1363–1369.

96. Olivares EC, Calos MP. Phage C31 integrase-mediated site-specic integration for gene therapy. *Gene Ther Regul* 2003;2(2):103–120.

97. Ogris M, Steinlein P, Kursa M, Mechtler K, Kircheis R, Wagner E. The size of DNA/transferrin–PEI complexes is an important factor for gene expression in cultured cells. *Gene Ther* 1998;5(10):1425–1433.

98. Luo D, Saltzman WM. Enhancement of transfection by physical concentration of DNA at the cell surface. *Nat Biotechnol* 2000;18(8):893–895.

99. Bunnell BA, Muul LM, Donahue RE, Blaese RM, Morgan RA. High-efficiency retroviral-mediated gene transfer into human and nonhuman primate peripheral blood lymphocytes. *Proc Natl Acad Sci USA* 1995;92(17):7739–7743.

100. Boussif O, Zanta MA, Behr JP. Optimized galenics improve *in vitro* gene transfer with cationic molecules up to 1000-fold. *Gene Ther* 1996;3(12):1074–1080.

101. O'Doherty U, Swiggard WJ, Malim MH. Human immunodeficiency virus type 1 spinoculation enhances infection through virus binding. *J Virol* 2000;74(21):10074–10080.

102. Huth S, Lausier J, Gersting SW, Rudolph C, Plank C, Welsch U, Rosenecker J. Insights into the mechanism of magnetofection using PEI-based magnetofectins for gene transfer. *J Gene Med* 2004;6(8):923–936.

103. Plank C, Schillinger U, Scherer F, Bergemann C, Remy JS, Krotz F, Anton M, Lausier J, Rosenecker J. The magnetofection method: using magnetic force to enhance gene delivery. *Biol Chem* 2003;384(5):737–747.

104. Chuck AS, Palsson BO. Consistent and high rates of gene transfer can be obtained using flow-through transduction over a wide range of retroviral titers. *Hum Gene Ther* 1996;7(6):743–750.

105. Williams AR, Bao S, Miller DL. Filtroporation: A simple, reliable technique for transfection and macromolecular loading of cells in suspension. *Biotechnol Bioeng* 1999;65(3):341–346.

106. Liu D, Knapp JE. Hydrodynamics-based gene delivery. *Curr Opin Mol Ther* 2001;3(2):192–197.

107. Isner JM. Myocardial gene therapy. *Nature* 2002;415(6868):234–239.

108. Herweijer H, Wolff JA. Progress and prospects: Naked DNA gene transfer and therapy. *Gene Ther* 2003;10(6):453–458.

109. Hodges BL, Scheule RK. Hydrodynamic delivery of DNA. *Expert Opin Biol Ther* 2003;3(6):911–918.

110. Koshkina NV, Agoulnik IY, Melton SL, Densmore CL, Knight V. Biodistribution and pharmacokinetics of aerosol and intravenously administered DNA–polyethyleneimine complexes: Optimization of pulmonary delivery and retention. *Mol Ther* 2003;8(2):249–254.

111. Rudolph C, Ortiz A, Schillinger U, Jauernig J, Plank C, Rosenecker J. Methodological optimization of polyethylenimine (PEI)-based gene delivery to the lungs of mice via aerosol application. *J Gene Med* 2004;7(1):59–66.

112. Klein TM, Arentzen R, Lewis PA, Fitzpatrick-McElligott S. Transformation of microbes, plants and animals by particle bombardment. *Biotechnology (NY)* 1992;10(3):286–291.

113. Bonadio J. Genetic approaches to tissue repair. *Ann NY Acad Sci* 2002;961:58–60.

114. Sharif F, Daly K, Crowley J, O'Brien T. Current status of catheter- and stent-based gene therapy. *Cardiovasc Res* 2004;64(2):208–216.

115. Eliaz RE, Szoka FC, Jr. Robust and prolonged gene expression from injectable polymeric implants. *Gene Ther* 2002;9(18):1230–1237.

116. Christman KL, Fang Q, Yee MS, Johnson KR, Sievers RE, Lee RJ. Enhanced neovasculature formation in ischemic myocardium following delivery of pleiotrophin plasmid in a biopolymer. *Biomaterials* 2005;26(10):1139–1144.

117. Scherer F, Schillinger U, Putz U, Stemberger A, Plank C. Nonviral vector loaded collagen sponges for sustained gene delivery *in vitro* and *in vivo*. *J Gene Med* 2002;4(6):634–643.

118. Bonadio J, Smiley E, Patil P, Goldstein S. Localized, direct plasmid gene delivery *in vivo*: Prolonged therapy results in reproducible tissue regeneration. *Nat Med* 1999;5(7):753–759.

119. Goldstein SA, Bonadio J. Potential role for direct gene transfer in the enhancement of fracture healing. *Clin Orthop* 1998 (355 Suppl):S154–S162.

120. Goldstein SA. *In vivo* nonviral delivery factors to enhance bone repair. *Clin Orthop* 2000 (379 Suppl):S113–S119.

121. Kyriakides TR, Hartzel T, Huynh G, Bornstein P. Regulation of angiogenesis and matrix remodeling by localized, matrix-mediated antisense gene delivery. *Mol Ther* 2001;3(6):842–849.

122. Tyrone JW, Mogford JE, Chandler LA, Ma C, Xia Y, Pierce GF, Mustoe TA. Collagen-embedded platelet-derived growth factor DNA plasmid promotes wound healing in a dermal ulcer model. *J Surg Res* 2000;93(2):230–236.

123. Berry M, Gonzalez AM, Clarke W, Greenlees L, Barrett L, Tsang W, Seymour L, Bonadio J, Logan A, Baird A. Sustained effects of gene-activated matrices after CNS injury. *Mol Cell Neurosci* 2001;17(4):706–716.

124. Pakkanen TM, Laitinen M, Hippelainen M, Hiltunen MO, Alhava E, Yla-Herttuala S. Periadventitial lacZ gene transfer to pig carotid arteries using a biodegradable collagen collar or a wrap of collagen sheet with adenoviruses and plasmid–liposome complexes. *J Gene Med* 2000;2(1):52–60.

125. Chandler LA, Gu DL, Ma C, Gonzalez AM, Doukas J, Nguyen T, Pierce GF, Phillips ML. Matrix-enabled gene transfer for cutaneous wound repair. *Wound Repair Regen* 2000;8(6):473–479.

126. Gu DL, Nguyen T, Gonzalez AM, Printz MA, Pierce GF, Sosnowski BA, Phillips ML, Chandler LA. Adenovirus encoding human platelet-derived growth factor-B delivered in collagen exhibits safety, biodistribution, and immunogenicity profiles favorable for clinical use. *Mol Ther* 2004;9(5):699–711.

127. Doukas J, Chandler LA, Gonzalez AM, Gu D, Hoganson DK, Ma C, Nguyen T, Printz MA, Nesbit M, Herlyn M, Crombleholme TM, Aukerman SL, Sosnowski BA, Pierce GF. Matrix immobilization enhances the tissue repair activity of growth factor gene therapy vectors. *Hum Gene Ther* 2001;12(7):783–798.

128. Takahashi A, Palmer-Opolski M, Smith RC, Walsh K. Transgene delivery of plasmid DNA to smooth muscle cells and macrophages from a biostable polymer-coated stent. *Gene Ther* 2003;10(17):1471–1478.

129. Klugherz BD, Jones PL, Cui X, Chen W, Meneveau NF, DeFelice S, Connolly J, Wilensky RL, Levy RJ. Gene delivery from a DNA controlled-release stent in porcine coronary arteries. *Nat Biotechnol* 2000;18(11):1181–1184.

130. Somiari S, Glasspool-Malone J, Drabick JJ, Gilbert RA, Heller R, Jaroszeski MJ, Malone RW. Theory and *in vivo* application of electroporative gene delivery. *Mol Ther* 2000;2(3):178–187.

131. Bloquel C, Fabre E, Bureau MF, Scherman D. Plasmid DNA electrotransfer for intracellular and secreted proteins expression: New methodological developments and applications. *J Gene Med* 2004;6(Suppl 1):S11–S23.

132. Reynolds PN, Nicklin SA, Kaliberova L, Boatman BG, Grizzle WE, Balyasnikova IV, Baker AH, Danilov SM, Curiel DT. Combined transductional and transcriptional targeting improves the specificity of transgene expression *in vivo*. *Nat Biotechnol* 2001;19(9):838–842.

133. Dhawan J, Rando TA, Elson SL, Bujard H, Blau HM. Tetracycline-regulated gene expression following direct gene transfer into mouse skeletal muscle. *Somat Cell Mol Genet* 1995;21(4):233–240.

134. Toniatti C, Bujard H, Cortese R, Ciliberto G. Gene therapy progress and prospects: Transcription regulatory systems. *Gene Ther* 2004;11(8):649–657.

135. Goodman R, Blank M. Insights into electromagnetic interaction mechanisms. *J Cell Physiol* 2002;192(1):16–22.

136. Guilhon E, Voisin P, de Zwart JA, Quesson B, Salomir R, Maurange C, Bouchaud V, Smirnov P, de Verneuil H, Vekris A, Canioni P, Moonen CT. Spatial and temporal control of transgene expression *in vivo* using a heat-sensitive promoter and MRI-guided focused ultrasound. *J Gene Med* 2003;5(4):333–342.

137. Ito A, Shinkai M, Honda H, Kobayashi T. Heat-inducible TNF-alpha gene therapy combined with hyperthermia using magnetic nanoparticles as a novel tumor-targeted therapy. *Cancer Gene Ther* 2001;8(9):649–654.

138. Binley K, Askham Z, Martin L, Spearman H, Day D, Kingsman S, Naylor S. Hypoxia-mediated tumour targeting. *Gene Ther* 2003;10(7):540–549.

139. Stacy DR, Lu B, Hallahan DE. Radiation-guided drug delivery systems. *Expert Rev Anticancer Ther* 2004;4(2):283–288.

140. Li Z, Ning W, Wang J, Choi A, Lee PY, Tyagi P, Huang L. Controlled gene delivery system based on thermosensitive biodegradable hydrogel. *Pharm Res* 2003;20(6):884–888.

141. Klibanov AL. Ultrasound contrast agents: Development of the field and current status. *Top Curr Chem* 2002;222:73–106.

142. Bekeredjian R, Chen S, Frenkel PA, Grayburn PA, Shohet RV. Ultrasound-targeted microbubble destruction can repeatedly direct highly specific plasmid expression to the heart. *Circulation* 2003;108(8):1022–1026.

143. Unger EC, Porter T, Culp W, Labell R, Matsunaga T, Zutshi R. Therapeutic applications of lipid-coated microbubbles. *Adv Drug Deliv Rev* 2004;56(9):1291–1314.

144. Tsutsui JM, Xie F, Porter RT. The use of microbubbles to target drug delivery. *Cardiovasc Ultrasound* 2004;2(1):23.

145. Zou SM, Erbacher P, Remy JS, Behr JP. Systemic linear polyethylenimine (L-PEI)-mediated gene delivery in the mouse. *J Gene Med* 2000;2(2):128–134.

146. Barron LG, Gagne L, Szoka FC, Jr. Lipoplex-mediated gene delivery to the lung occurs within 60 minutes of intravenous administration. *Hum Gene Ther* 1999;10(10):1683–1694.

147. Oh YK, Kim JP, Yoon H, Kim JM, Yang JS, Kim CK. Prolonged organ retention and safety of plasmid DNA administered in polyethylenimine complexes. *Gene Ther* 2001; 8(20):1587–1592.

148. Kircheis R, Schuller S, Brunner S, Ogris M, Heider KH, Zauner W, Wagner E. Polycation-based DNA complexes for tumor-targeted gene delivery *in vivo*. *J Gene Med* 1999;1(2):111–120.

149. Ogris M, Brunner S, Schuller S, Kircheis R, Wagner E. PEGylated DNA/transferrin-PEI complexes: Reduced interaction with blood components, extended circulation in blood and potential for systemic gene delivery. *Gene Ther* 1999;6(4):595–605.

150. Hood JD, Bednarski M, Frausto R, Guccione S, Reisfeld RA, Xiang R, Cheresh DA. Tumor regression by targeted gene delivery to the neovasculature. *Science* 2002; 296(5577):2404–2407.

151. Darji A, Guzman CA, Gerstel B, Wachholz P, Timmis KN, Wehland J, Chakraborty T, Weiss S. Oral somatic transgene vaccination using attenuated S. typhimurium. *Cell* 1997;91(6):765–775.

152. Niethammer AG, Xiang R, Ruehlmann JM, Lode HN, Dolman CS, Gillies SD, Reisfeld RA. Targeted interleukin 2 therapy enhances protective immunity induced by an autologous oral DNA vaccine against murine melanoma. *Cancer Res* 2001;61(16):6178–6184.

153. Niethammer AG, Xiang R, Becker JC, Wodrich H, Pertl U, Karsten G, Eliceiri BP, Reisfeld RA. A DNA vaccine against VEGF receptor 2 prevents effective angiogenesis and inhibits tumor growth. *Nat Med* 2002;4:4.

154. Reisfeld RA, Niethammer AG, Luo Y, Xiang R. DNA vaccines suppress tumor growth and metastases by the induction of anti-angiogenesis. *Immunol Rev* 2004;199:181–190.

155. Fu GF, Li X, Hou YY, Fan YR, Liu WH, Xu GX. Bifidobacterium longum as an oral delivery system of endostatin for gene therapy on solid liver cancer. *Cancer Gene Ther* 2004.

156. Zhao J, Lou Y, Pinczewski J, Malkevitch N, Aldrich K, Kalyanaraman VS, Venzon D, Peng B, Patterson LJ, Edghill-Smith Y, Woodward R, Pavlakis GN, Robert-Guroff M. Boosting of SIV-specific immune responses in rhesus macaques by repeated administration of Ad5hr-SIVenv/rev and Ad5hr-SIVgag recombinants. *Vaccine* 2003;21(25–26):4022–4035.

157. Fayad R, Zhang H, Quinn D, Huang Y, Qiao L. Oral administration with papillomavirus pseudovirus encoding IL-2 fully restores mucosal and systemic immune responses to vaccinations in aged mice. *J Immunol* 2004;173(4):2692–2698.

158. Kai E, Ochiya T. A method for oral DNA delivery with *N*-acetylated chitosan. *Pharm Res* 2004;21(5):838–843.

159. Hejazi R, Amiji M. Chitosan-based gastrointestinal delivery systems. *J Control Release* 2003;89(2):151–165.

160. Roy K, Mao HQ, Huang SK, Leong KW. Oral gene delivery with chitosan–DNA nanoparticles generates immunologic protection in a murine model of peanut allergy. *Nat Med* 1999;5(4):387–391.

161. Jones DH, Corris S, McDonald S, Clegg JC, Farrar GH. Poly(DL-lactide-co-glycolide)-encapsulated plasmid DNA elicits systemic and mucosal antibody responses to encoded protein after oral administration. *Vaccine* 1997;15(8):814–817.

162. Chen SC, Jones DH, Fynan EF, Farrar GH, Clegg JC, Greenberg HB, Herrmann JE. Protective immunity induced by oral immunization with a rotavirus DNA vaccine encapsulated in microparticles. *J Virol* 1998;72(7):5757–5761.

58   NUCLEIC ACID DELIVERY AND LOCALIZING DELIVERY WITH MAGNETIC NANOPARTICLES

163. Mittal SK, Aggarwal N, Sailaja G, van Olphen A, HogenEsch H, North A, Hays J, Moffatt S. Immunization with DNA, adenovirus or both in biodegradable alginate microspheres: Effect of route of inoculation on immune response. *Vaccine* 2000;19(2–3):253–263.

164. Moore RA, Walcott S, White KL, Anderson DM, Jain S, Lloyd A, Topley P, Thomsen L, Gough GW, Stanley MA. Therapeutic immunisation with COPV early genes by epithelial DNA delivery. *Virology* 2003;314(2):630–635.

165. Chang SF, Chang HY, Tong YC, Chen SH, Hsaio FC, Lu SC, Liaw J. Nonionic polymeric micelles for oral gene delivery *in vivo*. *Hum Gene Ther* 2004;15(5):481–493.

166. Howard KA, Li XW, Somavarapu S, Singh J, Green N, Atuah KN, Ozsoy Y, Seymour LW, Alpar HO. Formulation of a microparticle carrier for oral polyplex-based DNA vaccines. *Biochim Biophys Acta* 2004;1674(2):149–157.

167. Takamura S, Niikura M, Li TC, Takeda N, Kusagawa S, Takebe Y, Miyamura T, Yasutomi Y. DNA vaccine-encapsulated virus-like particles derived from an orally transmissible virus stimulate mucosal and systemic immune responses by oral administration. *Gene Ther* 2004;11(7):628–635.

168. Gander B. Trends in particulate antigen and DNA delivery systems for vaccines. *Adv Drug Deliv Rev* 2005;57(3):321–323.

169. Scherer F, Anton M, Schillinger U, Henke J, Bergemann C, Kruger A, Gansbacher B, Plank C. Magnetofection: Enhancing and targeting gene delivery by magnetic force *in vitro* and *in vivo*. *Gene Ther* 2002;9(2):102–109.

170. Alksne JF, Fingerhut AG. Magnetically controlled metallic thrombosis of intracranial aneurysms. A preliminary report. *Bull Los Angeles Neurol Soc* 1965;30(3):153–155.

171. Fingerhut AG, Alksne JF. Thrombosis of intracranial aneurysms. An experimental approach utilizing magnetically controlled iron particles. *Radiology* 1966;86(2):342–343.

172. Alksne JF, Fingerhut AG, Rand RW. Magnetically controlled focal intravascular thrombosis in dogs. *J Neurosurg* 1966;25(5):516–525.

173. Meyers PH, Nice CM, Jr., Meckstroth GR, Becker HC, Moser PJ, Goldstein M. Pathologic studies following magnetic control of metallic iron particles in the lymphatic and vascular system of dogs as a contrast and isotopic agent. *Am J Roentgenol Radium Ther Nucl Med* 1966;96(4):913–921.

174. Widder KJ, Senyel AE, Scarpelli GD. Magnetic microspheres: A model system of site specific drug delivery *in vivo*. *Proc Soc Exp Biol Med* 1978;158(2):141–146.

175. Fahlvik AK, Klaveness J, Stark DD. Iron oxides as MR imaging contrast agents. *J Magn Reson Imaging* 1993;3(1):187–194.

176. Carlin RL. *Magnetochemistry*. Heidelberg: Springer; 1986.

177. Weiss A, Witte H. *Magnetochemie: Grundlagen und Anwendungen*. Weinheim: Wiley/VCH; 1997.

178. Senyei AE, Reich SD, Gonczy C, Widder KJ. *In vivo* kinetics of magnetically targeted low-dose doxorubicin. *J Pharm Sci* 1981;70(4):389–391.

179. Widder KJ, Morris RM, Poore G, Howard DP, Jr., Senyei AE. Tumor remission in Yoshida sarcoma-bearing rts by selective targeting of magnetic albumin microspheres containing doxorubicin. *Proc Natl Acad Sci USA* 1981;78(1):579–581.

180. Widder KJ, Morris RM, Poore GA, Howard DP, Senyei AE. Selective targeting of magnetic albumin microspheres containing low-dose doxorubicin: Total remission in Yoshida sarcoma-bearing rats. *Eur J Cancer Clin Oncol* 1983;19(1):135–139.

181. Gupta PK, Hung CT. Comparative disposition of adriamycin delivered via magnetic albumin microspheres in presence and absence of magnetic field in rats. *Life Sci* 1990;46(7):471–479.

182. Gupta PK, Hung CT. Effect of carrier dose on the multiple tissue disposition of doxorubicin hydrochloride administered via magnetic albumin microspheres in rats. *J Pharm Sci* 1989;78(9):745–748.

183. Gupta PK, Hung CT, Rao NS. Ultrastructural disposition of adriamycin-associated magnetic albumin microspheres in rats. *J Pharm Sci* 1989;78(4):290–294.

184. Gupta PK, Hung CT. Magnetically controlled targeted micro-carrier systems. *Life Sci* 1989;44(3):175–186.

185. Gupta PK, Hung CT. Targeted delivery of low dose doxorubicin hydrochloride administered via magnetic albumin microspheres in rats. *J Microencapsul* 1990;7(1):85–94.

186. Kato T, Nemoto R, Mori H, Unno K, Goto A, Homma M. [An approach to magnetically controlled cancer chemotherapy. I. Preparation and properties of ferromagnetic mitomycin C microcapsules (author's transl)]. *Nippon Gan Chiryo Gakkai Shi* 1980;15(5):876–880.

187. Kato T, Nemoto R, Mori H, Abe R, Unno K, Goto A, Murota H, Harada M, Homma M. Magnetic microcapsules for targeted delivery of anticancer drugs. *Appl Biochem Biotechnol* 1984;10:199–211.

188. Lubbe AS, Bergemann C, Huhnt W, Fricke T, Riess H, Brock JW, Huhn D. Preclinical experiences with magnetic drug targeting: Tolerance and efficacy. *Cancer Res* 1996; 56(20):4694–4701.

189. Lubbe AS, Bergemann C, Riess H, Schriever F, Reichardt P, Possinger K, Matthias M, Dorken B, Herrmann F, Gurtler R, Hohenberger P, Haas N, Sohr R, Sander B, Lemke AJ, Ohlendorf D, Huhnt W, Huhn D. Clinical experiences with magnetic drug targeting: a phase I study with 4'-epidoxorubicin in 14 patients with advanced solid tumors. *Cancer Res* 1996;56(20):4686–4693.

190. Lübbe AS, Bergemann C. Selected preclinical and first clinical experiences with magnetically targeted 4-epidoxorubicin in patients with advanced solid tumors. In Häfeli U, Schütt W, Teller J, Zborowski M, editors. *Scientific and Clinical Applications of Magnetic Drug Carriers*. New York, London: Plenum Press; 1997, pp. 457–480.

191. Lemke AJ, Senfft von Pilsach MI, Lubbe A, Bergemann C, Riess H, Felix R. MRI after magnetic drug targeting in patients with advanced solid malignant tumors. *Eur Radiol* 2004;14(11):1949–1955.

192. Goodwin SC, Bittner CA, Peterson CL, Wong G. Single-dose toxicity study of hepatic intra-arterial infusion of doxorubicin coupled to a novel magnetically targeted drug carrier. *Toxicol Sci* 2001;60(1):177–183.

193. Goodwin SC, Peterson C, Hoh C, Bittner CA. Targeting and retention of magnetic targeted carriers (MTCs) enhancing intra-arterial chemotherapy. *J Magn Magn Mat* 1999;194:132–139.

194. Rudge SR, Kurtz TL, Vessely CR, Catterall LG, Williamson DL. Preparation, characterization, and performance of magnetic iron–carbon composite microparticles for chemotherapy. *Biomaterials* 2000;21(14):1411–1420.

195. Rudge S, Peterson C, Vessely C, Koda J, Stevens S, Catterall L. Adsorption and desorption of chemotherapeutic drugs from a magnetically targeted carrier (MTC). *J Control Release* 2001;74(1–3):335–340.

196. Johnson J, Kent T, Koda J, Peterson C, Rudge S, Tapolsky G. The MTC technology: A platform technology for the site-specific delivery of pharmaceutical agents. In *4th*

International Conference on the Scientific and Clinical Applications of Magnetic Carriers, 2002; Tallahassee: European Cells and Materials; 2002, pp. 12–15.

197. Lubbe AS, Alexiou C, Bergemann C. Clinical applications of magnetic drug targeting. *J Surg Res* 2001;95(2):200–206.

198. Widder KJ, Marino PA, Morris RM, Howard DP, Poore GA, Senyei AE. Selective targeting of magnetic albumin microspheres to the Yoshida sarcoma: Ultrastructural evaluation of microsphere disposition. *Eur J Cancer Clin Oncol* 1983;19(1):141–147.

199. Haim H, Steiner I, Panet A. Synchronized infection of cell cultures by magnetically controlled virus. *J Virol* 2005;79(1):622–625.

200. Wu J, Bratko D, Prausnitz JM. Interaction between like-charged colloidal spheres in electrolyte solutions. *Proc Natl Acad Sci USA* 1998;95(26):15169–15172.

201. Hughes C, Galea-Lauri J, Farzaneh F, Darling D. Streptavidin paramagnetic particles provide a choice of three affinity-based capture and magnetic concentration strategies for retroviral vectors. *Mol Ther* 2001;3(4):623–630.

202. Mah C, Fraites TJJ, Zolotukhin I, Song S, Flotte TR, Jon Dobson, Batich C, Byrne BJ. Improved method of recombinant AAV2 delivery for systemic targeted gene therapy. *Mol Ther* 2002;6(1):106–112.

203. Pandori MW, Hobson DA, Sano T. Adenovirus-microbead conjugates process enhanced infectivity: A new strategy to localized gene delivery. *Virology* 2002;299:204–212.

204. Raty JK, Airenne KJ, Marttila AT, Marjomaki V, Hytonen VP, Lehtolainen P, Laitinen OH, Mahonen AJ, Kulomaa MS, Yla-Herttuala S. Enhanced gene delivery by avidin-displaying baculovirus. *Mol Ther* 2004;9(2):282–291.

205. Chan L, Nesbeth D, Mackey T, Galea-Lauri J, Gaken J, Martin F, Collins M, Mufti G, Farzaneh F, Darling D. Conjugation of lentivirus to paramagnetic particles via nonviral proteins allows efficient concentration and infection of primary acute myeloid leukemia cells. *J Virol* 2005;79(20):13190–13194.

206. Krotz F, Wit C, Sohn HY, Zahler S, Gloe T, Pohl U, Plank C. Magnetofection—A highly efficient tool for antisense oligonucleotide delivery *in vitro* and *in vivo*. *Mol Ther* 2003;7(5):700–710.

207. Krotz F, Sohn HY, Gloe T, Plank C, Pohl U. Magnetofection potentiates gene delivery to cultured endothelial cells. *J Vasc Res* 2003;40(5):425–434.

208. Gersting SW, Schillinger U, Lausier J, Nicklaus P, Rudolph C, Plank C, Reinhardt D, Rosenecker J. Gene delivery to respiratory epithelial cells by magnetofection. *J Gene Med* 2004;6(8):913–922.

209. Schillinger U, Brill T, Rudolph C, Huth S, Gersting S, Krotz F, Hirschberger J, Bergemann C, Plank C. Advances in magnetofection—magnetically guided nucleic acid delivery. *J Magn Magn Mat* 2005;293:501–508.

210. Couto SS, Griffey SM, Duarte PC, Madewell BR. Feline vaccine-associated fibrosarcoma: Morphologic distinctions. *Vet Pathol* 2002;39(1):33–41.

211. Zborowski M, Fuh CB, Green R, Sun L, Chalmers JJ. Analytical magnetapheresis of ferritin-labeled lymphocytes. *Anal Chem* 1995;67(20):3702–3712.

212. Voltairas PA, Fotiadis DI, Michalis LK. Hydrodynamics of magnetic drug targeting. *J Biomech* 2002;35(6):813–821.

213. Babincova M, Babinec P. Possibility of magnetic targeting of drugs using magnetoliposomes. *Pharmazie* 1995;50(12):828–829.

214. Nagel S. *Theoretische und experimentelle Untersuchungen zum Magnetischen Drug Targeting.* Greifswald: Ernst-Moritz-Arndt-Universität Greifswald; 2004.

215. Babincova M, Babinec P, Bergemann C. High-gradient magnetic capture of ferrofluids: Implications for drug targeting and tumor embolization. *Z Naturforsch [C]* 2001;56(9–10):909–911.

216. Alexiou C, Diehl D, Henninger P, Iro H, Röckelein R, Schmidt W, Weber H. A high field gradient magnet for magnetic drug targeting. *IEEE T Appl Supercon* 2006;16(2):1527–1530.

217. Babincova M, Altanerova V, Lampert M, Altaner C, Machova E, Sramka M, Babinec P. Site-specific *in vivo* targeting of magnetoliposomes using externally applied magnetic field. *Z Naturforsch [C]* 2000;55(3–4):278–281.

218. Babincova M, Cicmanec P, Altanerova V, Altaner C, Babinec P. AC-magnetic field controlled drug release from magnetoliposomes: Design of a method for site-specific chemotherapy. *Bioelectrochemistry* 2002;55(1–2):17–19.

219. Schutt EG, Klein DH, Mattrey RM, Riess JG. Injectable microbubbles as contrast agents for diagnostic ultrasound imaging: The key role of perfluorochemicals. *Angew Chem Int Ed Engl* 2003;42(28):3218–3235.

220. Gramiak R, Shah PM, Kramer DH. Ultrasound cardiography: Contrast studies in anatomy and function. *Radiology* 1969;92(5):939–948.

221. Kremkau FW, Gramiak R, Carstensen EL, Shah PM, Kramer DH. Ultrasonic detection of cavitation at catheter tips. *Am J Roentgenol Radium Ther Nucl Med* 1970;110(1):177–183.

222. Mattrey RF, Wrigley R, Steinbach GC, Schutt EG, Evitts DP. Gas emulsions as ultrasound contrast agents. Preliminary results in rabbits and dogs. *Invest Radiol* 1994;29(Suppl 2):S139–S141.

223. El-Sherif DM, Wheatley MA. Development of a novel method for synthesis of a polymeric ultrasound contrast agent. *J Biomed Mater Res A* 2003;66(2):347–355.

224. Forsberg F, Lathia JD, Merton DA, Liu JB, Le NT, Goldberg BB, Wheatley MA. Effect of shell type on the *in vivo* backscatter from polymer-encapsulated microbubbles. *Ultrasound Med Biol* 2004;30(10):1281–1287.

225. Unger EC, McCreery T, Sweitzer R, Vielhauer G, Wu G, Shen D, Yellowhair D. MRX 501: A novel ultrasound contrast agent with therapeutic properties. *Acad Radiol* 1998;5(Suppl 1):S247–S249.

226. Unger EC, Hersh E, Vannan M, Matsunaga TO, McCreery T. Local drug and gene delivery through microbubbles. *Prog Cardiovasc Dis* 2001;44(1):45–54.

227. Klibanov AL. Targeted delivery of gas-filled microspheres, contrast agents for ultrasound imaging. *Adv Drug Deliv Rev* 1999;37(1–3):139–157.

228. Unger EC, Matsunaga TO, McCreery T, Schumann P, Sweitzer R, Quigley R. Therapeutic applications of microbubbles. *Eur J Radiol* 2002;42(2):160–168.

229. Chen S, Shohet RV, Bekeredjian R, Frenkel P, Grayburn PA. Optimization of ultrasound parameters for cardiac gene delivery of adenoviral or plasmid deoxyribonucleic acid by ultrasound-targeted microbubble destruction. *J Am Coll Cardiol* 2003;42(2):301–308.

230. Skyba DM, Price RJ, Linka AZ, Skalak TC, Kaul S. Direct *in vivo* visualization of intravascular destruction of microbubbles by ultrasound and its local effects on tissue. *Circulation* 1998;98(4):290–293.

231. Price RJ, Skyba DM, Kaul S, Skalak TC. Delivery of colloidal particles and red blood cells to tissue through microvessel ruptures created by targeted microbubble destruction with ultrasound. *Circulation* 1998;98(13):1264–1267.

232. Hynynen K, McDannold N, Vykhodtseva N, Jolesz FA. Noninvasive MR imaging-guided focal opening of the blood-brain barrier in rabbits. *Radiology* 2001;220(3):640–646.

233. Tata DB, Dunn F, Tindall DJ. Selective clinical ultrasound signals mediate differential gene transfer and expression in two human prostate cancer cell lines: LnCap and PC-3. *Biochem Biophys Res Commun* 1997;234(1):64–67.

234. Unger EC, McCreery TP, Sweitzer RH. Ultrasound enhances gene expression of liposomal transfection. *Invest Radiol* 1997;32(12):723–727.

235. Schratzberger P, Krainin JG, Schratzberger G, Silver M, Ma H, Kearney M, Zuk RF, Brisken AF, Losordo DW, Isner JM. Transcutaneous ultrasound augments naked DNA transfection of skeletal muscle. *Mol Ther* 2002;6(5):576–583.

236. Anwer K, Kao G, Proctor B, Anscombe I, Florack V, Earls R, Wilson E, McCreery T, Unger E, Rolland A, Sullivan SM. Ultrasound enhancement of cationic lipid-mediated gene transfer to primary tumors following systemic administration. *Gene Ther* 2000;7(21):1833–1839.

237. Teupe C, Richter S, Fisslthaler B, Randriamboavonjy V, Ihling C, Fleming I, Busse R, Zeiher AM, Dimmeler S. Vascular gene transfer of phosphomimetic endothelial nitric oxide synthase (S1177D) using ultrasound-enhanced destruction of plasmid-loaded microbubbles improves vasoreactivity. *Circulation* 2002;105(9):1104–1109.

238. Vannan M, McCreery T, Li P, Han Z, Unger E, Kuersten B, Nabel E, Rajagopalan S. Ultrasound-mediated transfection of canine myocardium by intravenous administration of cationic microbubble-linked plasmid DNA. *J Am Soc Echocardiogr* 2002;15(3):214–218.

239. Lawrie A, Brisken AF, Francis SE, Cumberland DC, Crossman DC, Newman CM. Microbubble-enhanced ultrasound for vascular gene delivery. *Gene Ther* 2000;7(23):2023–2027.

240. Frenkel PA, Chen S, Thai T, Shohet RV, Grayburn PA. DNA-loaded albumin microbubbles enhance ultrasound-mediated transfection *in vitro*. *Ultrasound Med Biol* 2002;28(6):817–822.

241. Shohet RV, Chen S, Zhou YT, Wang Z, Meidell RS, Unger RH, Grayburn PA. Echocardiographic destruction of albumin microbubbles directs gene delivery to the myocardium. *Circulation* 2000;101(22):2554–2556.

242. Song J, Tata D, Li L, Taylor J, Bao S, Miller DL. Combined shock-wave and immunogene therapy of mouse melanoma and renal carcinoma tumors. *Ultrasound Med Biol* 2002;28(7):957–964.

243. Miller DL, Pislaru SV, Greenleaf JE. Sonoporation: Mechanical DNA delivery by ultrasonic cavitation. *Somat Cell Mol Genet* 2002;27(1–6):115–134.

244. Danialou G, Comtois AS, Dudley RW, Nalbantoglu J, Gilbert R, Karpati G, Jones DH, Petrof BJ. Ultrasound increases plasmid-mediated gene transfer to dystrophic muscles without collateral damage. *Mol Ther* 2002;6(5):687–693.

245. Lu QL, Liang HD, Partridge T, Blomley MJ. Microbubble ultrasound improves the efficiency of gene transduction in skeletal muscle *in vivo* with reduced tissue damage. *Gene Ther* 2003;10(5):396–405.

246. Endoh M, Koibuchi N, Sato M, Morishita R, Kanzaki T, Murata Y, Kaneda Y. Fetal gene transfer by intrauterine injection with microbubble-enhanced ultrasound. *Mol Ther* 2002;5(5 Pt 1):501–508.

247. Fischer A, Hacein-Bey S, Cavazzana-Calvo M. Gene therapy of severe combined immun-odeficiencies. *Nat Rev Immunol* 2002;2(8):615–621.

248. Hacein-Bey-Abina S, von Kalle C, Schmidt M, Le Deist F, Wulffraat N, McIntyre E, Radford I, Villeval JL, Fraser CC, Cavazzana-Calvo M, Fischer A. A serious adverse event after successful gene therapy for X-linked severe combined immunodeficiency. *N Engl J Med* 2003;348(3):255–256.

249. Somia N, Verma IM. Gene therapy: Trials and tribulations. *Nat Rev Genet* 2000;1(2):91–99.

3

MAGNETIC NANOPARTICLES IN CANCER DIAGNOSIS AND HYPERTHERMIC TREATMENT

Robert H. Kraus, Jr. and Bradford Wright

3.1 INTRODUCTION

The publication of this book signifies the present and growing importance of nanotechnologies in the biosciences in general and medicine in particular. The diagnosis and treatment of tumors is one field in which an array of nanotechnologies are being applied. For example, nanoparticles can be targeted to tumors to enhance diagnosis and localization of tumors by many different imaging modalities [e.g., optical, magnetic resonance imaging (MRI), positron emission tomography (PET), single photon emission computed tomography (SPECT), computed tomography (CT), and X-ray]. Successful targeting of nanoparticles typically requires long resident times in the circulatory system to allow efficient uptake by a desired cell or cell receptor population. The rapid recognition and collection of most intravenously injected colloidal particles by Kupffer cells, for example, often impedes efficacy of a given colloidally based diagnostic or therapeutic. Moghimi et al. [15] have provided an extensive review of the current status of "stealth" technologies being developed to maximize the resident times of colloidal preparations.

Gold colloids and gold coatings have been used to study various cellular uptake and endocytosis [1–3] mechanisms. For example, gold nanoparticles have been conjugated to

Biomedical Applications of Nanotechnology. Edited by Vinod Labhasetwar and Diandra L. Leslie-Pelecky
Copyright © 2007 John Wiley & Sons, Inc.

antibodies [1], lipids, and DNA fragments [3] to study a variety of uptake mechanisms. In the early 1970s Faulk and Taylor [4] invented the immunogold staining procedure; since that time, researchers have been using gold nanoparticles to label proteins for techniques such as imaging, blotting, flow cytometry, and hybridization assays [5–7]. The most common method of coupling inorganic nanoparticles to biomolecules involves forming either a gold–thiol (Au–Sbiomolecule) or a metal–disulfide (Cd–S–Sbiomolecule) co-valent bond [8]. These types of metal sulfur bonds have been shown to be stable under aqueous conditions [8–12].

Gold coatings are particularly important for magnetic nanoparticle technology. The gold coating enhances the stealthiness of the nanoparticles by preventing macrophage recognition of the particles, and gold is an ideal surface for biomolecular conjuga-tion. Of particular importance to many magnetic materials, gold coatings also prevent oxidation and other chemical attack of the magnetic alloy, thereby reducing toxicity issues associated with the oxidation products of the magnetic alloy used to produce the nanoparticles.

A variety of nanotechnologies are used for tumor treatment from directing chemo-, gene-, or radiotherapeutic agents contained within nanoparticles to tumor sites to nanoparticles that enhance energy deposition in tissue from external energy sources (microwaves, light, magnetic, etc.). Mechanisms employed to direct nanoparticles to-ward tumors include molecular targeting, magnetic targeting, and even direct injection into the tumor. In this review, we will concentrate specifically on recent advances in the use of magnetic nanotechnology for diagnosing and treating tumors.

Magnetic nanoparticles in particular are used to diagnose, image, and treat cancer using different techniques based on a variety of physical mechanisms. We will review a selection of techniques using a framework of (1) concentrating the nanoparticles at the desired location, (2) localizing and/or imaging the concentrations of nanoparticles, and (3) inducing necrosis of the tumor tissue while minimizing damage of healthy tissue. While the synthesis and properties of the magnetic nanoparticles is an important and active area of research, it is beyond the scope of this chapter and the reader is referred to other reviews of the subject [13–15]. The concentration of nanoparticles at the tumor site is a prerequisite for both the imaging and treatment phases. In general, diagnostic imaging of the tumor is typically considered separately from the treatment; however, in recent years there has been a growing body of literature reporting multimodal and multifunction nanoparticles. Magnetic nanoparticles have been at the forefront of multifunctional applications by enabling the same materials to be used for both diagnosis and treatment. We will focus this chapter on the methods and materials used to image and treat tumors with magnetic nanotechnology. Targeting methods, toxicity, and magnetic material considerations are discussed in detail elsewhere in this book and will only be touched on as they relate specifically to the magnetic nanotechnology discussed here.

3.2 CONCENTRATING MAGNETIC MATERIAL AND THERAPY

Virtually all applications of magnetic nanoparticle technology to the diagnosis and treatment of tumors require the concentration of magnetic material at a site or sites of interest while minimizing uptake in healthy tissue and organs. The goal is to maximize the diagnostic signal-to-noise ratio (SNR) and maximize therapeutic effect while minimizing systemic or even local damage to healthy tissue. The total dose used for diagnostic techniques will depend on the limit of detection for any given method (total signal required) and the background or noise (the SNR). The primary noise source will be signals arising from the uptake of magnetic nanoparticles in healthy tissue and organs. The dose needed to treat tumors is determined by the nanoparticle load required in the affected tissue to attain therapeutic efficacy, while uptake in healthy tissue and organs may lead to undesirable tissue damage. Thus, as has been the case in cancer treatment for many years, the total dose needed for treatment depends on a balance between the quantity of material required for therapeutic efficacy and the deleterious effects of uptake in healthy tissue and organs.

A major focus of cancer research in recent years has been to target therapies to the tumors with a high degree of specificity that minimizes affecting healthy tissue. A variety of approaches that direct therapies to tumors are under active investigation including molecular targeting, mechanical targeting, and magnetic targeting. We define "mechanical targeting" as physical techniques that localize nanoparticles in the desired location. Two examples are direct injection or passive targeting [15] that relies on nanoparticle accumulation in porous vasculature common to many tumors. Direct injection may be useful for tumors that would not exhibit sufficient specificity or differentiation. In some cases, multiple targeting modalities may be combined, for example, by injecting a suspension of magnetic nanoparticles that are functionalized with monoclonal antibodies to target a specific tumor receptor. Many of these approaches are discussed elsewhere in this volume.

A growing number of therapies couple energy from an external source such as magnetic fields, microwaves, light (laser), and ultrasound to the tumor mediated by nanoparticles in the tissue. These approaches allow the nanoparticles to be preferentially targeted to the tumor *in vivo*, and the external energy source can be selectively applied to the nanoparticle concentrations that have been proven to be localized to the tumor.

3.3 LOCALIZING NANOPARTICLE CONCENTRATIONS

After the nanoparticles have been concentrated by means of molecular targeting, magnetic targeting, direct injection, and so on, it is important to map the foci of localized nanoparticle concentrations and the overall nanoparticle density distribution in tissue. Such a map provides a powerful diagnostic tool by detecting the abnormal localized presence of receptors expressed by tumor cells through molecular targeting [13, 16, 17], magnetic targeting [18, 19], mechanical targeting of compromised vasculature [15, 20], and new approaches being developed daily. Knowledge of the magnetic material distribution is also crucial for planning magnetic nanoparticle-mediated thermal energy deposition for cancer therapy. The distribution in tissue is extremely important for determining the dose required to realize the desired local temperature increase in the tumor while minimizing collateral damage. A wide range of techniques for localizing nanoparticles in tissue have been reported including radio-isotope imaging [21], optical methods

[22, 23], and X-ray [27]. Each of these approaches depends on using nanoparticles specifically designed for the imaging approach being applied. Imaging magnetic nanoparticles is most readily carried out by MRI susceptibility contrast approaches [25, 26]. It has also been suggested that superconducting quantum interference device (SQUID) sensor arrays could be used to localize concentrations of magnetic nanoparticles in tissue [27, 28]. We will examine these two approaches in the following sections.

3.3.1 Imaging Magnetic Nanoparticles with Magnetic Resonance Imaging

Magnetic resonance imaging has been shown to be very well suited for diagnostic cancer imaging as a result of the exceptional anatomical resolution of this modality [29, 30]. The basis of molecular MRI is generally based on the assumption that antibodies, peptides, or other targeting molecules, tagged with a magnetic contrast agent, binds to the target and produces a local magnetic field perturbation that results in an increased proton relaxation rate that is detectable by magnetic resonance techniques. Magnetic nanoparticles are a form of magnetic contrast agent in MRI. Para- and superparamagnetic agents such as Gd(III) and various forms of iron oxide in both molecular and nanoparticle form have been used in a broad range of MRI applications to enhance image contrast. This approach is only limited by the inherent sensitivity of MRI, and the specific pulse sequence chosen, to the presence and distribution of the magnetic contrast agent [25, 26, 31, 32].

The limit of detection for these contrast agents is determined by a variety of factors including the total quantity, local concentration or material distribution, and the material form (e.g., molecular, nanoparticle, oxide, etc.). One investigation reported that the limit of detection for paramagnetic gadolinium is approximately 15 μg Gd/g tissue [25]. Recent studies have used iron oxide nanoparticles conjugated to targeting molecules as the contrast agent [24, 33–35] both *in vitro* and *in vivo*. The larger local field perturbation caused by the nanoparticles, along with newer imaging methods (including image subtraction), has increased the detection sensitivity to significantly better than 2 μg Fe/g tissue [22]. The saturation magnetization of the contrast agent in the applied field is an important factor in maximizing sensitivity threshold [33, 36].

The recent interest in using MRI to directly measure neuronal currents has motivated developing new pulse sequences for maximizing sensitivity to local magnetic field perturbations. Bodurka et al. [37] suggested that neuronal currents in the brain will cause spatially varying magnetic field distributions on a very small scale. Such a model is very similar to what one would expect for magnetic nanoparticles in tissue. In an effort to determine the feasibility of detecting spatially varying magnetic fields on a microscopic scale, they suggested that the magnetic field perturbation resulting from local neuronal activity could be measured by a phase-shifted MRI technique. Bodurka originally demonstrated a detection threshold to local field inhomogeneities, using a gradient-recalled echo-planar imaging (EPI) sequence, of about 1.7 nT, in general agreement with theoretical estimates for an 8-mm^3 voxel. In a more recent work, Bodurka and Bandettini [38] demonstrated a sensitivity limit of ~200 pT (2×10^{-10} T) using a single-shot spin-echo (SE) echo-planar sequence. Finally, Konn et al. [80] reported a detection threshold of 1.1×10^{-10} T in a 2-mm × 3-mm × 3-mm voxel. The most important characteristics of the neuronal model that lead to the dephasing of the MRI signal is the variation of the magnetic field on a microscopic spatial scale. This characteristic is analogous to magnetic nanoparticles distributed in tissue.

We can now estimate the sensitivity of the SE EPI technique to detecting magnetic nanoparticles in tissue. If we assume a nominal voxel size of 2-mm $\times$ 2-mm $\times$ 2-mm, along with a conservative sensitivity limit of 3×10^{-10} T, we find that our detection limit is approximately 36 pg of γ-Fe_2O_3 magnetic nanoparticles ($\sim$1000–100 nm nanoparticles) distributed evenly in the 8-mm^3 voxel. We assumed a typical saturation magnetization, $J_s \approx 0.54$ T for γ-Fe_2O_3, and used an average volume magnetization in the voxel in keeping with previous MRI phantom studies. The saturation magnetization for such a material would be realized in virtually any standard MRI instrument. Recent advances in ultra-low-field (ULF) MRI may provide even greater sensitivity, particularly when using ferromagnetic nanoparticles. ULF MRI is particularly sensitive because of the extraordinarily narrow peak width [39–41]. Modeling in our own group has shown that such an approach would be sensitive to changes of less than 50 pT in a voxel. Given that nominal voxel size for our work today is about 2 mm $\times$ 2 mm $\times$ 2 mm, less than 200- to 100-nm nanoparticles could potentially be detected and imaged.

Finally, MR imaging of these localized concentrations of magnetic nanoparticles is essentially independent of the depth within the tissue (patient). This is readily understood when one considers that the effect of the nanoparticle magnetic fields is upon the spin population of the tissue in which the nanoparticles are distributed. MRI measures the dephasing effect of the nanoparticle fields on this spin population. The nanoparticle fields are not being directly measured at a distance by the MRI instrument.

3.3.2 Magnetic Nanoparticle Localization with SQUID Sensor Arrays

SQUIDs are the most sensitive magnetic field sensors known. SQUID sensors noninvasively measure the magnetic fields outside a given volume from which one deduces the magnetic sources inside the volume, such as in magnetoencephalography (MEG) [42]. The tremendous sensitivity of SQUID sensors led to the suggestion that they could be used to localize magnetic nanoparticles *in vivo* [27]. Indeed, the extraordinary sensitivity to detecting magnetic material *in vivo* has been demonstrated in a variety of early investigations of magnetic iron in the liver [43] and magnetic inclusions in the lungs [44]. Recently, Romanus and co-workers [28] reported on a system they built, and they demonstrated the detection of magnetic nanoparticles *in vitro* and *in vivo* in a mouse. They reported a detection limit for primarily γ-Fe_2O_3 magnetic nanoparticles of 2.4×10^{-8} g Fe_2O_3 (1.7×10^{-8} g Fe) at a distance of 2.2 cm (roughly equivalent to 7×10^5 100-nm nanoparticles). Unlike MRI, SQUID sensors detect the magnetic field of the nanoparticles directly. The field from a localized concentration of magnetic nanoparticles (such as measured by Romanus) falls off roughly with the cube of the distance between the nanoparticles and the sensor; thus with a constant sensitivity, the detection limit at 5-cm distance would be 12 times worse than at 2.2 cm.

While the extraordinary sensitivity of SQUIDs is unassailable, the value of using SQUID sensor arrays for detecting targeted magnetic nanoparticle concentrations *in*

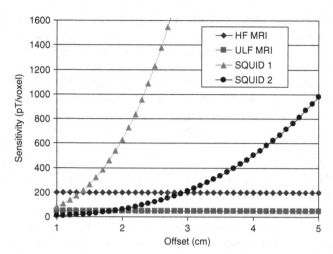

Figure 3.1. A plot of the anticipated sensitivity to the magnetization of a 2-mm × 2-mm × 2-mm volume as a function of distance from the sensor using MRI and direct magnetic field measurement by SQUID sensors. The high-field MRI (HF MRI, diamonds) sensitivity limit is 200 pT, and the expected ultra-low-field MRI (ULF MRI, squares) limit is 50 pT. Both MRI techniques are independent of sensor-source offset. The sensitivity limit for SQUID sensors with a nominal 0.1-pT and 0.01-pT noise floors are illustrated by the triangles and dots, respectively. The SQUID sensitivity curves reflect the $1/r^3$ dependence of a magnetometer to a dipolar magnetic field.

vivo over MR contrast methods is questionable for both technical and practical reasons. The most severe physical limitation is the need, in this approach, to directly measure the magnetic fields from nanoparticle concentrations at a given standoff. We illustrate this limitation in Figure 3.1, where we plot the anticipated sensitivity to the magnetization of a 2-mm × 2-mm × 2-mm voxel using MRI and direct magnetic field measurement by SQUID sensors. The sensitivities of high-field and ULF MRI are 200 pT and 50 pT, respectively, and independent of distance from the voxel, as described above. The sensitivity threshold to the magnetization of a 2-mm cube volume is plotted for two SQUID sensors with nominal field sensitivities of 0.1 pT (green triangles) and 0.01 pT (black dots). The curves reflect the $1/r^3$ dependence of a magnetometer to a dipolar magnetic field. The superconducting nature of SQUIDs requires that they be contained in a cryostat which immediately places the SQUID at least 1 cm from the magnetic source. The sensitivity would be further decreased if a gradiometer were coupled to the SQUID (commonly required to reduce ambient noise). While SQUIDs may outperform MRI techniques for surface sources, it appears that new MRI techniques will be more sensitive to magnetic nanoparticle concentrations localized 1–2 cm below the surface.

The second major technical drawback to direct measurement of magnetic nanoparticle concentrations is the inability to uniquely localize the source within the volume of interest. While years of work have developed constraints that have led to reliable solutions in functional brain imaging techniques using SQUID arrays (magnetoencephalography)

[45], the constraints for mapping localized concentrations of magnetic nanoparticles have yet to be explored. In contrast, MRI is a tomographic technique that uniquely *images* the contrast agent by encoding the signal in space using applied magnetic field gradients [46, 47].

The chief practical reasons that MR contrast approaches may be preferable to direct SQUID-based magnetic field measurement are the cost and availability. MRI instruments are widely available at virtually every hospital in the United States. In recent years, MRI instruments have been made available to smaller hospitals by installing them in specially designed trailers that are readily transported between local health care facilities. Finally, the cost of a new SQUID array is on the order of $1 million or more for an installed facility.

In conclusion, it is unclear that direct detection of the magnetic fields from *in vivo* magnetic nanoparticles with SQUID sensors will be any more sensitive than MRI contrast techniques, especially at ultra-low field. In addition, any direct field measurement approach will suffer from the ill-posed inverse problem as discussed above. Finally, MRI is extremely sensitive to magnetic inclusions (as we have shown), MR instrumentation is widely available, and MRI produces tomographic images that will unambiguously image concentrations of magnetic nanoparticles *in vivo*. Hence, MRI at high field in the near term and at ultra-low field in the future will likely provide the best approaches to image magnetic nanoparticle distributions *in vivo*.

3.4 ENERGY TRANSFER TO THE TUMOR

Cancer therapies include a broad array of options including surgery, chemotherapy, and radiotherapy. Another set of mechanisms include those by which energy is transferred from an external source into tissue, many of which are mediated by nanoparticles. Techniques such as nanoparticle-mediated laser energy absorption [48, 49], microwave absorption [50, 51], enhanced radiation absorption [52], and magnetic-field energy absorption [53] are a few examples of studies demonstrating nanoparticle-mediated energy transfer. Magnetic field energy absorption approaches use a wide variety of mediator materials from macroscopic "needles" to nanoparticles [54–56]. This review will focus on energy transfer mediated by magnetic nanoparticles. Coupling energy in the form of a time-varying magnetic field into tumors mediated by magnetic nanoparticles has been an active area of research for more than a decade. These mechanisms are included in the methodology broadly referred to as hyperthermia, which is used as either a primary therapy (directly killing tumor cells with heat) or an adjunctive therapy (using heat to increase the efficacy of chemo- or radiotherapy). More "exotic" therapies, such as the gamma-knife, can be used to treat inoperable tumors such as those in the brain where surgical intervention would risk damage to primary functional centers. Hyperthermia as primary therapy (directly killing cancer cells) requires heating the entire tumor by at least 10°C. Ideally, rapid energy deposition increasing the temperature by 25°C will minimize damage to nearby healthy tissue by limiting thermal diffusion and dissipation [57] while completely necroting tumor tissue. Perfusion in tissue has been found to increase thermal diffusion rates from ~20% to a factor of 2 near major vasculature [58].

3.4.1 Energy Transfer as Adjunctive Therapy

Success in using primary hyperthermic methods has been limited because of the inability to constrain collateral damage (phased-array microwave systems) or to reach even moderately deep tumors (laser therapy or microwave methods). Choosing the optimum therapy depends strongly on the type and location of a cancer. Therapeutic hyperthermia can be induced by a variety of methods with varying applications and efficacies. A comprehensive historical review was presented by Seegenschmiedt and Vernon [59]. Methods varied widely from laser-induced hyperthermia [60–62] to microwave-induced [63–66] and hysteretic (magnetic) heating using both large implanted "needles" or "rods" [54–56], or nanoparticles [53,67,68].

Hyperthermia, as a primary therapy, has been constrained by a variety of limitations. In some cases, antenna and emitters produced an extremely limited effective field of treatment resulting in necrosis. Other approaches (such as external microwave antennae) induced broader energy deposition that resulted in poorly focused energy deposition and extensive collateral damage of healthy tissue. Finally, energy transfer effected by magnetic materials either implanted or injected into the tissue resulted in thermal energy deposition insufficient to effect tumor necrosis. Consequently, most hyperthermia therapy, both systemic and targeted, has been reported largely as an adjunctive therapy (in support of the primary radio- and/or chemotherapies [53,59,66,69]). A novel approach combining thermal energy deposition from magnetic nanoparticles with dendritic cell (DC) therapy was also reported recently [70]. DCs have been found to play an important role in regulating the immune response in cancer. In this novel application, the thermal energy causes the expression of heat shock proteins which in turn activate the DCs. The method was tested on mice bearing a melanoma nodule and compared to control groups that received hyperthermia alone, DCs alone, and no therapy. The authors reported complete regression of the tumors in 60% of the mice receiving the combined therapy and no regression whatsoever in any of the other groups, an encouraging result indeed.

3.4.2 Energy Transfer as Primary Therapy

A review by Moroz et al. [53] refers to a significant body of work that found that maintaining elevated temperatures (5–10°C) for several minutes can induce apoptosis. The perspective of this review is that magnetically mediated hyperthermia (MMH) has the potential to address targeting and heat distribution shortcoming of other hyperthermia modalities. The MMH technique is broadly defined as localizing magnetic particles within tumor tissue and applying an external alternating magnetic field. The coupling of the external RF magnetic field to the magnetic particles result in energy being transferred to the tissue by a variety of well-known physical mechanisms. The greatest challenge to the efficacy of MMH is the balance between the rate of thermal energy deposition and the mechanisms responsible for thermal dissipation. The physical mechanisms responsible for energy transfer to tissue from an external RF magnetic field in all MMH reported to date are dominated by Neél relaxation and hysteretic heating. At the same time, the external RF field will cause global heating in all infused tissue by the ohmic (resistive) heating mechanism. The fastest *in vivo* energy transfer rates to tissue reported in the literature to date have been by Moroz et al. [71], where approximately $6.6 \text{ J}/(\text{sec}^{-1} \text{ g}_{\text{Fe}}^{-1})$ was realized using 75 mg of γ-Fe_2O_3 magnetic nanoparticles ($\sim$100 nm in size) directly injected into a hepatic tumor model in rabbits. More recently, Sonvico et al. [72] reported obtaining $13.8 \text{ J}/(\text{sec}^{-1} \text{ g}_{\text{Fe}}^{-1})$, but for an *in vitro* cell culture containing γ-Fe_2O_3 nanoparticles.

Let us examine whether it is feasible for magnetic nanoparticle-mediated energy transfer to be sufficiently fast to induce tumor necrosis as a primary therapy while minimizing collateral damage to healthy tissue, primarily due to thermal energy diffusion. There are three primary mechanisms of energy transfer to tissue impregnated with magnetic nanoparticles by time-varying magnetic fields. These are: (1) ohmic, caused by induced currents flowing in conducting media; (2) hysteretic, caused by traversing the magnetic hysteresis loop of magnetic materials; and (3) viscous, caused by physical motion of the magnetic nanoparticles in a viscous medium. Dielectric heating is negligible in the regime of interest here. Let us assume tissue with resistivity, R, containing magnetic nanoparticles with average magnetization, M, which is subjected to a spatially uniform (dipolar) magnetic field with strength, B_0, that is "switching" or *rotating* at a frequency, ω. The rate of energy deposition, dQ/dt, by these mechanisms can be simply written as

$$\frac{dQ_{\text{ohmic}}}{dt} \approx R(\omega B_0)^2, \tag{3.1}$$

$$\frac{dQ_{\text{hyst}}}{dt} \approx a^3 \omega B_0^2, \tag{3.2}$$

$$\frac{dQ_{\text{visc}}}{dt} = 8\pi \eta a^3 \omega^2, \tag{3.3}$$

In practice, hysteretic heating can be realized by *either* switching the external magnetic field back and forth (Figure 3.2, top row) or angularly rotating the field (Figure 3.2, bottom row). In contrast, the magnetic field must be varied by angular rotation to most efficiently transfer energy by the viscous mechanism. The maximum viscous energy transfer is realized at a critical rotation frequency:

$$\omega_{\text{crit}} = \frac{M B_0}{6\eta}, \tag{3.4}$$

where η is the viscosity of the fluid in which the nanoparticles are suspended. That maximum viscous energy transfer rate is then given by

$$\frac{dQ_{\text{visc}}}{dt}(\text{max}) = \frac{4}{3}\pi a^3 B_0 M \omega_{\text{crit}}, \tag{3.5}$$

The critical frequency, ω_{crit}, represents the maximal rotational frequency attainable for given nanoparticles magnetization, external field, and fluid viscosity. This occurs when

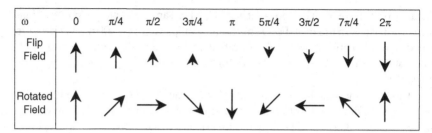

Figure 3.2. Vector representation of externally applied magnetic field as a function of ω. In the upper row, the field is switched or flipped (most common for hysteretic heating). In the bottom row the magnetic field is rotated (required for efficient viscous heating). The length of the arrows indicates field magnitude.

the drag on the rotating particle is equal to the coupling force between the external magnetic field and the nanoparticle.

A critical difference between the hysteretic and viscous heating rates is the presence of the nanoparticle magnetization in the energy transfer equation for viscous heating. Thus, when the nanoparticle moment exceeds the external drive field, the viscous mechanism will be greater (more efficient) than the hysteretic mechanism. In practice, B_0 will be limited to 10^{-2} to 10^{-1} T while the average magnetization for many magnetic materials can easily reach the order of 1 T. This illustrates that for readily attainable values of M and B_0, the viscous mechanism of magnetic nanoparticle-mediated energy transfer can be one to two orders of magnitude more efficient than the hysteretic mechanism.

Two properties of magnetic materials are particularly relevant to our discussion of energy deposition mechanisms: remnant magnetization, B_r, and coercivity, H_c. B_r is a measure of the magnetization of a given material that remains in the absence of any external field. H_c is a measure of the external magnetic field required to alter or reverse the magnetization of a material. One way to think of H_c (though somewhat of an oversimplification) is how strongly the magnetization vector is locked or pinned to the physical material.

The equations for energy deposition presented above assume specific properties of the magnetic materials that mediate the different mechanisms. Hysteretic heating is realized for generally paramagnetic and superparamagnetic materials where H_c is small or nonexistent; consequently, varying an external magnetic field (either by flipping or rotation) causes a portion of the material hysteresis loop to be traced. Viscous energy deposition is realized for materials where H_c is sufficiently large that it remains pinned to the lattice of the nanoparticle as the external magnetic field is rotated. If we begin with a magnetic nanoparticle with appreciable H_c (relative to the externally applied field) oriented parallel to the external magnetic field and then rotate the external magnetic field, the nanoparticle will experience a torque that will follow the direction of rotation. Friction between a nanoparticle and the viscous fluid provides the mechanism for energy transfer from the external magnetic field to the fluid mediated by the nanoparticle.

If we assume the nanoparticles are delivered to the tumor via a molecular targeting modality, the nanoparticle will most likely be coupled to a cell receptor by a specific molecule. In order for the viscous energy transfer to be successful, we must first break some molecular bond in order to free the nanoparticle to rotate. The energy associated with rotating the nanoparticle is given by

$$U_{\text{rot}} = \frac{8\pi}{3} a^3 B_0 M. \tag{3.6}$$

For typical values of M and B_0 (see example, below), we find that U_{rot} is 9.9×10^{-9} erg($\sim 6\,\text{keV}$) more than sufficient to break any molecular bond.

3.4.3 An Example of Viscous Energy Transfer

In the cgs system of units, the coefficient of viscosity has the dimensions of mass/(length × time) and is measured in poise (dyne sec/cm^2). The corresponding SI unit is the pascal second. The viscosity of water is close to 0.01 poise = 0.001 pascal seconds. Consider a spherical particle with diameter 0.5 μm and average magnetization 750 emu rotating in a fluid such as water. Let the amplitude of the rf field be 100 G. Then we have the following:

	Gaussian	MKS
M	750 emu	7.5×10^5 A/m
a	0.25×10^{-4} cm	0.25×10^{-6} m
B_0	100 G	0.01 T
η	0.01 poise	0.001 Pa · sec

And $\omega_{\text{crit}} = 1.25 \times 10^6\,\text{sec}^{-1}$ ($f_{\text{crit}} \sim 200\,\text{kHz}$). The maximum heating rate per particle is $dQ/dt_{\text{visc}}(\text{max}) = 6.1 \times 10^{-3}\,\text{erg/sec} = 6.1 \times 10^{-10}$ W. By comparison, we find that the hysteretic heating, $dQ/dt_{\text{hyst}} = 1.6 \times 10^{-5}\,\text{erg/sec} = 1.6 \times 10^{-12}$ W, more than two orders of magnitude less efficient than the viscous heating. The energy associated with rotating one 500-nm nanoparticle is, as presented above, $\sim 6\,\text{keV}$. We note that while nanoparticles down to $\sim 100\,\text{nm}$ will generate sufficient impulse to break molecular bonds, particles less than $\sim 50\,\text{nm}$ may be unable to break those bonds and thus will not contribute heating by the viscous mechanism.

3.4.4 A Viscous Energy Transfer Simulation

We developed a simple simulation to analyze the effectiveness of the viscous energy transfer mechanism. The simulation modeled a 1-cm^3 tumor containing 10^6 500-nm spherical magnetic nanoparticles. This represents less than one nanoparticle for every 10 cells in typical tissue, a nanoparticle density that one can realize by molecular targeting. We assumed that the nanoparticles had a net remnant magnetization of 0.9 T (similar to SmCo$_5$ or FePt magnetic materials). The tumor was centered inside a large volume with electrical conductivity equal to 0.5% saline to represent tissue. The entire construct was modeled inside a global *rotating* magnetic field. The result for one model

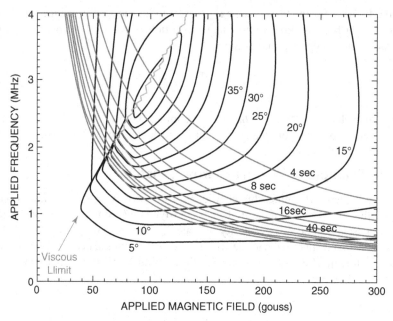

Figure 3.3. A plot of the computed temperature rise in tissue induced by viscous energy transfer as a function of the strength and rotational frequency of the applied magnetic field. The specifics of the model are described in the text.

analysis is presented in Figure 3.3, where the temperature increase is shown as white contours as a function of the amplitude of the applied magnetic field and the frequency of rotation. The "viscous limit" is where the viscous drag exceeds the force coupling the external magnetic field to the moment of the nanoparticles, hence the particle does not rotate fully but "oscillates," causing the efficiency of energy transfer to drop significantly. The model calculated both global ohmic and local viscous heating. The simulation was terminated when the global temperature rise ΔT_{global} is 3°C. The temperature contours in Figure 3.3 represent the temperature *increase* realized in the tumor volume at the time the simulation was terminated (e.g., $\Delta T_{global} = 3$°C). We see that the simulation predicts that increases in the tumor temperature, ΔT_{tumor} of at least 25°C can be realized, sufficient to necrote the tumor. Overlaid on the plot is the time required to realize the computed temperature increase. The short time, a minute or less, needed for the rotating nanoparticles to deposit sufficient energy in the tumor to raise the temperature to a level to effectively cause immediate cell necrosis will prevent significant thermal diffusion and minimize damage to nearby healthy tissue.

In addition to taking global ohmic heating into consideration, we must also examine the heating caused by the background magnetic nanoparticle concentration in healthy tissue. Current nanoparticle targeting has demonstrated *in vivo* specificity of better than 10:1, which would cause global heating of no more than 10% of the localized heating of the tumor. While we did not estimate this effect in our model, it is worth noting that one can reduce the global heating by using a rotating external magnetic field whose spatial distribution is peaked at the tumor site and falls off outside the tumor. Magnetic field engineering is a mature field, and such magnetic field patterns can be readily designed. Applying such a rotating field distribution to the tissue containing the magnetic nanoparticles will further localize energy deposition in the desired area.

While this extremely simple treatment of the viscous energy transfer mechanism assumes spherical nanoparticles in a uniformly viscous medium, the estimates are instructive. We have completed a more thorough treatment of (a) the types of particle motion that can be expected for given rf drive conditions and (b) the heating rates that can result [73]. We have developed a complete set of scaling principles and dimensionless parameters that govern magnetic particle behavior. A simple equation of motion was developed which incorporates a variety of significant factors including elliptical drive polarization, applied bias fields, particle–particle dipole coupling, finite particle inertia, and the effects of random external forces. Mathematical analysis of limiting cases provides insight on the classification and stability of various types of motion and establishes a framework for the interpretation of simulation results. The simulations (both single- and multi-particle) illustrate the onset of conditions less favorable to heating and demonstrate the utility of combining radio-frequency drive with a modest bias field. A significant consequence of this work is the suggestion that "rocker mode" operation, which combines a bias field with a single rf coil set, promises a simpler, but nearly as effective, rf drive technique in comparison with one giving full circular rotation.

3.4.5 Measuring the Temperature

Finally, knowledge of the temperature distribution in the treated tissue is crucial to controlling the therapy, determining the treatment outcome, and minimizing damage to healthy tissue and organs. The most common technique for thermal monitoring is to insert temperature probes in the affected region. Fiber-optic probes are commonly used because of their insensitivity to the time-varying magnetic field [71]. The greatest limitation of temperature probes is that they provide a single-point measurement with very limited information about the temperature distribution. Various teams have reported the development of thermochromic films based on thermal fluorescence [74] or thermally sensitive optical properties [75]. These techniques have demonstrated better than 1°C precision in laboratory tests, but the temporal response of these techniques is unclear. The current state-of-the-art in thermal imaging is tens of milli-Kelvins thermal accuracy with millisecond temporal resolution [76]. These systems are, however, extraordinarily expensive and the performance is perhaps more than required for hyperthermia monitoring. A recent development in thermal imaging was inspired by the realization that magnetic resonance spin relaxation is a temperature-dependent process [77, 78]. The current state-of-the-art for MRI thermometry is approaching 0.5°C temperature resolution [79] (MRI measures temperature change, not absolute temperature) with a temporal resolution on the order of tens of seconds. Because of the compatibility of MRI with RF fields, it may provide an ideal modality for providing both the RF fields for hyperthermia and the thermometry modality during the treatment.

3.5 CONCLUSION

Magnetic nanoparticle technology coupled with molecular targeting has, in recent years, enabled intriguing approaches to early cancer detection, imaging, and treatment options. Molecular targeting technology has successfully concentrated nanoparticles at tumor sites. As a consequence of recent developments in MRI technology, localized concentrations of magnetic nanoparticles may now be detected and imaged with extraordinary sensitivity that may soon enable the detection of submillimeter tumors. For more than two decades, magnetic nanotechnology has been used in cancer therapy to selectively deliver chemotherapeutic or radiological agents and induce heating to localized regions. Hyperthermia has historically been primarily an adjunctive therapy to chemo- or radiation therapy. Recent work suggests the possibility of a new and far more efficient energy transfer mechanism mediated by hard magnetic materials such as $SmCo_5$ or FePt nanoparticles. Efficient energy deposition capable of raising tumor tissue temperatures by $\Delta T_{tumor} = 25°C$ would provide a new and powerful primary therapy. Fast energy deposition would limit thermal diffusion and associated damage to healthy tissue. Finally, MRI is also proving a powerful thermal imaging tool for guiding hyperthermia treatment.

REFERENCES

1. Hurwitz E, Stancovski I, Sela M, Yarden Y. Suppression and promotion of tumor growth by monoclonal antibodies to ErbB-2 differentially correlate with cellular uptake. *Proc Natl Acad Sci USA* 1995;92:3353–3357.

2. Andreu EJ, Martin de Llano JJ, Moreno I, Knecht E. A rapid procedure suitable to assess quantitatively the endocytosis of colloidal gold and its conjugates in cultured cells. *J Histochem Cytochem* 1998;46:1199–1201.

3. Sandhu KK, McIntosh CM, Simard JM, Smith SW, Rotello VM. Gold nanoparticle-mediated transfection of mammalian cells. *Bioconjug Chem* 2002;13:6.

4. Faulk WP, Taylor GM. An immunocolloid method for the electron microscope. *Immunochemistry* 1971;8:1081–1083.

5. Hermanson GT. *Bioconjugate Techniques*. San Diego, CA: Academic Press; 1996, pp. 593–604.

6. Gee B, Warhol MJ, Roth J. Use of an anti-horseradish peroxidase antibody gold complex in the ABC technique. *J Histochem Cytochem* 1991;39:863–870.

7. Jackson P, Dockey DA, Lewis FA, Wells M. Application of 1-mm gold probes on paraffin wax sections for *in situ* hybridization histochemistry. *J Clin Pathol* 1990;43:810–812.

8. Niemeyer CM, Nanoparticles, proteins, and nucleic acids: Biotechnology meets materials science. *Angew Chem Int Ed* 2001;40:4128–4158.

9. Park S-J, Lazarides AA, Mirkin CA, Brazis PW, Kannewurf CR, Letsinger RL. The electrical properties of gold nanoparticle assemblies linked by DNA. *Angew Chem Int Ed* 2000;39:3845–3848.

10. Storhoff JJ, Lazarides AA, Mucic RC, Mirkin CA, Letsinger RL, Schatz GC. What controls the optical properties of DNA-linked gold nanoparticle assemblies? *J Am Chem Soc* 2000;122:4640–4650.

11. Park S-J, Lazarides AA, Mirkin CA, Letsinger RL. Directed assembly of periodic materials from protein and oligonucleotide-modified nanoparticle building blocks. *Angew Chem Int Ed* 2001;40:2909–2912.

12. Loweth CJ, Caldwell WB, Peng X, Alivisatos AP, Schultz PG. DNA-based assembly of gold nanocrystals. *Angew Chem Int Ed* 1999;38:1808–1812.

13. Gupta AK, Gupta M. Synthesis and surface engineering of iron oxide nanoparticles for biomedical applications. *Biomaterials* 2005;26:3995–4021.

14. Wilkinson JM. Nanotechnology applications in medicine. *Med Device Technol* 2003;14:29–31.

15. Moghimi SM, Hunter AC, Murray JC. Long-circulating and target-specific nanoparticles: theory to practice. *Pharmacol Rev* 2001;53:283–318.

16. Kim EE, Jackson EF. eds. *Molecular Imaging in Oncology*. Berlin: Springer-Verlag; 1999, p. 95ff, for example.

17. Zhang Y, Kohler N, Zhang M. Surface modification of superparamagnetic magnetite nanoparticles and their intracellular uptake. *Biomaterials* 2002;23:1553–1561.

18. Gupta AK, Wells S. Surface-modified superparamagnetic nanoparticles for drug delivery: preparation, characterization, and cytotoxicity studies. *IEEE Trans Nanobiosci* 2004;3:66–73.

19. Jain TK, Morales MA, Sahoo SK, Leslie-Pelecky DL, Labhasetwar V. Iron oxide nanoparticles for sustained delivery of anticancer agents. *Mol Pharm* 2005;2:194–205.

20. Brannon-Peppas L, Blanchette JO. Nanoparticle and targeted systems for cancer therapy. *Adv Drug Deliv Rev* 2004;56:1649–1659.

21. Rossin R, Pan D, Qi K, Turner JL, Sun X, Wooley KL, Welch MJ. ^{64}Cu-labeled folate-conjugated shell cross-linked nanoparticles for tumor imaging and radiotherapy: Synthesis, radiolabeling, and biologic evaluation. *J Nucl Med* 2005;46:1210–1218.

22. Sosnovik DE, Schellenberger EA, Nahrendorf M, Novikov MS, Matsui T, Dai G, Reynolds F. Magnetic resonance imaging of cardiomyocyte apoptosis with a novel magneto-optical nanoparticle. *Magn Reson Med* 2005;54:718–724.

23. Veiseh O, Sun C, Gunn J, Kohler N, Gabikian P, Lee D, Bhattarai N, Ellenbogen R, Sze R, Hallahan A, Olson J, Zhang M. Optical and MRI multifunctional nanoprobe for targeting gliomas. *Nano Lett* 2005;5:1003–1008.

24. Alexiou C, Arnold W, Hulin P, Klein RJ, Renz H, Parak FG, Bergemann C, Lübbe AS. Magnetic mitoxantrone nanoparticle detection by histology, X-ray and MRI after magnetic tumor targeting. *J Magn Magn Mater* 2001;225:187–193.

25. Allen M, Bulte JWM, Liepold L, Basu G, Zywicke HA, Frank JA, Young M, Douglas T. Paramagnetic viral nanoparticles as potential high-relaxivity magnetic resonance contrast agents. *Magn Reson Med* 2005;54:807–812.

26. Martina M-S, Fortin J-P, Ménager C, Clément O, Barratt G, Grabielle-Madelmont C, Gazeau F, Cabuil V, Lesieur S. Generation of superparamagnetic liposomes revealed as highly efficient MRI contrast agents for in vivo imaging. *J Am Chem Soc* 2005;127:10676–10685.

27. Andrä W, Nowak H, editors. *Magnetism in Medicine*. Berlin: Wiley-VCH; 1998.

28. Romanus E, Hückel M, Gross C, Prass S, Weitschies W, Bräuer R, Weber P. Magnetic nanoparticle relaxation measurement as a novel tool for *in vivo* diagnostics. *J Magn Magn Mater* 2002;252:387–389.

29. Smith BR, Johnson GA, Groman EV, et al. High resolution MRI and MRS: A feasibility study for the investigation of focal cerebral ischemia in mice. *NMR Biomed* 1998;11:423–429.

30. Ugurbil K, Hu X, Chen W, et al. Functional mapping in the human brain using high magnetic fields. *Philos Trans Roy Soc London Ser B: Biol Sci* 1999;354:1195–1213.

31. Lauffer RB. Paramagnetic metal complexes as water proton relaxation agents for NMR imaging: Theory and design. *Chem Revs* 1987;87:901–927.

32. Nunn AD, Linder KE, Tweedle MF. Can receptors be imaged with MRI agents? *J Nucl Med* 1997;41:155–162.

33. Zhao M, Beauregard DA, Loizou L, Davletov B, Brindle KM. Non-invasive detection of apoptosis using magnetic resonance imaging and a targeted contrast agent. *Nature Med* 2001;7:1241–1243.

34. Kim DK, Zhang Y, Voit W, Rao KV, Kehr J, Bjelke B, Muhammed M. Superparamagnetic iron oxide nanoparticles for biomedical applications. *Scripta Mater* 2001;44:1713–1717.

35. Johansson LO, Perez J, Weissleder R. Magnetic nanosensors for the detection of oligonucleotide sequences. *Angew Chem Int Ed* 2001;40:3204–3207.

36. Kim DK, Zhang Y, Kehr J, Klason T, Bjelke B, Muhammed M. Characterization and MRI study of surfactant-coated superparamagnetic nanoparticles administered into the rat brain. *J Magn Magn Mater* 2001;225:256–261.

37. Bodurka J, Jesmanowicz A, Hyde JS, Xu H, Estkowski L, Li S-J. Current-induced magnetic resonance phase imaging. *J Magn Reson* 1999;137:265–271.

38. Bodurka J, Bandettini PA. Toward direct mapping of neuronal activity: MRI detection of ultraweak, transient magnetic field changes. *Magn Reson Med* 2002;47:1052–1058.

39. McDermott R, Kelso N, Lee S-K, Mossle M, Muck M, Myers W, ten Haken B, Seton HC, Trabesinger AH, Pines A, Clarke J. SQUID-detected magnetic resonance imaging in microtesla magnetic fields. *J Low Temp Phys* 2004;135:793–821.

40. McDermott R, Trabesinger AH, Mück M, Hahn EL, Pines A, Clarke J. Liquid-state NMR and scalar couplings in microtesla magnetic fields. *Science* 2002;295:2247–2249.

41. Matlachov AN, Volegov PL, Espy MA, George JS, Kraus, Jr, RH. SQUID detected NMR in microtesla magnetic fields. *J Magn Reson* 2004;170:1–7.

42. Hämäläinen M, Hari R, Ilmoniemi RJ, Knuutila J, Lounasmaa OV. Magnetoencephalography —theory, instrumentation, and applications to noninvasive studies of the working human brain. *Rev Mod Phys* 1993;65:413–497.

43. Brittenham GM, et al. Magnetic susceptibility measurement of human iron stores. *N Engl J Med* 1982;307:1671–1675.

44. Cohen D. Ferromagnetic contamination in the lungs and other organs of the human body. *Science* 1973;173:745–748.

45. Baillet S, Mosher JC, Leahy RM. Electromagnetic brain mapping. *IEEE Sig Proc Mag* 2001;18:14–30.

46. Haacke EM, Brown RW, Thompson MR, Venkatesan R. *Magnetic Resonance Imaging—Physical Principles and Sequence Design*. New York: John Wiley & Sons; 1999.

47. Hu X, Norris DG. Advances in high-field magnetic resonance imaging. *Annu Rev Biomed Eng* 2004;6:157–184.

48. Zharov VP, Galitovskaya EN, Johnson C, Kelly T. Synergistic enhancement of selective nanophotothermolysis with gold nanoclusters: Potential for cancer therapy. *Lasers Surg Med* 2005;37:219–226.

49. Kam NWS, O'Connell M, Wisdom JA, Dai H. Carbon nanotubes as multifunctional biological

transporters and near-infrared agents for selective cancer cell destruction. *Proc Natl Acad Sci USA* 2005;102:11600–11605.

50. Jayasundar R, Hall LD, Bleehen NM. RF coils for combined MR and hyperthermia studies: I. Hyperthermia applicator as an MR coil. *Magn Reson Imaging* 2001;19:111–116.

51. Hiraoka M, Mitsumori M, Hiroi N, Ohno S, Tanaka Y, Kotsuka Y, Sugimachi K. Development of RF and microwave heating equipment and clinical applications to cancer treatment in Japan. *IEEE Trans Microwave Theory Tech* 2000;48:1789–1799.

52. Cho SH. Estimation of tumour dose enhancement due to gold nanoparticles during typical radiation treatments: A preliminary Monte Carlo study. *Phys Med Biol* 2005;50:N163–N173.

53. Moroz P, Jones SK, Gray BN. Magnetically mediated hyperthermia: Current status and future directions. *Int J Hyperthermia* 2002;18:267–284.

54. Strohbehn JW, Mechling JA. Interstitial techniques for clinical hyperthermia. In Hand J, James J, editors. *Physical Techniques in Clinical Hyperthermia.* Letchworth, Hertfordshine, UK: Research Studies Press; 1996, pp. 210–287.

55. van Wieringenyz N, Kottey ANTJ, van Leeuweny GMJ, Lagendijky JJW, van Dijkx JDP, Nieuwenhuysk GJ. Dose uniformity of ferromagnetic seed implants in tissue with discrete vasculature: A numerical study on the impact of seed characteristics and implantation techniques. *Phys Med Biol* 1998;43:121–138.

56. Case JA, Tucker RD, Park JB. Defining the heating characteristics of ferromagnetic implants using calorimetry. *J Biomed Mater Res* 2000;53:791–798.

57. Guiot C, Madon E, Allegro D, Piantá PG, Baiotto B, Gabriele P. Perfusion and thermal field during hyperthermia. Experimental measurements and modelling in recurrent breast cancer. *Phys Med Biol* 1998;43:2831–2843.

58. Kolios MC, Worthington AE, Sherar MD, Hunt JW. Experimental evaluation of two simple thermal models using transient temperature analysis. *Phys Med Biol* 1998;43:3325–3340.

59. Seegenschmiedt H, Fessenden P, Vernon CC., editors. A historical perspective on hyperthermia in oncology. *Thermoradiotherapy and Thermochemotherapy*, Vols. 1 and 2. Berlin: Springer; 1996.

60. Gutknecht N, Kanehl S, Moritz A, Mittermayer C, Lampert L. Effects of Nd:YAG-laser irradiation on monolayer cell cultures. *Lasers Surg Med* 1998;22:30–36.

61. Heisterkamp J, vanHillegersberg R, Ijzermans JNM. Critical temperature and heating time for coagulation damage: Implications for interstitial laser coagulation (ILC) of tumors. *Lasers Surg Med* 1999;25:257–262.

62. Dowlatshahi K. Percutaneous Interstitial laser therapy of a patient with recurrent hepatoma in a transplanted liver. *Surgery* 1992;112:603.

63. Feldmann HJ, et al. Hyperthermia in Eccentrically located pelvic tumors. Excessive heating of the perineal fat and normal tissue temperatures. *Int J Radiat Oncol Biol Phys* 1991;20:1017.

64. Lewin JS, Connell CF, Duerk JL, Chung YC, Clampitt ME, Spisak J, Gazelle GS, Haaga JR. Interactive MRI-guided radiofrequency interstitial thermal ablation of abdominal tumors: Clinical trial for evaluation of safety and feasibility. *J Magn Reson Imaging* 1998;8:40–47.

65. Myerson RJ, et al. Phantom studies and preliminary clinical experience with BSD-2000. *Int J Hyperthermia* 1991;7:937.

66. Hiraoka M, Mitsumori M, Hiroi N, Ohno S, Tanaka Y, Kotsuka Y, Sugimachi K. Development of RF and microwave heating equipment and clinical applications to cancer treatment in Japan. *IEEE Trans Microwave Theory Tech* 2000;48:1789–1799.

67. Jordan A, et al. Cellular uptake of magnetic fluid particles and their effects on human adenocarcinoma cells exposed to AC magnetic-field *in vitro*. *Int J Hyperthermia* 1996;12:705.

68. Jordan A, et al. Presentation of a new magnetic field therapy system for the treatment of human solid tumors with magnetic fluid hyperthermia. *J Magn Magn Mater* 2001;225:118.

69. Kong G, Braun RD, Dewhirst MW. Characterization of the effect of hyperthermia on nanoparticle extravasation from tumor vasculature. *Cancer Res* 2001;61:3027.

70. Tanaka K, Ito K, Kobayashi T, Kawamura T, Shimada S, Matsumoto K, Saida T, Honda H. Intratumoral injection of immature dendritic cells enhances antitumor effect of hyperthermia using magnetic nanoparticles. *Int J Cancer* 2005;116:624–633.

71. Moroz P, Jones SK, Gray BN. Tumor response to arterial embolization hyperthermia and direct injection hyperthermia in a rabbit liver tumor model. *J Surg Onco* 2002;80:149–156.

72. Sonvico F, Mornet S, Vasseur S, Dubernet C, Jaillard D, Degrouard J, Hoebeke J, Duguet E, Colombo P, Couvreur P. Folate-conjugated Iron oxide nanoparticles for solid tumor targeting as potential specific magnetic hyperthermia mediators: Synthesis, physicochemical characterization, and *in vitro* experiments. *Bioconjug Chem* 2005;16:1181–1188.

73. Wright B, Kraus RH, Jr. Dynamic modelling for magnetocarcinotherapy. *Phys Med Biol*; in press.

74. Amao Y, Okura I. Fullerene C60 immobilized in polymethylmethacrylate film as an optical temperature sensing material. *Analysis* 2000;28:847–849.

75. Smith CR, Sabatino DR, Praisner TJ. Temperature sensing with thermochromic liquid crystals. *Exp Fluids* 2001;30:190–201.

76. Maqueda RJ, Wurden GA, Terry JL, Stillerman JA. The new infrared imaging system on Alcator C-Mod. *Rev Sci Instrum* 1999;70:734–737.

77. Peller M, Löffler R, Baur A, Turner P, Abdel-Rahman S, Futschik G, Santl M, Hiddemann W, Reiser M, Issels R. MRI-controlled regional hyperthermia. *Radiologe* 1999;39:756–763.

78. Peller M, Kurze V, Loeffler R, Pahernik S, Dellian M, Goetz AE, Issels R, Reiser M. Hyperthermia induces T1 relaxation and blood flow changes in tumors. A MRI thermometry study *in vivo*. *Magn Reson Imaging* 2003;21:545–551.

79. Gellermann J, Wlodarczyk W, Feussner A, Fahling H, Nadobny J, Hildebrandt B, Felix R, Wust P. Methods and potentials of magnetic resonance imaging for monitoring radiofrequency hyperthermia in a hybrid system. *Int J Hyperthermia* 2005;21:497–513.

80. Konn D, Gowland P, Bowtell R, MRI detection of weak magnetic fields due to an extended current dipole in a conducting sphere: A model for direct detection of neuronal currents in the brain. *Magn Reson Med* 2003;50(1):40–49.

4

BROWNIAN MOTION IN BIOLOGICAL SENSING

Axel Hoffmann, Seok-Hwan Chung, Samuel D. Bader,
Lee Makowski and Liaohai Chen

4.1 INTRODUCTION

Albert Einstein developed a theoretical understanding of Brownian motion a century ago, during his *annus mirabilis*, by combining thermodynamics and mechanics [1]. In this ground-breaking work, Einstein showed that the random lateral motion of particles suspended in a liquid depends only on particle size, viscosity of the liquid, and temperature. The following year, Einstein generalized his ideas and predicted similar random rotational motion, which again is solely determined by the same three parameters [2]. The dependence on particle size makes Brownian motion attractive for biological sensing applications. This is because the Brownian motion changes due to an increase in particle size after particles with a well-defined affinity to a specific target bind with the target. The main problem that remains is to obtain an easily quantifiable measurement of the Brownian motion. Rotational Brownian motion of magnetic nanoparticles is particularly well-suited for such detection, since it can be readily measured, as we will see.

There has recently been an increased interest in fabricating magnetic nanoparticles with biologically relevant ligand coatings [3]. These particles are used for a variety of biological and medical applications, such as magnetic separation, targeted drug delivery,

Biomedical Applications of Nanotechnology. Edited by Vinod Labhasetwar and Diandra L. Leslie-Pelecky
Copyright © 2007 John Wiley & Sons, Inc.

hypothermal treatments, and contrast enhancement for magnetic resonance imaging. In addition, there are several approaches that utilize magnetic nanoparticles as labels for biological sensing. For most of these sensing schemes the basic idea is similar to other immunoassay techniques, which are based on labeling the targets with enzymes or with fluorescent or radioactive particles.

The magnetic labels can be fabricated with sizes as small as a few nanometers. This means that they can be prepared with high surface-to-volume ratios and that their mobility in solution can be very high. In addition, the magnetic labels have several properties that make them distinct from other commonly used labels. Most biological systems do not exhibit ferromagnetism and typically have only small magnetic susceptibilities due to their dia- or paramagnetism. Therefore, the magnetic properties can be detected with little noise interference from the biological environment, which is advantageous in comparison to fluorescence or optical based detection schemes. Furthermore, it is possible to apply significant forces onto the magnetic nanoparticles using magnetic field gradients. This can be used to guide their motion or to separate the magnetic nanoparticles from a liquid [4]. The shelf life of magnetic particles can be long compared to that for fluorescence, radioactivity, or enzyme-based labels. Also, there is a wide variety of magnetic materials that are nontoxic and environmentally benign in contrast to radioactive labels. The absence of toxicity of many magnetic labels means that they are suitable for both *in vitro* and *in vivo* applications. Most importantly, as is the main focus of this chapter, magnetic labels distinguish themselves by the fact that the binding itself may modify the dynamic magnetic response. This means that for the sensing applications described in this chapter there is no need for washing away unbound labels.

The chapter is organized as follows: In Section 4.2 we discuss general properties of rotational Brownian motion and present sensing schemes based on this effect, which do not utilize magnetic particles. Section 4.3 presents the magnetic relaxation of ferrofluids and how it relates to rotational Brownian motion. Section 4.4 discusses the application of ferrofluid magnetic relaxation to biomagnetic sensing schemes and compares them to alternative nonrelaxation-based biomagnetic sensing approaches. Section 4.5 provides an outlook to future developments, and Section 4.6 concludes with a brief summary.

4.2 SENSING APPLICATIONS UTILIZING ROTATIONAL BROWNIAN MOTION

Translational Brownian motion is readily accessible even through optical imaging provided that the particles suspended in liquid solution are large enough [5]. However, in order to have an appreciable change of the Brownian motion for sensing purposes, the particle size should be comparable to the target size, since the relative size change upon binding determines the sensitivity. This means that if the targets have sizes of a few nanometers, such as molecules or proteins, the particle size should be, at most, in the submicron range. However, the detection of translational Brownian motion for such small particles is very challenging [6–8]. Hence, an alternative approach is to use rotational Brownian motion for the signal transduction of biochemical binding events. In one of the earliest measurements, the random rotational motion of a wire in a liquid was detected by light reflected from a mirror attached to the wire [9]. Obviously, the detection of rotational Brownian motion for spherical particles is more complicated and requires an anisotropy to break the spherical symmetry. Examples of symmetry-breaking are permanent electric dipole moments, whose motion can be detected with inelastic light scattering [10], or anisotropic optical properties, which leave a distinct signature of the rotational motion in dynamical light-scattering experiments [11]. Clearly, a permanent magnetic moment will also break the spherical symmetry; its use for the detection of rotational Brownian motion is discussed in Section 4.3.2. An alternative approach for the detection of Brownian relaxation is nuclear magnetic resonance, since the relaxation time for nuclear spins is modified by the rotational motion of the particles [12].

In addition to using intrinsic anisotropies, it is also possible to add anisotropy artificially. For example, fluorescent particles can be coated on one side with an optically opaque medium such as gold, so that the fluorescence can only be observed from one side [13, 14]. The rotational motion of these half-coated fluorescent particles gives rise to a characteristic "blinking." The intensity of the fluorescent signal can be analyzed in terms of an autocorrelation function [14], which decays exponentially in time with the Brownian relaxation time τ_B [15] given by [16, 17]:

$$\tau_B = \frac{\kappa V_H \eta}{2k_B T},$$ (4.1)

where V_H is the hydrodynamic volume of the particle, η is the viscosity of the liquid, $k_B = 1.38 \times 10^{-23}$ J/K is Boltzmann's constant, T is temperature, and κ is a shape factor, which is for spherical particles κ is equal to 6 [15].

Values of τ_B are tabulated in Table 4.1 for different solvents and sphere diameters. As can be seen, the relaxation times can vary over a large timescale ranging from microseconds to days. The dependence of τ_B on viscosity has been employed to determine

Table 4.1. Calculatd Relaxation Time for Rotational Brownian Motion for Spherical Particles in Pure Water and 98% Glycerol Solution

Sphere Diameter (μm)	Water 37°C (0.7 mPa·sec)	Water 20°C (1.0 mPa·sec)	Glycerol 20°C (957 mPa·sec)
13.	9.8 min	14 min	9.3 day
4.4	22 sec	32 sec	8.7 hr
2.	2.1 sec	3.1 sec	48 min
1.	270 msec	380 msec	6.1 min
0.3	7.2 msec	10 msec	9.9 sec
0.1	270 μsec	380 μsec	370 msec
0.05	34 μsec	48 μsec	46 msec

Adapted from Ref. 18.

local rheological properties in biological systems using half-coated fluorescent particles [14].

The simple expression of Eq. (4.1) describes most of the experimentally observed relaxation times reasonably well and independent of the particle concentration, since interactions between particles due to hydrodynamic forces are typically very weak. However, at sufficiently high particle concentrations, small deviations from Eq. (4.1) have been observed [11]. The interactions give rise to a rescaling of the temperature in Eq. (4.1) with a concentration-dependent effective temperature. That means that even at high concentrations, precise relative measurements of viscosities or particle sizes are possible as long as the concentration stays constant.

As mentioned above, there are several fluorescence and optical approaches for the detection of rotational Brownian motion. Using a magnetism-based approach, as discussed in the following section, has several distinct advantages. While biological systems may have sizable fluorescence of their own, they generally present only a very low background for the detection of magnetic properties. More importantly, the motion of magnetic particles can be directly driven by externally applied magnetic fields. This allows measurements of Brownian motion directly in the frequency domain (see Section 4.3.2.), which in turn makes it easier to detect the motion of smaller particles with short relaxation times.

4.3 MAGNETIC RELAXATION IN FERROFLUIDS

Ferromagnetic particles suspended in a liquid carrier are generally referred to as ferrofluids [19]. The dynamic behavior of these ferrofluids is governed by two very different magnetization relaxation pathways. Either the magnetization can change within each magnetic particle (Néel relaxation) or the whole particle can rotate in the liquid (Brownian relaxation). We will first discuss each mechanism individually and then turn to systems with size dispersions, where both relaxation mechanisms can be present simultaneously.

4.3.1 Internal Magnetic Relaxation

In the following we will neglect magnetic interactions between magnetic particles, which may arise from dipolar coupling. This assumption is nontrivial, but in the case of biologically functionalized magnetic nanoparticles, the ligand coating helps to separate the individual particles independent of their concentration. The important parameter for assessing the importance of interactions is the ratio of interaction to thermal energies $\lambda = E_{int}/k_B T$. E_{int} depends on both the saturation magnetization, particle volume, and interparticle distance (concentration of suspension). As an example, for magnetite (Fe_3O_4) particles with 10-nm diameter, λ becomes negligibly small ($\lambda < 0.1$) for ligand coatings with a thickness exceeding 8 nm [20]. In any case, interactions are more important for Néel (internal) relaxation and less so for Brownian relaxation [21].

The magnetization in each particle is coupled to the orientation of the particle through magnetic anisotropies, which may arise from the crystal structure or the overall shape of the particles. In nanoparticles the magnetic anisotropy is typically increased compared to bulk materials due to additional surface contributions [22]. These anisotropies create energy barriers ΔE between different orientations of the magnetization relative to the particle. In the simplest case of a uniaxial anisotropy (characterized by an anisotropy constant K) this energy barrier is given by $\Delta E = KV_m$, where V_m is the magnetic volume of the particle. In this case the Néel relaxation follows a simple Arrhenius law [23]:

$$\tau_N = \tau_0 e^{KV_m/k_BT}. \tag{4.2}$$

The pre-exponential factor τ_0 is generally assumed to be between 10^{-8} and 10^{-12}, see Ref. 24. Since the Néel relaxation only depends on parameters specific for the given magnetic materials (K, V_m, and τ_0), it is not modified by targets bound to the ligand coating of the magnetic particles.

4.3.2 Magnetic Relaxation Through Mechanical Motion

When magnetic particles are suspended in a liquid, the magnetization can also relax via rotational Brownian motion in addition to Néel relaxation [19]. The relaxation time associated with rotational motion is identical with the Brownian relaxation time τ_B in Eq. (4.1) and thus is independent of magnetic properties. As is shown in Figure 4.1, the two relaxation times τ_N and τ_B depend very differently on particle size. Notice that the particle diameter d in Figure 4.1 refers in one case (τ_N) to the magnetic diameter, while in the other case (τ_B) it indicates the hydrodynamic diameter. In general the two are not the same due to the surfactant coating of the magnetic particles.

Since for most particle diameters the timescales for the two relaxation mechanisms are very different, the two processes are generally independent of each other [25, 26].

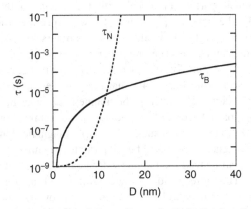

Figure 4.1. Néel τ_N (dashed line) and Brownian τ_B (solid line) relaxation time as a function of particle diameter D. D indicates the magnetic diameter for τ_N, while it indicates the hydrodynamic diameter for τ_B. For the calculations based on Eqs. (4.1) and (4.2) we used bulk magnetic properties of magnetite (Fe_3O_4) and the viscosity of water at room temperature.

In this case, one can define an effective relaxation time $\tau^{-1} = \tau_N^{-1} + \tau_B^{-1}$. From this expression it is clear that the shorter of the two relaxation times determines the dominant relaxation mechanism. As can be verified from Figure 4.1, below a critical particle size (determined by $\tau_N = \tau_B$) the relaxation is mainly Néel type, while above a critical particle size, it is mainly Brownian. Magnetic particles that are small enough, such that their relaxation is dominated by Néel relaxation, are also referred to as superparamagnetic or unblocked particles.

In principle the relaxation time can be determined directly by measuring the time-dependent response of the magnetization to a change in magnetic fields [27]. However, it is difficult to measure the time-dependent magnetization of ferrofluids with a time resolution below 1 msec. This means that the Brownian relaxation of submicron ferromagnet particles in aqueous solution cannot be determined by time-resolved measurements (see Table 4.1).

In order to get information about the relaxation mechanisms in ferrofluids it is often more convenient to determine the frequency-dependent susceptibility, instead of measuring the direct time dependence of the magnetization [28]. The complex susceptibility as a function of frequency ω can be expressed as

$$\chi(\omega) = \chi'(\omega) - i\chi''(\omega), \tag{4.3}$$

$$\chi'(\omega) = \frac{\chi_0}{1 + (\omega\tau)^2}, \tag{4.4}$$

$$\chi''(\omega) = \frac{\chi_0 \omega \tau}{1 + (\omega\tau)^2}, \tag{4.5}$$

where τ is the effective relaxation time and χ_0 is the static susceptibility. As a function of frequency, the real part χ' decreases monotonically with a step at the inverse relaxation time, while the imaginary part χ'' has a peak with its maximum at the inverse relaxation time [29]. Generally, the peak in χ'' occurs at low frequencies (up to 100 kHz) for Brownian relaxation [see Eq. (4.1)], while Néel relaxation (see Eq. (4.2) typically gives rise to a peak at higher frequencies (i.e., megahertz to gigahertz range).

In order to determine pure Brownian motion from magnetic susceptibility measurements, we have seen from Figure 4.1 that the particles need to have a minimum size to avoid Néel relaxation. However, if the magnetic particles become too large, the magnetization in each particle ceases to be homogeneous (single domain) and an inhomogeneous magnetization structure, such as multiple magnetic domains, can develop [30]. In this case the magnetic response of each particle may no longer be dominated by Brownian relaxation, but instead the magnetization can change internally—that is, via domain-wall motion. Therefore, besides a lower limit there is also an upper limit for the suitable size of ferromagnetic particles for which their Brownian rotational motion can readily be detected via susceptibility measurements [31]. The maximum single-domain size depends on materials-specific parameters, such as exchange stiffness, magnetic anisotropy, and saturation magnetization [30]. We show in Figure 4.2 suitable particle sizes for stable single-domain magnetic particles (large enough to avoid Néel relaxation,

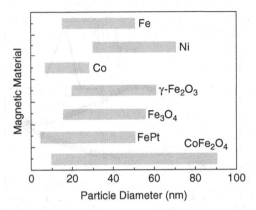

Figure 4.2. Particle diameters for stable single domain magnetic nanoparticles. (Adapted from Ref. 31.)

small enough to avoid multiple domains) for various magnetic materials. For almost all materials the useful particle size is of the order of tens of nanometers.

The frequency dependence of the complex magnetic susceptibility can be experimentally determined several ways. A small ac magnetic field applied to the sample results in a time-varying magnetic moment in accordance with Eq. (4.4) ($\chi = dM/dH$, where H is the applied magnetic field and M is the magnetic moment). By placing the sample into a pick-up coil, the time-dependent moment induces a current in the pick-up coil, thus allowing measurement of χ without sample motion. Alternatively, the pick-up coil itself can be used as the source of the alternating magnetic field and the susceptibility can be determined directly from the impedance of the coil. For measurements at very high frequencies (>0.1 GHz) a transmission line can be used instead of a pick-up coil [32].

In Figure 4.3 we show an example of Brownian rotational relaxation measured via the frequency dependence χ''. The solid symbols show χ'' at 300 K for avidin-coated Fe_3O_4 particles suspended in a phosphate buffered saline (PBS, pH = 7) solution with a concentration $\approx 2 \times 10^{15}$ particles/ml. For these particles, the magnetic diameter is ≈ 10 nm and the hydrodynamic diameter is ≈ 50 nm. A peak is observed at 210 Hz. The measurement was repeated at 250 K (open symbols in Figure 4.3), which is below the freezing point of the PBS solution. In this case the peak in χ'' disappears, verifying that the peak at 210 Hz in the 300 K data is indeed due to rotational Brownian motion [31]. The solid line in Figure 4.3 shows a fit of the data to Eqs. (4.1) (4.6) and (4.7) using a size dispersion of $\pm 12\%$ (see also Section 4.3.3).

4.3.3 Systems with Size Distributions

Any real ferrofluid system will generally have a distribution of sizes for the ferromagnetic particles, which may influence the experimentally observed susceptibility. In fact, it has been shown that in some cases the Néel relaxation peak cannot be understood completely in terms of a size distribution alone, but a distribution of effective magnetic anisotropies is also required [33]. However, since in this review we are mostly concerned with

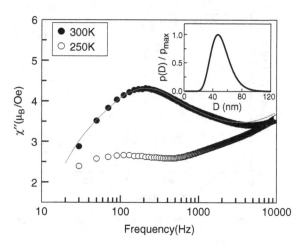

Figure 4.3. Imaginary part χ'' of the ac magnetic susceptibility as a function of frequency for avidin-coated magnetite nanoparticles suspended in a PBS buffer solution. Solid symbols are measured at 300 K, open symbols at 250 K, which is below the freezing point of the liquid. Solid line is a fit using Eqs. (4.1), (4.6), and (4.7) and utilizing the 250 K data as background. The inset shows the log-normal size distribution [Eq. (4.7)] used for the best fit, with D being the hydrodynamic diameter of the particles. (Adapted from Ref. 20).

Brownian relaxation, we can ignore any dispersion in the anisotropy. In this case the measurable susceptibility can be represented as an integral of Eq. (4.5) over the particle size distributions [20]:

$$\chi''(\omega) = \frac{1}{3}\hat{\chi}_0 \int_0^\infty dr_m \int_0^\infty dr_h P(r_m)P(r_h)\frac{r_m^6}{4}\frac{\omega\tau(r_m, r_h)}{1 + [\omega\tau(r_m, r_h)]^2}, \qquad (4.6)$$

where $P(r_m)$ is the size distribution of the magnetic radius, and $P(r_h)$ is the size distribution of the hydrodynamic radius. Both size distributions are required, since the effective relaxation time τ depends through τ_N and τ_B on r_m and r_h, respectively.

Typically the size distribution will follow a log-normal distribution [34] given by

$$P(r) = \frac{1}{\sigma\sqrt{2\pi}}\exp\left[-\frac{(\ln r - \ln \bar{r})^2}{2\sigma^2}\right], \qquad (4.7)$$

where $\bar{r}$ is the median size and σ is the logarithm of the standard deviation. A typical log-normal size distribution is shown in the inset of Figure 4.3, which is the distribution we used in Figure 4.3 for fitting χ''. We will base our discussion of size distributions on Eq. (4.7) even though other distributions are possible. In order to understand a given system, it is generally best to directly determine the size distribution or to obtain it from magnetization measurements [35, 36].

In Figure 4.4 we show susceptibilities obtained by using a size distribution of the magnetic core [according to Eq. (4.7)] and a fixed shell thickness. The calculations in

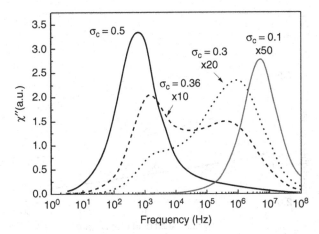

Figure 4.4. Imaginary part χ'' of ac susceptibility as a function of frequency calculated for different size dispersions σ_c for the magnetic core of magnetic nanoparticles. (Adapted from Ref. 20.)

Figure 4.4 were based on magnetite (Fe_3O_4) nanoparticles with 10-nm median magnetic core diameter and 20-nm shell thickness suspended in aqueous solution [20]. As can be seen, depending on the width of the size distribution, given by σ, there can be one or two peaks in χ'' at low and/or high frequencies. This indicates that due to the well-separated τ_N and τ_B as a function of size (see Figure 4.1) the size distribution becomes effectively bisected at a size given by $\tau_N = \tau_B$, such that the particles show either only pure Néel or Brownian relaxation. As a consequence the positions of the Néel and Brownian relaxation peaks become dependent on the distribution width.

An important question is how size dispersion of the magnetic core affects the detectability of an increase in the hydrodynamic radius, which serves as a basis for some of the Brownian sensing scheme under discussion (see Section 4.4.2). For a vanishing size dispersion one expects from Eqs. (4.1) and (4.5) that the peak frequency changes as $\omega_B \propto 1/r_h^3$. However, as is shown in Figure 4.5, this dependence becomes slower with increased size distribution. This means that with increased size dispersion a modified hydrodynamic radius will be detected with reduced sensitivity [20].

Furthermore, as can be seen from Figures 4.4 and 4.5, the Brownian relaxation peak shifts to lower frequencies with increased size distribution. This is a disadvantage for dynamic susceptibility measurements, since for most detection methods the signal amplitude is proportional to the frequency, and thus the shift to lower frequencies requires a larger number of magnetic particles for successful detection [37].

4.4 SENSING BASED ON MAGNETIC RELAXATION

As discussed in the previous section rotational Brownian motion of magnetic nanoparticles can be easily detected via measurements of the complex magnetic susceptibility. There are two approaches: one where biological targets are detected via complete

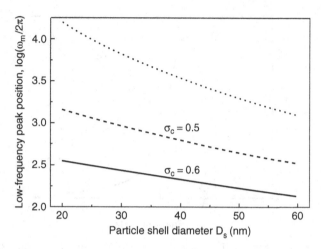

Figure 4.5. Evolution of the low-frequency peak position of the imaginary part χ'' of the dynamical susceptibility as a function of shell thickness. (Adapted from Ref. 20).

suppression of the Brownian relaxation (Section 4.4.1), and the other based on a modification of the Brownian relaxation (Section 4.4.2). After presenting each approach, we will compare them to other magnetic-biomedical sensing schemes in Section 4.4.3.

4.4.1 Suppression of Brownian Relaxation

When magnetic particles in a ferrofluid are immobilized, their magnetization can only relax via Néel relaxation. For sufficiently large particles that implies that the relaxation time will be significantly increased compared to the Brownian relaxation time (see Figure 4.1). One example from Grossman et al. [38] is shown in Figure 4.6, where 50-nm-diameter γ-Fe_2O_3 particles coated with antibodies against *Listeria monocytogenes* are used. Using a superconducting quantum interference device (SQUID) magnetometer, the magnetization is measured after applying a 0.4-mT magnetic field for 1 sec. As can be seen in Figure 4.6 when *L. monocytogenes* is added to the magnetic nanoparticles, their relaxation is slowed compared to the free particles. Also by using *Escherichia coli* as a control, it is shown that, indeed, nonspecific binding can be avoided. Using this approach, Grossman et al. [38] demonstrated the detection of $(1.1 \pm 0.2) \times 10^5$ *L. monocytogenes* bacteria in a 20-μl sample volume.

If the targets are large enough, such as bacteria (see the above example) or cells, then binding to the targets directly leads to immobilization of the magnetic particles. For smaller targets, where the binding event itself does not completely suppress Brownian relaxation, there are two alternative approaches to sensing via immobilization [39]. One possibility is a sandwich detection, where the target is first labeled by the magnetic nanoparticle and the target is then subsequently immobilized—that is, by binding to a substrate. Alternatively, one can use a competitive detection, where the nonferromagnetic analyte and the label compete for the same binding to a substrate. In this case the presence of the target material is indicated by continued Brownian relaxation. It should also be

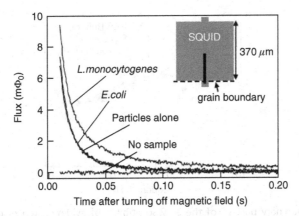

Figure 4.6. Magnetization relaxation of γ-Fe_2O_3 nanoparticles conjugated with monoclonal mouse anti-biotin antibodies. Samples with *L. monocytogenes* show a slower relaxation compared to samples with the free particles or with *E. coli*. (Adapted from Ref. 38).

mentioned that in the rare case when the target itself is ferromagnetic, direct detection is possible by immobilizing the target, without the need for additional labels.

Most sensing schemes based on suppression of Brownian relaxation proceed by first applying a magnetic field to saturate the magnetization of the magnetic nanoparticles, and then, after the field is removed, the remanent magnetization and its time-dependent decay is measured with high-sensitivity magnetic field sensors, such as a SQUID [38, 40–43]. The distinct advantage of this approach is that the actual magnetization measurements are performed without the presence of an applied magnetic field. Therefore any magnetic background from para- or diamagnetism of the environment is minimized.

The amount of reacted magnetic nanoparticles can be quantified via either an increase in remanence [40, 42, 43] or an increase in the relaxation time [41, 42]. Using the later approach, a lower detection limit of $(5 \pm 2) \times 10^4$ 30-nm diameter magnetite particles has been demonstrated [42]. This compares favorably to the well-established enzyme linked immunosorbent assay (ELISA) technique, which typically has a lower limit of $\approx 10^5$ labeled antigens. It is also possible to determine binding rates with successive relaxation measurements since there is no need to wash away unbound labels [38]. Furthermore, magnetic relaxation measurements can potentially be used for *in vivo* diagnostics [44].

4.4.2 Modification of Brownian Relaxation

An alternative approach to biomedical sensing monitoring the relaxation time τ_B for rotational Brownian motion provides for. Valberg *et al.* [45] used the dependence of τ_B on viscosity η [see Eq. (4.1)] in order to determine cytoplasmic viscosities. Connolly and St. Pierre were the the first to suggest that the increase in hydrodynamic size, due to the binding of the target materials to the magnetic nanoparticles, would also give rise to a detectable change in the magnetic susceptibility [46]. This latter approach is particularly well-suited to smaller targets—that is, individual molecules or proteins.

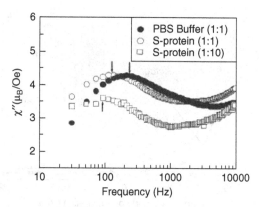

Figure 4.7. Imaginary part χ'' of the ac susceptibility of avidin-coated magnetic particles before (solid) and after (open) binding to biotinylated S protein. Different concentrations of S protein (open circles and squares) result in similar frequency shifts. (Adapted from Ref. 31.)

An example of this approach is shown in Figure 4.7, were the imaginary part of the magnetic susceptibility was measured for 10-nm magnetite (Fe_3O_4) nanoparticles coated with avidin and suspended in a PBS solution [31]. The hydrodynamic diameter of these samples is approximately 50 nm due to the avidin coating. We used biotinylated S-protein as a target, since the interaction between biotin and avidin has been well characterized in the literature with a large affinity constant on the order of a femtomole [47]. As can be seen in Figure 4.7, the relaxation peak is at 210 Hz for the particles without the target, and it shifts to 120 Hz after the target is added to the solution. We expect that the peak frequency changes as $\omega_B \propto 1/r_h^3$ from Eqs. (4.1) and (4.5). Using the initial diameter of the avidin-coated magnetite particles of about 50 nm, this relation implies that the particle diameter after S-protein binding is about 60 nm. This increase corresponds well to the known size of S-protein (4 nm, estimated from the crystal structure of its "parent" protein, ribonuclease A). In the measurements shown in Figure 4.7 the sensor is overdosed, which means that there is more target material than binding sites on the magnetic nanoparticles. This is indicated by the fact that further addition of S-protein does not continue to change the peak frequency. Since the sample gets more diluted with the addition of more analyte, the fixed peak in χ'' also shows that the result is concentration independent. This indicates that interparticle interactions are negligible and that the peak shift is not caused by changes of the viscosity, which could be concentration-dependent. It should also be pointed out that viscosity changes due to changes in temperature can be ignored, since the estimated frequency shift is only ≈ 0.1 Hz for a 1°C temperature change of an aqueous solution near room temperature [46].

Besides biotinylated S-protein, we also added biotinylated T7 bacteriophage to the avidin-coated magnetite particles. In this case the relaxation peak was completely suppressed, indicating that the binding to the T7 bacteriophage immobilized the magnetite particles, similar to the measurements discussed in Section 4.4.1. In our example, each T7 bacteriophage had 415 copies of biotin, which permits an effective cross-linking of the avidin-coated magnetite particles. This result indicates that the sensing scheme based on modification of the Brownian motion can discriminate between different targets, even when the biochemical affinity is the same. The reason for this is that while chemical affinity is a first step to detect the desired target, the frequency shift contains additional information about the size of the detected targets. Another advantage is that since the detection is in a solution medium, changing the activity of the biomolecules can be readily achieved. Furthermore, the magnetic particles remain in solution during the detection. Thus, after detection, the magnetic particles, together with the targets, can be removed from solution using high-gradient magnetic filters [4].

In the example shown in Figure 4.7 the magnetic nanoparticles were overdosed. Astalan et al. [37] investigated how the Brownian relaxation peak in χ'' changes in the case of a partial dosage. This study used magnetite dextran composite particles coated with antiPSA, which can be used to detect binding to prostate-specific antigen (PSA). Depending on the target concentration, the maximum in χ'' shifted from 255 Hz (without target) to 195 Hz (saturation), such that $\Delta\omega \propto 1/c$, where $\Delta\omega$ is the frequency shift relative to the frequency at saturation and c is the relative coverage of the nanoparticles by the target. This shows that the modified Brownian motion approach can yield quantitative information. A lower limit of 0.3-pmol sensitivity can be obtained from the concentration-dependent data, which are comparable to conventional ELISA. It should be noted that this sensitivity can be expected to increase significantly by narrowing the size distribution of the magnetic nanoparticles (see Section 4.3.3) and by miniaturizing the ac susceptibility measurement system.

4.4.3 Alternative Biomagnetic Sensing Schemes

The main distinction of the biological sensing approaches discussed in the previous two sections is that the binding of the target to the magnetic nanoparticles changes the time-dependent magnetic properties via either suppressing or modifying the Brownian relaxation. This means that these approaches can be used in a substrate-free mode (mix-and-measure), where the analyte is simply combined with the magnetic nanoparticles and no additional steps are required, such as washing away unbound particles. Furthermore, in the case of modified Brownian motion, there is a useful magnetic signature from the nanoparticles both with and without the target present. This inherent integrity check makes this approach especially suitable for continuous monitoring.

There has also been much work recently on biological sensors which emulate more traditional immunoassay techniques by using magnetic nanoparticles as a label with a readily detectable, magnetic moment. One of the earliest examples is the bead array counter (BARC). BARC uses a magnetic field sensor to detect the stray field of a magnetic particle bound to the target that is also bound to the sensor. The latter is accomplished via affinity selective binding, such as with DNA–DNA, antibody–antigen, or ligand–receptor interactions [48, 49]. The magnetic field sensor can be based on a variety of detection methods, such as via giant magnetoresistance [48–51], spin valves [52, 53], anisotropic magnetoresistance [54], Hall effect [55], planar Hall effect [56], or SQUIDs made of either low transition temperature (T_c) [41, 57] or high T_c [42, 43]

superconductors. Interestingly, in spite of the variety of techniques, the sensitivity for detecting biologically functionalized magnetic particles is fairly comparable. Using giant magnetoresistive sensors, Schotter *et al.* [51] demonstrated the possibility to detect about 200 labels, based on 0.86-μm-diameter magnetic particles. This sensitivity is about two orders of magnitude higher than for assays based on fluorescent labels. Note that the size of the magnetic particles is substantially larger than in the Brownian motion case. This is because the magnetic particles have to be large enough to produce a significant magnetic field. For example, while it has been demonstrated that with spin valve sensors a single 2.8-μm-diameter magnetic bead can be detected [53], there is an insufficient signal-to-noise ratio for smaller particles, such that even for a larger number of beads the smallest detectable bead size has a diameter of 0.25 μm [52]. Similarly, for sensors based on Hall and planar Hall effects, single 2-μm beads have been detected [55,56], but there are no comparable data available for smaller particles. However, SQUID-based sensors have a better sensitivity and for the best available sensing systems a detection limit of $\approx$4000 35-nm-diameter particles has been demonstrated [58]. But one of the drawbacks of SQUID-based sensors is that they require a cryogenic operating environment.

Besides high sensitivity, there are other advantages of using a substrate-based stray-field detection technique for biomedical sensing. In comparison to fluorescent-based approaches, the shelf-life of magnetic markers is virtually unlimited. Also, the electronic sensor-based detection of the magnetic stray field permits a direct electrical read-out of the binding information. Furthermore, substrate-based sensors can be integrated with microfluidic devices for delivering the analyte and the magnetic markers to the active sensing area [50]. In addition, the location of magnetic beads can be manipulated by field gradients of current lines integrated in the substrate [59]. This can be utilized for moving and concentrating the magnetic beads near the field sensors. However, one of the distinct disadvantages is that two binding events are always required (magnetic marker to target and target to field sensor) and that unbound labels need to be removed in order to avoid false positive signals. In Table 4.2 we present a brief comparison between the

Table 4.2. Comparison Between Sensing Schemes Based on Detection of Stray Magnetic Fields After Magnetically Tagging the Target and Based on Either Suppression or Modification of Brownian Relaxation

	Detection via		
	Stray Field	Suppression of Brownian Relaxation	Modification of Brownian Relaxation
Binding events required	2	1 or 2	1
Magnetic particle size	$\approx$1 μm	10 nm to 1 μm	10–100 nm
Signal present without target	·No	Maybe	Yes
Sensitivity to size of target	No	No	Yes
Separation of target	No	Maybe	Yes
In vivo application	No	Maybe	Maybe

different requirements and possibilities of stray-field-based and Brownian-relaxation-based sensing techniques.

4.5 OUTLOOK

4.5.1 Open Questions

The development of Brownian relaxation approaches for biomedical sensing applications is experiencing a promising start. In order to harness the full potential of this method, there are still many open questions that need to be addressed. Most of the measurements—in particular, ones based on modified Brownian relaxation—are proof-of-concepts [20,31] and therefore have not optimized the apparatus in order to determine the ultimate achievable sensitivity. There are two major pathways to increase the sensitivity of the detection based on modified Brownian relaxation. Narrowing the size distribution of the magnetic particles used will result in better resolution for detecting changes of the relaxation time, and it will also provide for a better signal, since the Brownian relaxation signal can be detected at higher frequencies. At the same time, a miniaturization of the susceptibility measurement system will allow us to investigate smaller sample volumes.

For the modification of Brownian motion there are first results indicating the possibility to quantify concentration levels [37]. However, it is unclear if there is a universal relation between the relative target-to-nanoparticle concentration and the shift of the relaxation peak. It might be necessary to analyze the full shape of the relaxation peak. For example, with a partial dosage of targets to the magnetic nanoparticles, one expects that the peak width would increase due to a broadening of the particle size distribution.

Since the modification of Brownian motion approach permits one to infer information about the target size, it has been shown that it can discriminate between targets with similar biochemical affinity [31]. Another question is: can this discrimination still be possible in heterogeneous mixtures of potential targets? If it is possible, what are the limiting factors? That is, what is the smallest detectable size difference?

One of the unique opportunities for relaxational detection of magnetic nanoparticle is its potential for *in vivo* applications. Magnetic nanoparticles can have low toxicity, and measurements are, in principle, possible within a living body. There is no need for optical access. In fact, Néel relaxation of magnetic nanoparticles injected into mice has been observed [44]. But an open question that remains is whether useful information can also be obtained from the Brownian relaxation. Are the hydrodynamic properties in the living body (i.e., in the bloodstream) homogeneous enough to allow conclusions to be drawn from the Brownian relaxation time? If not, is it still possible to use the suppression of Brownian motion for detection purposes?

Clearly, many questions remain before biomedical sensing based on Brownian rotation of magnetic nanoparticles achieves its full potential. There are many opportunities for interdisciplinary approaches to realize the promises of early laboratory tests. Finally, we will outline below one such opportunity, which might significantly enhance the usefulness of the sensing approaches discussed in this chapter. It is based on magnetic viruses.

4.5.2 Magnetic Viruses

In Section 4.3.3 we showed that the sensing scheme based on modification of Brownian relaxation can tolerate a broad size distribution. However, a narrow size distribution is desirable, since it increases the sensitivity to small targets. Also, functionalizing the magnetic nanoparticles for the desired affinity can be challenging. Both issues can be addressed by our recent development of a magnetic virus [60].

Combining biological self-organization with inorganic materials is a promising way to create a wide variety of new functional materials [61]. In particular, several biological systems have been used as templates to fabricate inorganic nanoparticles with well-defined sizes. An example is ferritin, which is composed of 24 nearly identical subunits. Ferritin encapsulates iron(III) oxy-hydroxide, ferrihydrite, whose biological function is to store iron [62]. Apoferritin is the same protein without the native iron core; its 8-nm cavity can be used for templating different magnetic particles [63].

Viruses are particularly interesting as templates for inorganic materials synthesis. Viruses are systems in which nucleic acid strands are confined within a nano-sized ($\approx$10–500 nm) compartment, which is referred to as the virus capsid shell. Virus capsids come in a variety of sizes and shapes that are suitable for templating [64]. Virus capsids are also mechanically stable [65] and give rise to a narrow size distribution. For example, synthetic virus capsids have been used to template (4.1 ± 1.1)-nm-diameter γ-Fe_2O_3 nanoparticles [66].

Using a natural virus instead of synthetic virus capsids has the advantage that their biochemical functionality can be readily selected through phage display libraries. Phage display technology permits the display of suitable ligands on the virus capsid for any desired target [67,68]. In addition, the recombinant viral particles are highly stable and can be amplified (grown).

We demonstrated the feasibility to fabricate biofunctionalized magnetic nanoparticles using viruses selected through phage display with T7 bacteriophage. T7 offers a capsid with about a 40-nm inner diameter [69]. This size is ideally suited for sensing applications based on the modification of Brownian relaxation (see Figure 4.2).

As shown in Figure 4.8 the basic idea of the magnetic virus is first to remove the DNA from the intact virus (see Figure 4.9a for a transmission electron microscopy image of T7 bacteriophage) and then to replace it with a suitable magnetic material. The first step of removing the DNA generates an empty virus capsid or ghost virus. Treating T7 phage with an alkali buffer produced ghost viruses with a high yield (>98%), as shown in Figure 4.9b. Finally, using biomineralization, the inorganic material can be synthesized inside the empty virus capsids with a narrow size distribution (42 ± 2 nm). This is shown in Figure 4.9c for iron oxide particles. The preparation of ferromagnetic Co-filled T7 bacteriophage has also been demonstrated [60].

Most importantly, we also demonstrated that the magnetic viruses maintained the original biological recognition functionality of the original viruses selected by phage display. Thus, besides uniform magnetic properties due to well-controlled size, these

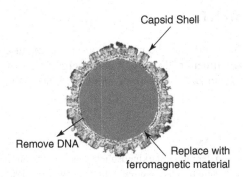

Figure 4.8. Schematic of a magnetic virus. The DNA is removed from the virus interior, which results in a rigid empty protein shell. This shell can be used as a template for ferromagnetic nanoparticles [Ref. 60].

novel hybrid inorganic/biological particles also have uniformity of their biofunctionality. For example, in our case each magnetic virus had 415 copies of 15 amino acid (aa) S-tag displayed on its capsid.

4.6 SUMMARY

A century after the first theoretical explanation of Brownian motion by Einstein there are now several approaches to utilize this information for biomedical sensing applications. The dependence of Brownian motion on particle size is a powerful tool for signal transduction when it is combined with the determination of rotational Brownian motion enabled by the use of magnetic nanoparticles. We have reviewed the recent approaches to these applications and have shown how they compare to other biomagnetic sensing schemes. In particular, the simplicity of mix-and-measure, the ability to obtain information beyond the biochemical affinity, and the potential for *in vivo* measurements indicate that further development for actual practical applications of these ideas are highly desirable. We are thus confident that ultimately this research will lead to useful technologies.

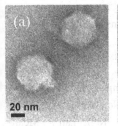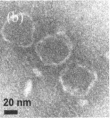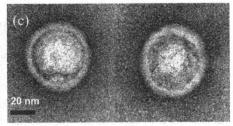

Figure 4.9. TEM images of T7 bacteriophage. (a) Normal phage, (b) ghost particles after osmotic shock, and (c) phage particles with iron oxide nanoparticles at their center.

ACKNOWLEDGMENTS

Our research in this field and thus the preparation of this manuscript were financially supported the U.S. Department of Energy, Basic Energy Sciences, under Contract No. DE-AC02-06CH11357 and DARPA, under Contract No. 8C67400.

REFERENCES

1. Einstein A. Über die von der molekularkinetischen Theorie der Wärme geforderte Bewegung von in ruhenden Flüssigkeiten suspendierten Teilchen. *Ann Phys* 1905;17:549–560.

2. Einstein A. Zur Theorie der Brownschen Bewegung. *Ann Phys* 1906;19:371–381.

3. Pankhurst QA, Connolly J, Jones SK, Dobson J. Applications of magnetic nanoparticles in biomedicine. *J Phys D: Appl Phys* 2003;36:R167–R181.

4. Šafařík I, Šafaříková M. Use of magnetic techniques for the isolation of cells. *J Chromatogr B* 1999;722:33–53.

5. Brown R. A brief account of microscopical observations made in the months of June, July and August,1827, on the particles contained in the pollen of plants; and on the general existence of active molecules in organic and inorganic bodies. *Edinburgh Philos J* 1828;5:358–371.

6. Schätzel K, Neumann W-G, Müller J, Materzok B. Optical tracking of single Brownian particles. *Appl Opt* 1992;31:770–778.

7. Bar-Ziv R, Meller A, Tlusty T, Moses E, Stavans J, Safran SA. Localized dynamic light scattering: Probing single particle dynamics at the nanoscale. *Phys Rev Lett* 1997;78:154–157.

8. Khim KD, Banerjee A, Choi CK, Takagi T. Near-wall hindered Brownian diffusion of nanoparticles examined by three-dimensional ratiometric total internal reflection fluorescence microscopy (3-D R-TRIM). *Exp Fluids* 2004;37:811–824.

9. Gerlach W, Lehrer E. Über die Messung der rotatorischen Brownschen Bewegung mit Hilfe einer Drehwage. *Naturwissenschafen* 1927;15:15.

10. McConnell JR. *Rotational Brownian Motion and Dielectric Theory.* London: Academic Press; 1980.

11. Piazza R, Degiorgio V, Corti M, Stavans J. Rotational and translational self-diffusion of interacting spherical Brownian particles. *Phys Rev B* 1990;42:4885–4888.

12. Estève D, Urbina C, Goldman M, Frisby H, Raynaud H, Strzelecki L. Direct observation of rotational Brownian motion of spheres by NMR. *Phys Rev Lett* 1984;52:1180–1183.

13. Choi J, Zhao Y, Zhang D, Chien S, Lo Y-H. Patterned fluorescent particles as nanoprobes for the investigation of molecular interactions. *Nano Lett* 2003;36:995–1000.

14. Behrend CJ, Anker JN, Kopelman R. Brownian modulated optical nanoprobes. *Appl Phys Lett* 2004;84:154–156.

15. Valberg PA, Butler JP. Magnetic particle motions within living cells. Physical theory and techniques. *Biophys J* 1987;52:537–550.

16. Debye PJW. *Polar Molecules.* New York: The Chemical Catalog Company, Inc.; 1929.

17. Einstein A. *Investigations on the Theory of the Brownian Movement.* New York: Dover; 1956.

18. Behrend CJ, Anker JN, McNaughton BH, Brasuel M, Philbert MA, Kopelman R. Metal-capped Brownian and magnetically modulated optical nanoprobes (MOONs): Micromechanics in chemical and biological microenvironments. *J Phys Chem B* 2004;108:10408–10414.

19. Shliomis MI. Magnetic fluids. *Sov Phys-Usp* 1974;17:153–169.

20. Chung S-H, Hoffmann A, Guslienko K, et al. Biological sensing with magnetic nanoparticles using Brownian relaxation (invited). *J Appl Phys* 2005;97:10R101.

21. Zhang J, Boyd C, Luo W. Two mechansims and a scaling relation for dynamics in ferrofluids. *Phys Rev Lett* 1996;77:390.

22. Bødker F, Mørup S, Linderoth S. Surface effects in metallic iron nanoparticles. *Phys Rev Lett* 1994;72:282–285.

23. Néel L. Théorie du traînage magnétique des ferromagnétiques en grains fins avec applications aux terres cuites. *Ann Géophys* 1949;5:99–136.

24. Dickson DPE, Reid NMK, Hunt C, Williams HD, El-Hilo M, O'Grady K. Determination of f_0 for fine magnetic particles. *J Magn Magn Mater* 1993;125:345–350.

25. Shliomis MI, Stepanov VI. Theory of the dynamic susceptibility of magnetic fluids. *Adv Chem Phys* 1994;87:1–30.

26. Coffey WT, Fannin PC. Internal and Brownian mode-coupling effects in the theory of magnetic relaxation and ferromagnetic resonance in ferrofluids. *J Phys: Condens Matter* 2002;14:3677–3692.

27. Kötitz R, Fannin PC, Trahms L. Time domain study of Brownian and Néel relaxation in ferrofluids. *J Magn Magn Mater* 1995;149:42–46.

28. Fannin PC, Scaife BKP, Charles SW. Relaxation and resonance in ferrofluids. *J Magn Magn Mater* 1993;122:159–163.

29. Fannin PC. Investigating magnetic fluids by means of complex susceptibility measurements. *J Magn Magn Mater* 2003;258–259:446–451.

30. Kittel C. Physical theory of ferromagnetic domains. *Rev Mod Phys* 1949;21:541–583.

31. Chung SH, Hoffmann A, Bader SD, et al. Biological sensors based on Brownian relaxation of magnetic nanoparticles. *Appl Phys Lett* 2004;85:2971–2973.

32. Fannin PC, Relihan T, Charles SW. Investigation of ferromagnetic resonance in magnetic fluids by means of the short-circuited coaxial line technique. *J Phys D: Appl Phys* 1995;28:2003–2006.

33. Fannin PC, Relihan T, Charles SW. Experimental and theoretical profiles of the frequency-dependent complex susceptibility. *Phys Rev B* 1997;55:14423–14428.

34. Granqvist CG, Buhrman RA. Ultrafine metal particles. *J Appl Phys* 1976;47:2200–2219.

35. Berkov DV, Görnet P, Buske N, et al. New method for the determination of the particle magnetic moment distribution in a ferrofluid. *J Phys D: Appl Phys* 2000;33:331–337.

36. Fischer B, Huke B, Lücke M, Hempelmann R. Brownian relaxation of magnetic colloids. *J Magn Magn Mater* 2005;289:741–777.

37. Astalan AP, Ahrentorp F, Johansson C, Larsson K, Krozer A. Biomolecular reactions studied using changes in Brownian rotation dynamics of magnetic particles. *Biosens Bioelectron* 2004;19:945–951.

38. Grossman HL, Myers WR, Vreeland VJ, et al. Detection of bacteria in suspension by using a superconducting quantum interference device. *Proc Natl Acad Sci USA* 2004;101:129–134.

39. Kriz CB, Rådevik K, Kriz D. Magnetic permeability measurements in bioanalysis and biosensors. *Anal Chem* 1996;68:1966–1970.

40. Kötitz R, Matz H, Trahms L, et al. SQUID based remanence measurements for immunoassays. *IEEE Trans Appl Supercond* 1997;7:3678–3681.

41. Haller A, Hartwig S, Matz H, et al. Magnetic nanoparticles measured by a low-T_c SQUID system. *Supercond Sci Technol* 1999;12:956–958.

42. Chemla YR, Grossman HL, Poon Y, et al. Ultrasensitive magnetic biosensor for homogeneous immunoassay. *Proc Natl Acad Sci USA* 2000;97:14268–14272.

43. Enpuku K, Minotani T. Biological immunoassay with high-T_c superconducting quantum interference device (SQUID) magnetometer. *IEICE Trans Electron* 2001;84:43–48.

44. Romanus E, Hückel M, Groß C, et al. Magnetic nanoparticle relaxation measurement as a novel tool for in vivo diagnostics. *J Magn Magn Mater* 2002;252:387–389.

45. Valberg PA, Feldman HA. Magnetic particle motions within living cells. Measurements of cytoplasmic viscosity and motile activity. *Biophys J* 1987;52:551–561.

46. Connolly J, St. Pierre TG. Proposed biosensors based on time-dependent properties of magnetic fluids. *J Magn Magn Mater* 2001;225:156–160.

47. Green NM. Avidin. *Adv Protein Chem* 1975;29:85–133.

48. Baselt DR, Lee GU, Natesan M, Metzger SW, Sheehan PE, Colton RJ. A biosensor based on magnetoresistance technology. *Biosens Bioelectron* 1998;13:731–739.

49. Edelstein RL, Tamanaha CR, Sheehan PE, et al. The BARC biosensor applied to the detection of biological warfare agents. *Biosens Bioelectron* 2000;14:805–813.

50. Miller MM, Sheehan PE, Edlestein RL, et al. A DNA array sensor utilizing magnetic microbeads and magnetoelectronic detection. *J Magn Magn Mater* 2001;225:138–144.

51. Schotter J, Kamp PB, Becker A, et al. A biochip based on magnetoresistive sensors. *IEEE Trans Magn* 2002;38:3365–3367.

52. Ferreira HA, Graham DL, Freitas PP, Cabral JMS. Biodetection using magnetically labeled biomolecules and arrays of spin valve sensors (invited). *J Appl Phys* 2003;93:7281.

53. Li G, Joshi V, White RL, et al. Detection of single micron-sized magnetic bead and magnetic nanoparticles using spin valve sensors for biological applications. *J Appl Phys* 2003;93:7557–7559.

54. Miller MM, Prinz GA, Cheng S-F, Bournak S. Detection of a micron-sized magnetic sphere using a ring-shaped anisotropic magnetoresistance-based sensor: A model for a magnetoresistance-based biosensor. *Appl Phys Lett* 2002;81:2211–2213.

55. Besse PA, Boero G, Demierre M, Pott V, Popovic R. Detection of a single magnetic microbead using a miniaturized silicon Hall sensor. *Appl Phys Lett* 2002;80:4199–4201.

56. Ejsing L, Hansen MF, Menon AK, Ferreira HA, Graham DL, Freitas PP. Planar Hall effect sensor for magnetic micro- and nanobead detection. *Appl Phys Lett* 2004;84:4729–4731.

57. Weitschies W, Kötitz R, Bunte T, Trahms L. Determination of relaxing or remanent nanoparticle magnetization provides a novel binding-specific technique for the evaluation of immunoassays. *Pharm Pharmacol Lett* 1997;7:5–8.

58. Lee S, Myers WR, Grossman HL, Cho H-M, Chemla YR, Clarke J. Magnetic gradiometer based on a high-transition temperature superconducting quantum intereference device for improved sensitivity of a biosensor. *Appl Phys Lett* 2002;81:3094–3096.

59. Graham DL, Ferreira H, Bernardo J, Freitas PP, Cabral JMS. Single magnetic microsphere placement and detection on-chip using current line designs with integrated spin valve sensors: Biotechnological applications. *J Appl Phys* 2002;91:7786–7788.

60. Liu C, Chung S-H, Jin Q, et al. Magnetic viruses via nano-capsid templates. *J Magn Magn Mater* 2006;302:47–51.

61. Sarikaya M, Temrler C, Jen AK-Y, Schulten K, Baneyx F. Molecular biomimetic: nanotechnology through biology. *Nat Mater* 2003;20:577–585.

62. Harrison PM, Artymiuk PJ, Ford GC, et al. Ferritin: Function and structural design of an iron-storage protein. In Mann S, Webb J, Williams RJP, editors. *Biomineralization: Chemical and Biochemical Perspectives*. Weinheim, Germany: VCH; 1989, pp. 257–294.

63. Meldrun FC, Heywood BR, Mann S. *in vitro* synthesis of a novel magnetic protein. *Science* 1992;257:522–523.

64. Douglas T, Young M. Host–guest encapsulation of materials by assembled virus protein cages. *Nature (London)* 1998;393:152–155.

65. Ivanovska IL, Pablo PJ, Ibarra B, et al. Bacteriophage capsids: Tough nanoshells with complex elastic properties. *Proc Natl Acad Sci USA* 2004;101:7600–7605.

66. Allen M, Willitis D, Mosolf J, Young M, Douglas T. Protein cage constrained synthesis of ferrimagnetic iron oxide nanoparticles. *Adv Mater* 2002;14:1562–1565.

67. Rodi DJ, Makowski L. Phage-display technology—Finding a needle in a vast molecular haystack. *Curr Opin Biotech* 1999;10:87–93.

68. Kay BK, Kasanov J, Yamabhai M. Screening phage-displayed combinatorial peptide libraries. *Methods* 2001;24:240–246.

69. Studier FW. Bacteriophage T7. *Science* 1972;176:367–376.

5

DENDRIMERS AND HYPERBRANCHED POLYMERS FOR DRUG DELIVERY

Rangaramanujam M. Kannan, Omathanu P. Perumal and Sujatha Kannan

5.1 INTRODUCTION

Significant breakthroughs in molecular biology and material science have led to new possibilities of delivering therapeutic agents at an appropriate length scale to the biological system (Figure 5.1). In this process, drug delivery systems have metamorphosed from "macro" (delivery to an organ) to "micro" (to tissues) to "nano" scale lengths (to cells). One of the key challenges in the design of a drug delivery system is to deliver the drug effectively and efficiently to the target site with minimal systemic exposure. Although conventional polymers can achieve spatial and temporal control of drug release, they are somewhat limited by their ability to carry a high payload. Recent advances in supramolecular chemistry have resulted in the emergence of new architectures including dendritic polymers [1]. Dendritic polymers are tree-like polymers that can be classified into two main types based on their branching architecture as "perfectly branched" (dendrimers) and "imperfectly branched" (hyperbranched polymers or HBP). The tree-like architecture of these polymers is probably one of the most pervasive topologies observed universally throughout the biological system [2]. The concept of dendrimers

Biomedical Applications of Nanotechnology. Edited by Vinod Labhasetwar and Diandra L. Leslie-Pelecky
Copyright © 2007 John Wiley & Sons, Inc.

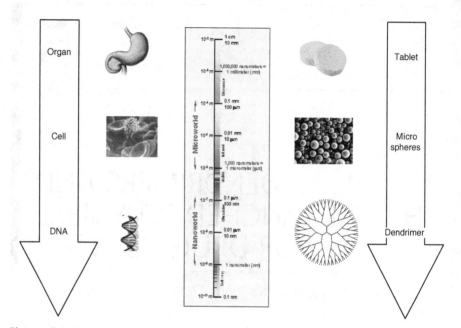

Figure 5.1. Progress of drug delivery from "macro" systems that interact at the organ level to "micro" systems that interact at the cellular level to "nano" systems that interact at the cellular level. Length scale has a significant influence on drug delivery in terms of reaching the target site, modifying the biodistribution of the drug, and enhancing the efficacy of the drug. See insert for color representation of this figure.

was first introduced by Tomalia in the 1980s with the poly(amidoamine) (PAMAM) family of dendrimers [3].

The dendritic polymers are characterized by a multifunctional central core, branching units (generation or "G"), and multiple functionalities at the periphery (nanoscale container or nanoscaffold). One of the distinct aspect of dendritic polymers is their nanoscale dimension varying in size from 2 to 15 nm, mimicking a diverse class of biopolymers from the smallest protein insulin (G3 dendrimer) to large proteins like histone (G7 dendrimer) [2]. Dendritic polymers are unique compared to linear polymers in a number of ways: (i) Dendrimers have controlled structures with discrete number of functionalities in a single molecule; (ii) they are a uniform and monodisperse system; (iii) they have a high density of functional groups and are water-soluble; (iv) functional groups are placed in close proximity without any chain entanglements; (v) they allow easy accessibility of functional groups that can result in cooperative substrate binding to receptors; (vi) the globular structure is amenable to host–guest entrapment and affords better stability to the encapsulated drug molecules; (vii) drugs can be loaded in multiple ways such as encapsulation, complexation, and conjugation, as well as in a matrix or a nanogel structure; (viii) dendrimers exhibit ease of modifying the size, charge, type, and number of functional groups. Most importantly, dendrimers can carry a number of drug molecules, targeting ligands, solubilizing agents, and imaging agents in one single molecule to perform multiple functions in the biological system.

This chapter discusses the design of novel dendritic nanodevices for various therapeutic areas, interfaced with material science and biology, using suitable examples. Since there is a large body of work, an attempt is made to illustrate specific applications with examples. After a brief introduction to the synthesis of dendrimers and hyperbranched polymers, their applications for intracellular drug delivery, their biodistribution, and future prospects are discussed. PAMAM and polypropyleneimine dendrimers are commercially available and have been studied extensively for drug delivery applications. Therefore, most of the applications covered in this chapter would deal with these two classes of dendrimers.

5.2 SYNTHESIS OF DENDRIMERS AND HYPERBRANCHED POLYMERS

The well-defined architecture of dendrimers arises from the stepwise synthetic procedure consisting of a series of coupling and activation steps. The synthesis of dendrimers follow either a "divergent" or a "convergent" growth strategy [4–7]. In the "divergent" approach (Figure 5.2a), the dendrimer grows outwards from the core, each generation at a time. Divergent growth is useful for the production of large quantities of dendrimers, since the molar mass is doubled after each generation. However, incomplete growth and side reactions may lead to somewhat imperfect samples, where purification of the end products could be tedious. Tomalia [3] used this approach to synthesize "star-burst" polyamidoamine (PAMAM) dendrimers starting from an ammonia core. The PAMAM dendrimers are commercially available. On the other hand, "convergent" approach (Figure 5.2b) growth begins with what will be the end-surface groups, and it works inward by attaching more monomer units. When the growing wedges are large enough, they are brought together using a suitable core. Even though the purification steps are simple, the yields of the reactions are low for larger dendrimers. To improve the synthesis methodology, "hypercores" and "branched monomers" have been used as starting points, and improved growth strategies such as "double exponential" and "mixed growth" have been explored. Because of the meticulous stepwise synthesis, dendrimers are nearly monodisperse, as opposed to the typically used linear polymers. Since the building blocks for the dendritic polymers are covalent bonds (as opposed to "specific interactions" in the cases of liposomes and micelles), the structure is stable and is not dependent on thermodynamic factors, as in the case of liposomes.

In contrast to the "structurally perfect" dendrimers prepared by multi-step synthesis, somewhat less-perfect "hyperbranched" polymers (Figure 5.3a, b) can be synthesized in one-step reactions. These are being considered as possible, inexpensive alternatives for applications where a well-defined structure is not a prerequisite. Typically, hyperbranched polymers are prepared by condensation of AB_2-type monomers. Since there is no specific focus point for the growth of the molecule (as in the case of a dendrimer

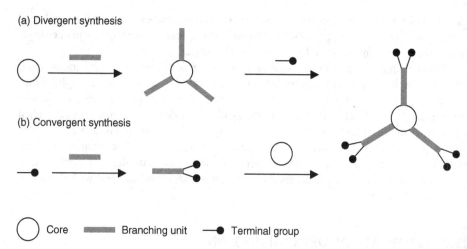

Figure 5.2. Schematic representation of synthesis of dendrimers using (a) divergent growth approach and (b) convergent growth approach. The details are described in the text, and further information can be gained from Refs. 3 and 7.

core), the polydispersities are high ($\sim$3–8) [8]. The polydispersity of hyperbranched polymers have been significantly reduced recently, using controlled polymerization reactions starting from AB_2-type monomers, where the B monomers become reactive only after the reaction of A groups. This has been improved further through the use of anionic polymerization and slow monomer addition conditions [9, 10]. Hyperbranched polyglycerols prepared using such methods have improved polydispersities [10] and are significantly less expensive, making them amenable for large-scale *in vivo* trials and bringing highly branched polymers as candidates for drug delivery of even common drugs such as ibuprofen.

5.3 INTRACELLULAR DRUG DELIVERY

Most of the therapeutic targets exist at the cellular level either at the cell membrane, in the cytoplasm, or in the nucleus [11]. Transport of drug molecules across the cell membrane occurs by a variety of processes *such as* passive diffusion, endocytosis, or active transport mechanisms, depending on the physicochemical properties of the drug and/or the carrier. The cell entry dynamics of dendritic polymers is influenced by size, charge, and the surface functionality. Using fluorescent (fluoroisothiocyanate; FITC)-labeled dendrimers, the cellular uptake and fate of dendrimers have been understood with a variety of techniques including flow cytometry, UV–visible spectroscopy, confocal microscopy, electron microscopy, and atomic force microscopy.

Typically, dendrimers appear to enter the cells rapidly within a few minutes to hours. In lung epithelial cell lines [12], it was found that most of the PAMAM G4 NH_2 dendrimers entered the cells within 1 hr and is suggested by the decrease in fluorescence intensity of the cell supernatant and corresponding increase in the intracellular

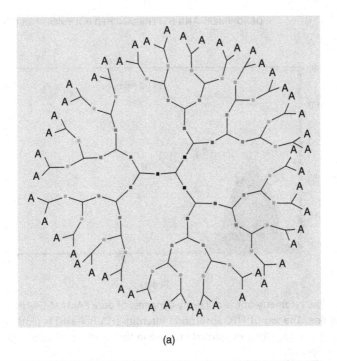

(a)

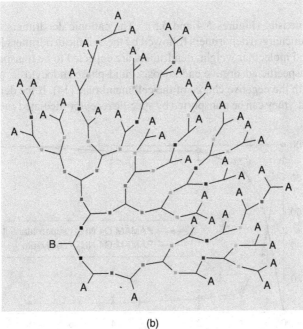

(b)

Figure 5.3. Schematic representation of (a) perfectly branched dendrimer structure and (b) imperfectly branched hyperbranched polymer. A and B represent the different monomer units in the polymer. The functional groups at the periphery A can be used to link drug molecules or targeting or imaging ligands by covalent or noncovalent bonds, while the interior of the dendritic polymer can be used for encapsulating drug molecules.

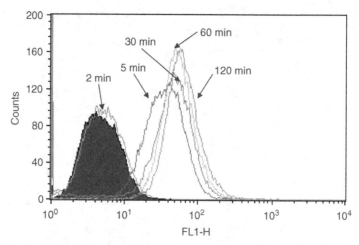

Figure 5.4. Flow cytometry of the cell entry dynamics of pure PAMAM-G4-NH$_2$ in A549 lung epithelial cell line. The log of FITC absorption intensity (FL1-H x-axis) is plotted against the number of cells (counts). The exponential increase in the cellular uptake of polymers within few minutes is evident.

fluorescence intensity (Figures 5.4 and 5.5). The cationic dendrimers enter the cells faster than the uncharged dendrimers followed by the anionic dendrimers [12,13]. Based on their size and molecular weight, dendrimers are expected to be transported by endo-cytosis, by nonspecific adsorptive endocytosis, fluid-phase endocytosis, or electrostatic interactions with the negative charge of the cell membrane [14]. If the dendrimers carry targeting ligands, they can be transported by specific receptor-mediated endocytosis [15].

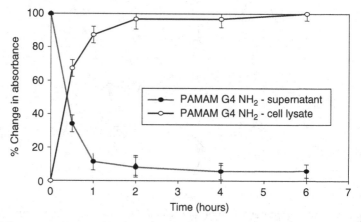

Figure 5.5. The normalized absorbance (A) values versus time (t). For the supernatant [$A(t)-A(\text{final})$]/[$A(\text{initial}) - A(\text{final})$] is used. For the cell lysate, [$A(t)/A(\text{final})$] is used. It can be seen that the increase in the normalized absorbance of the cell lysate correlates very well with the decrease in the absorbance of cell supernatant.

The cationic (amine terminated) dendrimers are transported by electrostatic interaction with the negatively charged cell membrane. Studies using supported lipid bilayers and atomic force microscopy have shown that the amine-terminated PAMAM dendrimers form holes of 15–40 nm in diameter by interacting with the lipid bilayer [16]. However, upon variation of the surface functionality by acetylating the amine group, the dendrimer did not form any holes in the lipid bilayer, but adsorbed to the periphery of the existing defects in the bilayer. In another study with an endothelial cell line, it was found that the cationic dendrimers are transported into the cell by interacting with the cholesterol in the cell membrane [17]. Cholesterol is involved in the cellular trafficking including signaling and endocytosis processes. Uncharged dendrimers such as PAMAM hydroxyl-terminated dendrimers are transported by nonspecific adsorptive and fluid-phase endocytosis [18]. When the temperature of the cell culture medium was reduced from 37°C to 10°C, the cellular uptake of FITC-labeled PAMAM G4OH dendrimer decreased significantly (80% reduction in comparison to the control). Similarly, when a high concentration of sucrose was used in the cell culture medium, the uptake was reduced to 65% of the control [19]. The results imply that the dendrimers enter the cell by energy-dependent process such as adsorptive and fluid-phase endocytosis. In addition to the surface functionality, the rate of cellular uptake of dendrimers is also dependent on the generation number, concentration, and incubation time. Generally, the lower-generation dendrimers are transported somewhat faster than the higher-generation ones. Cellular transport of dendrimer increases linearly with the increase in concentration of the dendrimer and incubation time. The linear increase in transport as a function of concentration is consistent with the endocytosis uptake process [20].

Once endocytosed, the dendrimers are transported through endosomes to the lysosomes and then released into the cytoplasm. Using confocal microscopy, it was observed that the dendrimer localized in the cytosol (Figure 5.6a, b). Since most of the therapeutic targets are located in the cytoplasm dendrimers can be suitable vehicles to deliver a high drug payload to the cytosol. The presence of dendrimer in the lysosome can be visualized using a lysosomal marker (lysotracker dye), where co-treatment of FITC-labeled dendrimer with the lysostracker dye results in co-localization in the lysosomes (Figure 5.6c, d). Although hyperbranched polymers (HBP) are imperfectly branched, they are also transported by a mechanism similar to that of dendrimers [21]. Polyol-a hyperbranched polymer and PAMAM G4 OH showed similar cell entry dynamics [12, 21]. Cationic dendrimers can interact with the lysosomes or endosomes by increasing the pH of the lysosome and escape by permeabilizing the lysosomal membrane [17]. This assumes importance in gene delivery, where the dendrimer–gene complex can bypass the lysosomal breakdown and deliver the intact DNA into the cytoplasm for further transport into the nucleus.

5.3.1 Anti-Inflammatory Drugs

Inflammation is a cell-mediated response occurring through the synthesis of pro-inflammatory mediators such as prostaglandins in the cytoplasm. Anti-inflammatory agents act by inhibiting the cyclooxygenase enzyme involved in the synthesis of prostaglandins [22]. The anti-inflammatory drugs generally enter the cell by passive

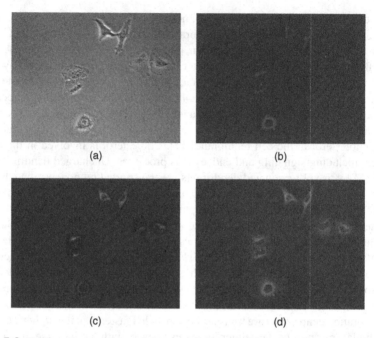

Figure 5.6. Confocal images showing lung epithelial cells. (a) Under phase contrast. (b) Localization of FITC-labeled PAMAM G4 hydroxyl terminated dendrimer in the cytoplasm. (c) Localization of lysosomal marker (lysotracker) in the lysosomes. (d) Co-localization of FITC-labeled dendrimer and lysostracker in the lysosomes. Images were captured 30 min after treatment. See insert for color representation of this figure.

diffusion and are transported in and out of the cell as a function of concentration gradient inside and outside of the cell membrane. This would result in a reduction in the effective drug concentration at the target site inside the cell, which prolongs the time required to produce a pharmacological response. Since passive diffusion is nonspecific, the free drug gives rise to a number of side effects due to the systemic drug exposure. When the drug is loaded on a dendrimer, the drug is transported by a more specific endocytosis process, resulting in a high drug concentration locally at the target site.

When conjugated to PAMAM-G4-OH dendrimer, ibuprofen, a nonsteroidal anti-inflammatory drug was found to be transported into the cell by endocytosis rapidly within a few hours and was localized in the cytoplasm (Figure 5.7a, b). The drug released from the dendrimer conjugate inhibited the prostaglandin synthesis as early as 30 min, while there was no inhibition for the free drug. However, at later time points, there was no difference in the activity of free and dendrimer-conjugated drug [23]. Similarly the HBP conjugate showed activity within 30 min (Figure 5.8). Among the HBP, polyglycerol ($M_w \approx 5$ kDa) showed higher activity than the polyol ($M_w \approx 15$ kDa) conjugate. The difference can be attributed to the difference in the rate of cell entry and drug release from the conjugate due to the difference in their molecular weights [21]. As the polymer conjugates carry a high drug payload (42–70% w/w), they produce a high local drug

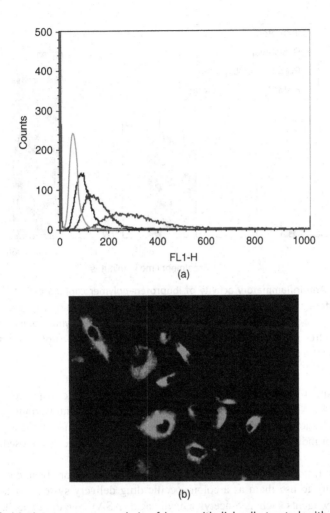

Figure 5.7. (a) Flow cytometry analysis of lung epithelial cells treated with FITC-labeled ibuprofen–PAMAM G4 OH dendrimer conjugate at different time points. The shift in intracellular fluorescence intensity indicates the rapid cellular uptake of the conjugate. Key: Red, 0 min; green, 5 min; black, 30 min; blue, 60 min; brown, 240 min (b) Confocal image showing the localization of FITC-labeled ibuprofen–dendrimer conjugate in the cytoplasm after 2 hr of treatment. See insert for color representation of this figure.

concentration resulting in a rapid pharmacological response. Under *in vivo* conditions, this would translate into reduced dose and decrease in frequency of dosing, since high drug concentration can be sustained for longer time periods at the target site in the cell.

The skin permeation of indomethacin, a hydrophobic drug, was enhanced on complexing with NH$_2$-terminated PAMAM dendrimer. Increase in permeation of indomethacin was due to the increase in drug solubility in the presence of the dendrimer,

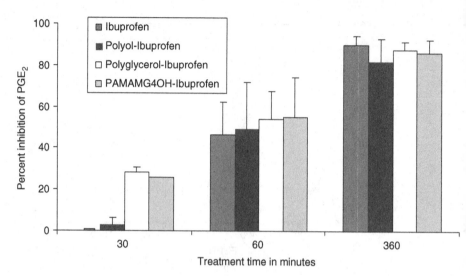

Figure 5.8. Anti-inflammatory activity of ibuprofen–polymer conjugates in lung epithelial cells with different treatment time. Percent inhibition of prostaglandin synthesis compared to control with no treatment is shown on the y-axis. Ibuprofen–polymer conjugates are compared against free ibuprofen. The concentration of ibuprofen was 10 μg/ml both for the free ibuprofen and the ibuprofen–polymer conjugates.

and this resulted in higher blood levels and pharmacodynamic response *in vivo* in a rat model [24]. When the same complexes were administered intravenously to arthritic rats, the dendrimer complex was passively targeted to the inflamed tissue and high concentrations of indomethacin were found at the inflammation site as opposed to the free drug [25].

The absorption of PAMAM dendrimer is low when given orally, and this provides an opportunity to use them as a colon-specific drug delivery system for topical antiinflammatory agents such as 5-acetyl salicylic acid (ASA). The drug was linked to the dendrimer through two different spacers (p-amino hippuric acid and p-amino benzoic acid) through an azo bond [26]. The azo bond can be specifically cleaved by the bacterial flora in the colon to release the free drug at the target site in the colon. Release of the drug from the dendrimer was found to be dependent on the spacer, and it was slower than the commercially available prodrug of 5-ASA [26]. Hence dendrimer can be used for sustained delivery of drugs specifically to the colon.

Steroidal drugs such as methylprednisolone are used for lung inflammation, but are rapidly cleared when given by conventional formulations. In an effort to achieve sustained intracellular delivery, PAMAM-G4-OH dendrimer or polyol was conjugated to methylprednisolone through a spacer (glutaric acid or succinic acid) and the conjugates showed comparable activity to the free drug [18, 27]. Release of free methylprednisolone from the conjugate was dependent on the branching architecture and the spacer. Dendrimer conjugate prepared with succinic acid showed higher activity than the one containing glutaric acid as a spacer due to a higher rate of drug release from

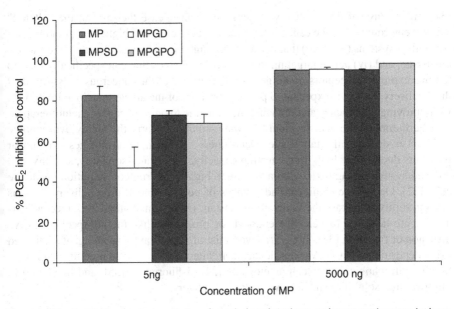

Figure 5.9. Anti-inflammatory activity of methylprednisolone–polymer conjugates in lung epithelial cells after treatment for 4 hr. Percent inhibition of prostaglandin synthesis compared to control with no treatment is shown on y-axis. Methylprednisolone-PAMAM G4 OH conjugate linked using either glutaric acid or succinic acid spacer and the methylprednisolone-polyol conjugate was linked using glutaric acid spacer. **Key:** MP, methyl prednisolone; MPGD, methylprednisolone–glutaricacid–PAMAMG4 OH dendrimer conjugate; MPSD, methylprednisolone–succinicacid–PAMAMG4 OH dendrimer conjugate; MPGPO, methylprednisolone–glutaric acid–polyol conjugate.

the former in the lysosomes [28]. On the other hand, with the glutaric acid spacer, the imperfectly branched polyol conjugates showed a higher activity than the dendrimer conjugate (Figure 5.9). From the results it can be inferred that the imperfect architecture provides an easy access for the lysosomal enzymes to release the drug, while the dendrimer conjugate can sustain the release for longer time periods [29]. Therefore it is possible to modulate the intracellular drug release by suitable choice of branching architecture and spacer.

5.3.2 Cancer Chemotherapy

Selective intracellular drug delivery assumes great importance in cancer chemotherapy, because most of the anti-cancer drugs are cytotoxic to normal cells. To achieve selective drug targeting, drug delivery scientists have been exploiting the differences in cell physiology of normal and cancerous cells [30]. Various approaches based on these differences include (i) the use of high-molecular-weight polymer–drug conjugates that preferentially accumulate in solid tumor tissue by enhanced permeability and retention (EPR) effect

[31], (ii) the use of drug delivery systems that can release the drug in the "altered" cellular environment of the cancer tissue such as acidic pH and higher temperature, (iii) particulate systems (<50 nm) that permeate through the leaky vasculature of the cancerous tissue, and (iv) active targeting methods using ligands that can specifically bind to the overexpressed receptors or antigens on the surface of the cancerous cells. An ideal drug delivery system is expected to perform or more of the above-mentioned functions for improving the efficacy and reducing the systemic toxicity of cancer chemotherapeutics. Furthermore, the vehicle should be water-soluble to carry the highly hydrophobic anti-cancer drugs to the target site. Dendrimers offer unique advantages over linear polymers due to their high drug-carrying capacity, good water-solubility, stability, and reduced systemic drug toxicity by achieving a high local drug concentration inside the cells [32]. One of the significant advantages of polymer-based drug delivery systems is their ability to bypass the multi-drug-resistant (MDR)-mediated efflux mechanisms as they are taken up by selective endocytosis process, unlike the nonspecific passive diffusion of free drug [33]. Table 5.1 gives a list of various anti-cancer agents delivered using dendritic polymers. Various dendritic architectures have been used to encapsulate highly hydrophobic drugs such as etopside [34], 5-fluorouracil [35], and taxol [36] to improve their solubility and achieve sustained delivery.

Table 5.1. Dendritic Polymer-Based Drug Delivery Systems For Anti-Cancer Agents

Anti-Cancer Agent	Dendritic Polymer	Remarks	Ref.
Methotrexate (MTX)	Conjugated with PAMAM-G5-NH$_2$, COOH, OH, NH$_2$COCH$_3$	FA and FITC attached for targeting and imaging cells, respectively, which overexpress folate receptors. Acetamide conjugate showed 10 to 100-fold activity than free drug.	15
	Conjugated with polyether dedrimer (G1 and G2) through hydrazide bond	Water-soluble conjugate with 12.6 folate residues and 4.7 MTX per G2 dendrimer.	37
	PAMAM-(G2-G4)-NH$_2$ dendrimer encapsulated in liposomes.	Entrapment of MTX in the liposomes improved in the presence of dendrimer.	44
5-Fluorouracil (5-FU)	PAMAM-(G4)-NH$_2$ dendrimer and PEG were attached.	PEGylated dendrimer increased drug encapsulation, reduced hemolytic toxicity, and increased drug circulation time.	35
Doxorubicin (DOX)	Polyester dendritic scaffold used for linking DOX through hydrazone bond.	The drug release was higher in acidic pH, improved drug circulation time and reduced cytotoxicity.	33

Anti-Cancer Agent	Dendritic Polymer	Remarks	Ref.
Cisplatin	PAMAM-(G3.5)-NH$_2$ dendrimer was conjugated to the drug.	The drug payload was 25% w/w of the conjugate, and the conjugate was water-soluble. The conjugate preferentially accumulated in the tumor and showed improved efficacy and reduced toxicity in comparison to free drug.	32
Etopside	PAMAM core was used and inner block arm was lipophilic (polycaprolactone) and the outer block was hydrophilic (PEG).	Drug loading was 22% and the encapsulated drug was nontoxic in *in vitro* cell culture studies.	34
Taxol	Polylgylcerol dendrimer was used for complexation and encapsulation.	Increased water solubility of taxol by 10,000-fold and also can achieve controlled release. The polymer was biodegradable.	36
5-aminolevulinic acid (ALA)	ALA dendrons conjugated to periphery by ester linkages of G1 and G2 dendrimers of di- or tripodent aromatic or tripodent aliphatic core.	High concentration of ALA inside the tumorigenic keratinocyte cells and showed greater potency than free ALA.	43
Boron (^{10}B)	PAMAM-(G3)-NH$_2$ dendrimer used for conjugation and boron was attached through PEG.	FA residues attached for targeting to folate receptors on cancer cells. Showed selective tumor uptake *in vitro* and *in vio* in mice.	41
	PAMAM-(G5)-NH$_2$ dendrimer used for conjugation and monoclonal antibody Cetuximab was attached.	Targeted to EGFR of brain tumor in rats compared to normal brain	42
J591 anti-PSMA	PAMAM-(G5)-NH$_2$ dendrimer used for conjugation and FITC or 6-TAMRA was used as an imaging agent.	A high-payload antibody conjugate that specifically targeted to PC3 cell line with overexpressed PSMA on its surface.	40
60B	PAMAM-(G5)-NH$_2$ dendrimer used for conjugation and FITC were used as an imaging agent.	A high-payload antibody conjugate which specifically bound to CD14 expressed on human myeloblastic leukemia cell lines.	39

Key: PAMAM, poly(amido amine); PEG, polyethylene glycol; FITC, fluoroisothiocyanate; FA, folic acid; PSMA, prostate-specific membrane antigen; 6-TAMRA, carboxytetramethyl rhodamine succinimidyl ester; EGFR, endothelial growth factor receptor.

Malik and Duncan [32] have shown superior anti-tumor efficacy of cisplatin–dendrimer conjugate by achieving a high drug payload (25% w/w). The conjugate showed preferential accumulation in a tumor mice model by EPR effect (size of the conjugate was 40 nm) and showed good water-solubility, reduced cytotoxicity (3.5 times less toxicity than cisplatin), and improved efficacy compared to the free drug. A polyester dendritic system was developed for the delivery of doxorubicin, where the drug was conjugated to the dendrimer through acid-labile hydrazone linkage [33]. The drug showed higher release at an acidic pH (pH < 6) *in vitro*, which is consistent with the acidic environment of the tumor. An additional advantage of the pH-sensitive linker is that the release of the drug attached by acid-labile linkage is less affected by the type of the drug than in the case of enzyme cleavable linkages [33]. The polymer conjugate was water-soluble, while showing longer circulation time. Moreover, the cytotoxicity of doxorubicin was reduced significantly (compared to 80–98% of the free drug) and also did not accumulate in any of the vital organs.

More specific and selective targeting of anti-tumor agents can be accomplished by attaching folic acid (FA) molecules to the dendrimer for binding with the overexpressed folate receptors on the surface of the tumor cell. By using a combination of molecular modeling techniques and *in vitro* cell uptake studies, a multifunctional PAMAM dendrimer-based nanodevice was designed [15] containing the drug (methotrexate), targeting ligand (FA), and an imaging agent (FITC). Studies with folate-containing amine-terminated PAMAM dendrimer showed poor cellular uptake in KB cells expressing high-affinity folate receptors. There was also nonspecific uptake into cells that lacked high-affinity folate receptors by nonspecific electrostatic interaction of the amine groups with the negatively charged cell membrane. Computer modeling of three different surface modifications (carboxyl, hydroxyl, and acetamide) revealed that the folic acid molecules were better accessible for receptor interaction from an acetamide surface. When tested *in vitro* using KB cells and monitored using confocal microscopy, the cellular uptake was rapid for acetamide surface and continued to accumulate upto 4–6 hr of incubation in contrast to other surface modifications. Methotrexate (MTX) was linked to the dendrimer through an ester or amide bond. The ester-linked conjugate was cytotoxic similar to the free drug in cancer cells, while the amide linked conjugate did not show any activity. Folate-targeted MTX dendrimer conjugate was significantly more active than the free drug. Other dendritic polymers such as polyarylethers and poly(L -glutamic acid) have also been studied for targeting using folic acid molecules [37,38].

Alternatively, dendrimers can be designed to carry other targeting ligands such as monoclonal antibodies that bind to specific surface antigens expressed on the specific tumors including prostate-specific membrane antigen (expressed on prostate cancer cells) and CD14 (expressed on myeloblastic leukemia cell lines) [39, 40] for selective drug delivery and imaging. The dendrimer-based immunoconjugates overcome some of the limitations of linear polymeric immunoconjugaes including decreased solubility and reduced binding efficacy [40].

Dendritic polymers can serve as high payload vehicles for carrying radioisotopes for radiation therapy like boron nuclear capture therapy. The therapeutic effectiveness of radiation therapy can be further enhanced by targeting to specific tumor cells in different organs using folate residues (for lung and other sarcomas) or antibodies (for brain glioma cells) on the dendrimer surface [41, 42]. Similarly, dendrimers have been used as high-payload water-soluble vehicles in photodynamic therapy for delivering high concentrations of 5-aminolevulinic acid to tumirogenic keratinocytes [43].

5.3.3 Gene Therapy

The success of gene therapy is governed by the efficiency of the delivery vehicle to protect the gene expression system from premature degradation in the extracellular biological milieu and to achieve delivery to a target cell (Figure 5.10). Several cationic lipids and polymers have been used for transfecting cells *in vitro*. However, their use is limited by low water-solubility, low transfection efficiency, and cytotoxicity [45]. Cationic amine-terminated PAMAM dendrimers have the inherent ability to associate, condense, and efficiently transport DNA into a wide variety of cell types without significant cytotoxicity [46]. Such complexes are termed "dendriplexes," and the electrostatic interaction takes place between the positive amino groups of the dendrimer and the negatively charged phosphate groups of the nucleic acid.

Stable complex formation and transfection efficiency are influenced by a number of factors including dendrimer/DNA molar ratio, size of the complex, generation,

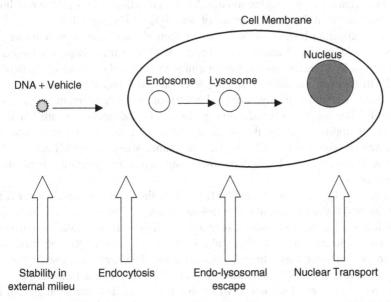

Figure 5.10. Schematic representation of various barriers involved in gene delivery to the nucleus using polymeric vehicles. The polymer–gene complex is taken up by cells through endocytosis and is transported to endosomes and lysosomes. Cationic polymers can escape from the lysosomes, thus preventing the degradation of DNA inside the lysosomes. The DNA has to be transported into the nucleus using specific signals to transfect at the target site in the nucleus.

and concentration of the dendrimer [45, 47]. Complexes of PAMAM dendrimer have been reported with various forms of nucleic acids including short single-stranded oligonucelotides, circular plasmid DNA, linear RNA, and various sizes of double-stranded DNA [45–48]. In general, the larger the nucleic acid the lower the dendrimer concentration required to generate high-density complexes. Mean size of DMA–dendrimer complexes has been shown to below 200 nm and monodisperse at various ratios of dendrimer to DNA [45]. Uniform particles of dendriplexes are formed when very low DNA concentration (1–10 ng/ml) are used [46]. As the ratio of dendrimer is increased, the size of the condensed particles decreases, while the transfection efficiency significantly increases [45].

In a recent study by Braun et al. [47], they demonstrated that for dendriplexes within an optimal size/charge ratio, a small size complex (<150 nm in diameter) and a positively charged surface are important for cell association and endocytosis process. After being entrapped within the endosome, release of the complex into the cytosol is essential before acidic or enzymatic degradation of DNA within the endosomal–lysosomal cavity (Figure 5.10). Cationic dendrimers promote the complex release from the endosomes by buffering the acid pH in the endosome and disrupting the membrane barrier of the organelle [49]. Once in the cytosol, the DNA translocates into the nucleus within 30 min of post-transfection. It has been suggested that the dendriplex may directly translocate into the nucleus, although the exact mechanism is not clearly known [46]. Hence, the endosomal release of the complex and translocation into the nucleus are the rate-limiting processes rather than the cellular uptake of dendriplexes (Figure 5.10).

The dendriplexes have been shown to be effective transfecting agents in a variety of cell systems [46], and the transfection efficiency has been further improved by attaching cyclodextrins, hydrophobic fluorophor, or glutamic acid on the surface of the dendrimer [50–52]. In this regard, Tang et al. [53] have observed that the transfection activity of dendrimer was dramatically enhanced by 50-fold upon heat treatment in a variety of solvents. The enhanced transfection of fractured dendrimers was attributed to the increased flexibility, enabling the dendrimer to be compact when complexed and to swell when released from DNA. On the other hand, Wang et al. [54] enhanced gene transfer and transfection of DNA–dendrimer complex into murine cardiac implants using electroporation.

Antisense oligonucleotides (ODN) for endothelial growth factor receptor (EGFR) was conjugated with a novel pentaerythritol-based anionic dendrimer [55]. The conjugate showed four-fold higher cellular uptake and 80% cell growth inhibition in comparison to the free ODN. Greater activity of the dendrimer conjugate is due to the protection of ODN from exo- and endonucleases. In addition, the anionic dendrimers are less cytotoxic than the cationic dendrimers. Recently, Marano et al. [56] have designed a novel lipid–lysine dendrimer for delivery of sense ODN for inhibition vascular endothelial growth factor (VEGF) expression in the eye. The lipid–lysine dendrimer facilitates transmembrane transportation, acts as lipid solubilizer, and protects the labile ODN from nuclease digestion. Time course studies in a rat model showed that the dendrimer–ODN complex remain active for up to 2 months, indicating that the dendrimer provides protection against the effects of nuclease.

5.4 BIODISTRIBUTION AND SAFETY

It is essential that the polymeric carrier is nontoxic, biocompatible, and nonimmunogenic for *in vivo* applications. The biodistribution and toxicity profile of dendritic polymers is influenced by a number of factors including dose, generation, and surface charge. Malik et al. [57] have carried out a systematic study on the *in vitro* toxicity and *in vivo* biodistribution of PAMAM and poly(propylene imine) dendrimers. Cationic dendrimers bearing amine functionality were found to display concentration- and generation-dependent toxicity. These dendrimers showed an increased hemolytic effect with increase in the generation of dendrimer, even at low concentrations for higher generations (10 μg/ml). On the other hand, anionic dendrimers were neither hemolytic nor cytotoxic toward a variety of cell lines *in vitro* at concentrations up to 1 mg/ml. In general, cationic dendrimers were more cytotoxic and their IC$_{50}$ values varied from 5 to 300 μg/ml depending on the cell type and generation of the dendrimer. Kannan et al. [12] have shown that the amine-terminated dendrimer (G4) showed significant cytotoxicity at concentrations ∼1 mg/ml, while the neutral hydroxyl terminated dendrimer and polyol were not cytotoxic even up to 1 mg/ml in A549 lung epithelial cells (Figure 5.11).

On IV administration, radiolabeled cationic dendrimers were rapidly cleared from the circulation, but the anionic dendrimers circulated for a longer time [57].

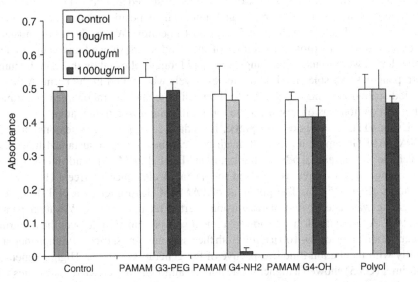

Figure 5.11. Comparative cytotoxicity profile of various dendritic polymers as a function of polymer concentration on lung epithelial cells. The cytotoxicity was estimated using MTT assay where the absorbance on the y-axis is proportional to the number of viable cells after treatment for 72 hr. Methylthiazol diphenyl tetrazolium (MTT) bromide is a yellow substrate that is cleaved by live mitochondria cells to form dark blue formazan crystals and is measured at 570 nm. **Key:** PEG, poly(ethyleneglycol).

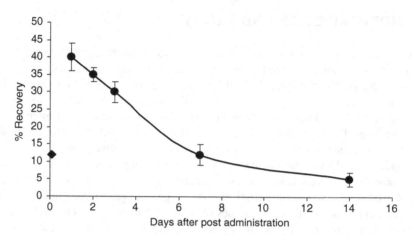

Figure 5.12. Lung disposition of free and PAMAM (G4 NH₂) dendrimer-conjugated FITC after intranasal administration in mice. Values are mean of three measurements. The square represents free FITC, while the closed circles represent FITC- labeled dendrimer.

Lower-generation dendrimers circulate longer than the higher generation dendrimers. For both classes of dendrimers, the blood levels correlated with the extent of liver uptake. In a tumor mice model, Nigavekar et al. [58] observed that the cationic and neutral dendrimers localized in major organs and tumor. The deposition peaked at 1 hr in most organs and stabilized from 24 hr to 7 days post injection. Within 24 hr post injection, the excretion was maximal and neither of the dendrimers showed any acute toxicity. Higher levels were found in the lungs, liver and kidney followed by those in the tumor, heart, pancreas and spleen, while the lowest levels were found in the brain. Although, the biodistribution trend was similar for both cationic and neutral charged dendrimers, higher disposition was observed for the former in comparison to the latter.

Using FITC as a fluorescent probe, Khandare et al. [59] have demonstrated that the PAMAM G4 amine terminated dendrimer can be used as a sustained lung delivery vehicle. On intranasal administration, FITC-labeled PAMAM dendrimer is retained in the lung until 14 days as opposed to the rapid clearance of free FITC within a few hours (Figure 5.12). The potential of PAMAM dendrimer as a oral drug delivery vehicle was investigated *in vitro* using everted rat intestine by Wiwattanapatapee et al. [60]. Anionic dendrimers showed a linear serosal transfer rate over the dendrimer concentration range of 10–100 µg/ml. Both the tissue and serosal accumulation increased linearly with incubation time irrespective of the generation number. Higher-generation dendrimer (G5.5) showed a higher tissue uptake, although the serosal transfer rates were quite similar to the lower-generation dendrimers (G2.5 and G3.5). Overall, the serosal transfer rate was higher than the tissue uptake. In sharp contrast, for cationic dendrimers the tissue uptake was higher than the serosal transfer rate, probably due to the nonspecific adsorption of cationic dendrimers to the mucosal surface.

Furthermore, using cationic dendrimers on Caco-2 cells, El-Sayed et al. [20] have found that the mannitol (a permeability marker) permeability, transepitheleal electrical resistance, and the leakage of lactate dehydrogenase increased as a function of donor concentration, incubation time, and generation number (G0–G4). Lower generations (G0–G2) were nontoxic to Caco-2 cells. Dendrimers have been suggested to be transported by paracellular transport and through adsorptive endocytosis or fluid-phase endocytosis by transcellular transport [14]. Florence and Hussain [61] studied the *in vivo* oral absorption of a lipid dendrimer (G4, 2.5 nm) in rats and observed that the dendrimer is preferentially taken up through the Peyer's patches in the small intestine. The maximum uptake was observed at 6 hr after administration from all parts of the GI tract. At 24 hr, only negligible amounts of dendrimer could be observed in the GI tract, possibly due to excretion or absorption into the systemic circulation.

Various surface modifications on the dendrimer have been shown to alter the biodistribution and improve the safety profile of amine terminated PAMAM dendrimers. Conjugation of lauroyl chains [13], polyglutamic acid [38], and PEG [62] to cationic dendrimers reduced their cytotoxicity significantly. Very recently, Schatzlein et al. [63] used a novel propyleneimine dendrimer to achieve preferential delivery of DNA to the liver. Quaternary amine group in the dendrimer improved DNA binding and tolerability of the vehicle *in vivo*. Most importantly, quaternisatoin reduced the particle aggregation of the complex (as particulate aggregates tend to localize in the lungs) and thus diverting the complex from the lung to the liver. Similar biodistribution has also been achieved by attaching cyclodextrin molecules on the surface of PAMAM dendrimers [64].

5.5 FUTURE PROSPECTS AND CONCLUSIONS

Since their introduction 20 years ago, dendrimers have received considerable attention because of their unique structure. Now, there are more than 50 families of dendrimers, each with unique properties [65]. Many potential applications of dendrimers are based on their unparalleled molecular uniformity, multifunctional surface, and the presence of internal cavities. Dendrimers have dual properties to function as endo- or exo-receptor [2]. They function as unimolecular endoreceptors (nanocontainer) manifesting noncovalent chemistry reminiscent of traditional regular and inverse micelles or liposomes. Alternatively, the dendrimers can function as nanoscaffold exhibiting a high propensity to cluster or complex in an exoreceptor fashion with a wide variety of biological polymers (e.g. DNA or protein) or metals.

Dendrimers mimic the hydrophobic and hydrophilic core shell topology of regular micelles. They can be viewed as stable unimolecular micelles, unlike the dynamic polymeric micelles. This results in a better control of drug delivery *in vivo* where the conventional micelles can dissociate into monomers on dilution. On the other hand, the dendrimer surface can be modified to induce "unimolecular encapsulation" behavior mimicking inverse micelles or liposomes. Jansen et al. [66] designed a "dendritic box" by modifying the surface of propyleneimine dendrimer with amino acids. Depending on the available internal diameter cavities, a variety of guest molecules can be encapsulated. The "dendritic box" can be opened to release either all or some of the entrapped guest molecules, depending on the immediate microenvironment of the dendrimer [67].

Dendrimers can be made to release the drug in response to specific stimuli such as temperature or pH by surface modification with various polymers and linkers [33,68]. A variety of architectures can be developed from the dendrimers through modification of either the core or the surface with block copolymers (copolymer). It would be pertinent to mention that a "perfect" branching architecture may not be always necessary for drug delivery applications. The "imperfectly" branched HBP may also be able to carry a similar drug payload and perform similar functions to a "perfectly" branched dendrimer. Therefore, HBP can be used as cheaper alternative to dendrimer for therapeutic areas, where the high cost of the dendrimer would not be justifiable. Further application of HBP as delivery vehicles is dependent on developing better synthetic procedures to make them more monodisperse.

Novel vehicles can be designed by using biomolecules to generate dendrimer or by attaching biomolecules on the dendrimer surface to make them more biocompatible or interact with specific biomolecules or bioassemblies. Glycodendrimers incorporate carbohydrate units on the surface of PAMAM dendrimer [69]. These systems are useful to study the protein–carbohydrate interactions that are implicated in many intercellular recognition events and to develop new therapies. Bayele et al. [70] have designed a new family of cationic lipidic polylysine dendron structures known as "dendrisomes" by synergistically combining the exceptional features of dendrimers and liposomes [71]. Dendrisomes serve as functional nonviral vector systems characterized by positively charged (lysine) molecules for interaction with nucleic acids and neutral lipidic moieties for membrane lipid bilayer transit [70].

The payload of dendrimers can be remarkably amplified by innovative integration of "chemistry" and "biology." Very recently, Choi et al. [72] have developed nanoclusters of PAMAM dendrimer by linking them together using complementary DNA oligonucleotides. In the proof of principle study, amine-terminated PAMAM G5 dendrimers were conjugated to different bifunctional moieties (FITC and FA), which were then further attached to a single strand of oligonucleotide. Hybridization of these oligonucleotide conjugates led to self-assembled nanoclusters of FITC- and FA-conjugated dendrimers. *In vitro* studies showed that the nanoclusters specifically bind to KB cells expressing high-affinity folate receptors and could be visualized through confocal microscopy. This strategy overcomes some of the synthetic limitations when multiple copies of small molecules are conjugated to a single dendrimer including limited number of subunits, decreased water-solubility, and low yield due to steric hindrance. Furthermore, these nanoclusters open up new possibilities of developing combinatorial therapeutics (different drugs, imaging agents, radiotherapeutic agents).

To conclude, dendritic polymers are exceptionally versatile vehicles that not only mimic the other nano- drug delivery systems (Table 5.2), but also offer some distinct advantages including design, type, and amount of payload it can carry. At the same time, some of the issues such as complexity of the preparation method, structure–property relationships, and biocompatibility need to be resolved before the dendritic nanodevices can become a "clinical" reality. Thus, dendritic polymers can be "smarter" (with regard to its ability to target, treat, and track the progression of therapy) as well as "smaller" (ability to interact with the biological system at nanoscale dimensions) offering exciting possibilities in the future.

Table 5.2. Comparative Performance Characteristics of Various Nanodelivery Systems

Nanosystem	Smallest Size	Drug Loading	Sustained Release	Endolysosomal Escape	Targeting	Stimuli Responsive	In Vivo Stability	Biocompatibility	Low Cost/ Complexity
Liposomes	+	+	+	++	++	++	+	+++	++
Nanoparticles	++	++	+++	++	++	++	++	++	+++
Micelles	++	++	++	+	+	++	+	++	++
Dendrimers	+++	+++	++	++	+++	++	+++	+	+

The comparison is based on available literature reports and is compared on a relative scale from low to high. **Key:** Low (+); moderate (++); high (+++).

REFERENCES

1. Mathews OA, Shipway AN, Stoddart JF. Dendrimers branching out from curiosities into new technologies. *Prog Polym Sci* 1998;23:1–56.

2. Esfand R, Tomalia DA. Poly (amidoamine) (PAMAM) dendrimers: From biomimicry to drug delivery and biomedical applications. *Drug Discov Today* 2001;6:427–436.

3. Tomalia DA. A new class of polymers: starbust dendritic macromolecules. *Polym J* 1985;17: 117–132.

4. Tomalia DA, Durst HD. Genealogically directed synthesis—Starburst cascade dendrimers and hyperbranched structures. *Top Curr Chem* 1993;165:193–313.

5. Frechet JMJ. Functional polymers and dendrimers—Reactivity, molecular architecture, and interfacial energy. *Science* 1994;263:1710–1715.

6. Newkome GR, Moorefield CN, Vogtle F. *Dendritic Macromolecules: Concepts, Synthesis, Perspectives.* Weinheim, Germany: VCH; 1996.

7. Liu M, Frechet JMJ. Designing dendrimers for drug delivery. *Pharm Sci Tech Today* 1999; 2:393–401.

8. Jikei M, Kakimoto, MA. Hyperbranched polymers: A promising new class of materials. *Prog Polym Sci* 2001;26:1233–1285.

9. Voit BI. Hyperbranched polymers: A chance and a challenge. *C R Chimie* 2003;6:821–832.

10. Frey H, Haag R. Dendritic polyglycerol: A new versatile biocompatible material. *Rev Mol Biotechnol* 2002;90:257–267.

11. Drews J. Drug discovery: A historical perspective. *Science* 2000;287:1960–1964.

12. Kannan S, Kolhe P, Raykova V, Glibatec M, Kannan RM, Lieh-Lai M, Bassett D. Dynamics of cellular entry and drug delivery by dendritic polymers into human lung epithelial carcinoma cells. *Biomater Sci Polym Ed* 2004;15:311–330.

13. Jevprasephant R, Penny J, Atwood D, McKeown NB, Emanuele AD. Engineering of dendrimer surfaces to enhance transepithelial transport and reduce cytotoxicity. *Pharm Res* 2003;20:1543–1550.

14. Kolhe P, Misra E, Kannan RM, Sujatha K, Lieh-Lai M. Drug complexation, in vitro release and cellular entry of dendrimers and hyperbranched polymers. *Int J Pharm* 2003;259:143–160.

15. Quintana A, Raczka E, Piehler L, Lee I, Myc A, Majoros I, Patri AK, Thomas T, Mule J, Baker JR, Jr. Design and function of a dendrimer based therapeutic nanodevice targeted to tumor cells through the folate receptor. *Pharm Res* 2002;19:1310–1316.

16. Hong S, Bielinska AU, Mecka A, Keszler B, Beals JL, Shi X, Balogh L, Orr BG, Baker JR Jr, Holl MMB. Interaction of poly (amidoamine) dendrimers with supported lipid bilayers and cells: Hole formation and the relation to transport. *Bioconjug Chem* 2004;15:774–782.

17. Manuta M, Tan PH, Sagoo P, Kasjafo K, George AJT. Gene delivery by dendrimers operates via a cholesterol dependent pathway. *Nucleic Acid Res* 2004;32:2730–2739.

18. Khandare J, Kolhe P, Pillai O, Kannan S, Lieh-Lai M, Kannan RM. Synthesis, cellular transport and activity of PAMAM dendrimer–methylprednisolone conjugates. *Bioconjug Chem* 2005;16:330–337.

19. Pillai O, Khandare J, Kannan S., Kannan RM. Cell entry dynamics of dendrimers. *Pharm Res* (in preparation).

20. El-Sayed M, Ginski M, Rhodes C, Ghandehari H. Transepithelial transport of poly (amidoamine) dendrimers across Caco-2 cell monolayers. *J Control Release* 2002;81:355–365.

21. Kolhe P, Khandare J, Pillai O, Sujatha K, Lieh-Lai M, Kannan RM. Hyperbranched polymer-drug conjugates with high drug payload for enhanced cellular delivery. *Pharm Res* 2004;21:2185–2195.

22. McAdam BF, Lawson FC, Mardini IA, Kapoor S, Lawson JA, Fitzgerald GA. Systemic biosynthesis of prostacyclin by cyclooxygenase (COX) -2: The human pharmacology of a selective inhibitor of COX-2. *Proc Natl Acad Sci USA* 1999;96:272–277.

23. Kolhe P, Khandare J, Pillai O, Sujatha K, Lieh-Lai M, Kannan RM. Design and evaluation of dendritic nanodevices with high drug payload for enhanced cellular delivery. *Biomaterials* 2006;27:660–669.

24. Chauhan AS, Sridevi S, Chalasani KB, Jain AK, Jain SK, Jain NK, Diwan PV. Dendrimer mediated transdermal delivery: Enhanced bioavailability of indomethacin. *J Control Release* 2003;90:335–343.

25. Chauhan AS, Jain NK, Diwan PV, Khopade AJ. Solubility enhancement of indomethacin with poly(amidoamine) dendrimers and targeting to inflammatory regions of arthritic rats. *J Drug Target* 2004;12:575–583.

26. Wiwattanapatapee R, Luelak L, Saramunee K. Dendrimer conjugates for colonic delivery of 5-amino salicylic acid. *J Control Release* 2003;88:1–9.

27. Kannan S, Pillai O, Khandare J, Kohle P, Lieh-Lai M, Kannan RM. Enhanced delivery of anti-inflammatory drugs using nanopolymers. *Crit Care Med* 2004;32(12) suppl: A142.

28. Pillai O, Khandare J, Kolhe P, Kannan S, Leih-Lai M, Kannan RM. Dendritic nanodevice of methylprednisolone-succinate conjugate for drug delivery to lung epithelial cells. *J Control Release* (in preparation).

29. Khandare J, Pillai O, Kolhe P, Kannan S, Leih-Lai M, Kannan RM. Synthesis and evaluation of polyol–methylprednisolone conjugate for drug delivery to lung epithelial cells. *J Control Release* (in preparation).

30. Duncan R. The dawning era of polymer therapeutics. *Nat Rev Drug Discov* 2003;2:347–360.

31. Duncan R. Polymer conjugates for tumor targeting and intracytolplasmic delivery. The EPR effect as a common gateway. *PSTT* 1999;2:441–449.

32. Malik N, Duncan R, inventors; Dow Chemical Company, assignee. Dendritic–platinate drug delivery system. US Patent 6,585,956. 2003 July 1.

33. DeJesus OLP, Ihre HR, Gagne L, Frechet JM, Szoka FC, Jr. Polyester dendritic systems for drug delivery applications: *In vitro* and *in vivo* evaluation. *Bioconjug Chem* 2002;13:453–461.

34. Wang F, Bronich TK, Kabanov AV, Rauh RD, Rovers J. Synthesis and evaluation of a star amphiphilic block co-polymer from poly(caprolactone) poly(ethylene glycol) as a potential drug delivery carrier. *Bioconjug Chem* 2005;16:397–405.

35. Bhadra D, Bhadra S, Jain S, Jain NK. A PEGylated dendritic nanoparticlate carrier of fluorouracil. *Int J Pharm* 2003: 257:111–124.

36. Ooya T, Lee J, Park K. Effects of ethylene glycol-based graft, star-shaped and dendritic polymers on solubilization and controlled release of paclitaxel. *J Control Release* 2003;93:121–127.

37. Kono K, Liu M, Frechet JM. Design of dendritic macromolecular containing folate or methotrexate residues. *Bioconjug Chem* 1999;10:1115–1121.

38. Tansey W, Ke S, Cao XY, Pasuelo MJ, Wallace S, Li C. Synthesis and characterization of branched poly(L-glutamic acid) as a biodegradable drug carrier. *J Control Release* 2004;94:39–51.

39. Patri AK, Myc A, Beals J, Thomas TP, Bander NH, Baker JR, Jr. Synthesis and in vitro testing of J591 antibody–dendrimer conjugates targeted for prostate cancer therapy. *Bioconjug Chem* 2004;15:1174–1181.

40. Thomas TP, Patri AK, Myc A, Myaing MT, Ye JY, Norris TB, Baker JR, Jr. *In vitro* targeting of synthesized antibody-conjugated dendrimer nanoparticles. *Biomacromolecules* 2004;5:2269–2274.

41. Shukla S, Wu G, Chatterjee M, Yang W, Sekido M, Diop LA, Muller R, Sudimack JJ, Lee RJ, Barth RF, Tjarus W. Synthesis and biological evaluation of folate receptor targeted boronated PAMAM dendrimers as potential agents for neutron capture therapy. *Bioconjug Chem* 2003;14:158–167.

42. Wu G, Barth RF, Yang W, Chatterjee M, Tjarus W, Ciesielski MJ, Fenstermaker RA. Site specific conjugation of boron containing dendrimers to anti-EGF receptor monoclonal antibody cetuximab (IMC-C225) and its evaluation as a potential delivery agent for neutron capture therapy. *Bioconjug Chem* 2004;15:185–194.

43. Battah SH, Chee CE, Nakanishi H, Gerscher S, MacRobert AJ, Edwards C. Synthesis and biological studies of 5-aminolevulinic acid containing dendrimers for photodynamic therapy. *Biconjug Chem* 2001;12:980–988.

44. Khopade AJ, Caruso F, Tripathi P, Nagaich S, Jain NK. Effect of dendrimer on entrapment and release of bioactive from liposomes. *Int J Pharm* 2002;31:157–162.

45. Tomlinson E, Rolland AP. Controllable gene therapy: Pharmaceutics of non-viral gene delivery systems. *J Control Release* 1996;39:357–372.

46. Eichman JD, Bielinska AU, Kukowska-Latallo JF, Baker JR, Jr. The use of PAMAM dendrimers in the efficient transfer of genetic material into cells. *Pharm Sci Technol Today* 2000;3:232–245.

47. Braun CS, Vetro JA, Tomalia DA, Koe GS, Koe JG, Middaugh CR. Structure/function relationships of polyamidoamine/DNA dendrimers as gene delivery vehicles. *J Pharm Sci* 2005;94:423–436.

48. Bielinska A, Kukowska-Latallo JF, Johnson J, Tomalia DA, Baker JR, Jr. Regulation of *in vitro* gene expression using antisense oligonucleotides or antisense expression plasmids transfected using starburst PAMAM dendrimers. *Nucleic Acid Res* 1996;24:2176–2182.

49. Hansler J, Szoka FC, Jr. Polyamidoamine cascade polymers mediate efficient transfection of cells in culture. *Bioconjug Chem* 1993;4:372–379.

50. Wada K, Arima H, Tsutsumi T, Hirayama F, Uekama K. Enhancing effects of galactosylated dendrimer/α-cyclodextrin conjugates on gene transfer efficiency. *Biol Pharm Bull* 2005;28:500–505.

51. Yoo H, Juliano RL. Enhanced delivery of antisense oligonucleotides with fluorophore-conjugated PAMAM dendrimers. *Nucleic Acid Res* 2000;28:4225–4231.

52. Choi JS, Nam K, Park J, Kim J, Lee J, Park J. Enhanced transfection efficiency of PAMAM dendrimer by surface modification with L-arginine. *J Control Release* 2004;99:445–456.

53. Tang MX, Redemann CT, Szoka FC, Jr. *In vitro* gene delivery by degraded polyamidoamine dendrimers. *Bioconjug Chem* 1996;7:703–714.

54. Wang Y, Bai, Y, Price C, Boros P, Qin L, Bielinska AU, Kukowska-Latallo JF, Baker JR, Jr, Bromberg JS. Combination of electroporation and DNA/dendrimer complexes enhances gene transfer into murine cardiac transplantats. *Am J Transplant* 2001;1:334–338.

55. Hussain M, Shchepinov M, Sohail M, Benter IF, Hollins AJ, Southern EM, Akhtar S. A novel anionic dendrimers for improved cellular delivery of antisense oligonucleotides. *J Control Release* 2004;99:139–155.

56. Marano RJ, Wimmer N, Kearns DS, Thomas BG, Toth I, Brankov M, Rakoezy PE. Inhibition of in vitro VEGF expression and choroidal neovascularization by synthetic dendrimer. Peptide mediated delivery of a sense oligonucleotides. *Exp Eye Res* 2004;79:525–535.

57. Malik N, Wiwattanapateppe R, Klopsch R, Lorenz K, Frey H, Weener JW, Meijer EW, Paulus W, Duncan R. Dendrimers: Relationship between structure and biocompatibility *in vitro* and preliminary studies on the biodistribution of ^{125}I-labeled polyamidoamine dendrimers *in vivo*. *J Control Release* 2000;65:133–148.

58. Nigavekar SS, Sung LY, Llanes M, El-Jawahri A, Lawrence TS, Becker CW, Balogh L, Khan MK. 3H dendrimer nanoparticle organ/tumor distribution. *Pharm Res* 2004;21:476–483.

59. Khandare J, Kolhe P, Pillai O, Kannan S, Kannan RM, Basett D, Leih-Lai, M. Dendritic drug delivery systems with high drug payload for asthma. Proceedings of AAPS Pharmaceutics and Drug Delivery Conference, 2004 Jun 7–9; Philadelphia, p. 70.

60. Wiwattanapatpee R, Laueno-Gomez B, Malik N, Duncan R. Anionic PAMAM dendrimers rapidly cross adult rat intestine *in vitro*: A potential oral delivery system. 2000;17:991–998.

61. Florence AT, Hussain N. Transcytosis of nanoparticle and dendrimer delivery systems:evolving vistas. *Adv Drug Deliv Rev* 2001;50:569–589.

62. Chen H, Neuman MF, Parrish AR, Simanek EE. Cytotoxicity, hemolysis and acute in vivo toxicity of dendrimers based on melamine, candidate vehicles for drug delivery. *J Am Chem Soc* 2004;126:10044–10048.

63. Schatzlein AG, Zinselmeyer BH, Elowzi A, Dufes C, Chim YTA, Roberts CJ, Davies MC, Munro A, Gray AT, Uchegbu IF. Preferential liver gene expression with polypropyleneimine dendrimers. *J Control Release* 2005;101:247–258.

64. Kihara T, Arima H, Tsutsumi T, Hirayama F, Uekama K. *In vitro* and *in vivo* gene transfer by an optimized α-cyclodextrin conjugate with polyamidoamine dendrimer. *Bioconjug Chem* 2003;14:342–350.

65. Klajnert B, Bryszewska M. Dendrimers: Properties and applications. *Acta Biochim Pol* 2001;48:199–208.

66. Jansen JFGA, deBrabander-van den Berg EMM, Meijer EW. Encapsulation of guest molecules into a dendritic box. *Science* 1994;266:1226–1229.

67. Jansen JFGA, Meijer EW, deBrabander-van den Berg EMM. The dendritic box: Shape-selective liberation of encapsulated guests. *J Am Chem Soc* 1995;117:4417–4418.

68. Namazi H, Adeli M. Dendrimers of citric acid and poly (ethylene glycol) as the new drug delivery agents. *Biomaterials* 2005;26:1175–1183.

69. Cloninger MJ. Biological applications of dendrimer. *Curr Opin Chem Biol* 2002;6:742–748.

70. Bayele HK, Sakthivel T, O'Donell M, Pasi KJ, Wilderspin AF, Le CA, Toth I, Florence AT. Versatile peptide dendrimers for nucleic acid delivery. *J Pharm Sci* 2005;94:446–457.

71. Al-Jamal KJT, Sakthivel T, Florence AT. Dendrisomes: Cationic lipidic dendron vesicular assemblies. *Int J Pharm* 2003;254:33–36.

72. Choi Y, Thomas T, Kotlyar A, Islam MT, Baker JR Jr. Synthesis and functional evaluation of DNA-assembled polyamidoamine dendrimer clusters for cancer cell-specific targeting. *Chem Biol* 2005;12:35–43.

6

NANOGELS: CHEMISTRY TO DRUG DELIVERY

Murali Mohan Yallapu, Maram K. Reddy, and
Vinod Labhasetwar

6.1 INTRODUCTION

Macro-, micro-, and nanogels are composed of soft materials (polymeric gels) that are of significant interest for applications in various areas of research including polymer and material science, physical and pharmaceutical sciences, medical sciences, and biotechnology. Hydrogels or macrogels are composed of hydrophilic polymeric chains that exhibit distinct property of good swelling capacity in aqueous media. Almost all water-soluble polymers containing hydrophilic functional groups such as –OH, –COOH, –NH$_2$, –CONH$_2$, –CONH, –SO$_2$H, and so on, in their macromolecular chains are capable of producing hydrogels. The excellent properties of hydrogels such as high swelling capacity, hydrophilicity, and biocompatibility have enhanced their widespread applications in different fields including in agriculture and horticulture, health, bioengineering, pharmaceutical, food industry, and so on [1–13].

Microgels and nanogels are a special class of polymeric gel family that varies in size range from micrometer to submicrometer scale and exists as a three-dimensional network of hydrophilic and/or hydrophobic polymers. Microgels were first synthesized by Staudinger and Husemann in 1935 [14] and are defined as a disperse phase of

Biomedical Applications of Nanotechnology. Edited by Vinod Labhasetwar and Diandra L. Leslie-Pelecky
Copyright © 2007 John Wiley & Sons, Inc.

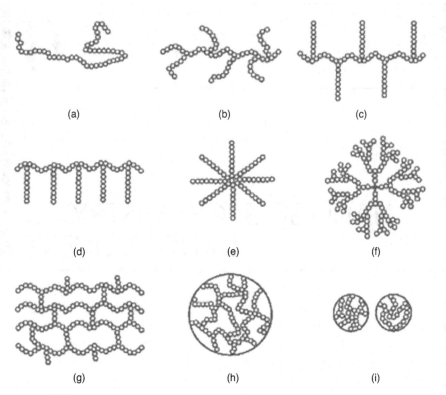

Figure 6.1. Molecular structures of (a) linear polymer, (b) branched polymer, (c) graft polymer, (d) comb polymer, (e) star polymer, (f) dendri-polymer, (g) cross-linked polymer or gel, (h) microgel, and (i) nanogels.

discrete polymeric gel particles typically in the size range of 1 mm to 1 μm [15, 16]. Different names and definitions have been used to define and describe nanogels; however, based on the available concepts, these are defined as gel macromolecules in the size range of tens to hundreds in nanometers. These are formed either through covalent bonds for stable and insoluble three-dimensional networks or unstable (physical) gels via hydrogen bonds, van der Waals forces, and chain entanglements or through formation of crystalline regions [17–19]. Figure 6.1 summarizes the variations in the molecular structure of nanogels beyond traditional polymeric chains and gel structures.

Nanogels have a broad spectrum of applications in biomimetics, biosensors, artificial muscles, chemical separation, biomaterials, cell culture systems, catalysis, photonics, and drug delivery systems [15, 20–25]. Recently, a few novel biomimetic nanogels have been evaluated in biomachines due to their self-oscillating property [26–29]. These nanogels undergo an autonomic and periodic coil-globular oscillation or swelling and de-swelling oscillation without any external stimuli in a closed system. Ryan and co-workers [30,31] and Varga et al. [32] have developed pulsating temperature and pH responsive nanogels.

Because of their inherent rapid swelling and de-swelling nature in response to external stimuli such as temperature, solvent composition, pH, light, pressure, magnetic field, specific chemicals, electric fields, and so on, nanogels have attracted attention as functional smart materials for biotechnological and biomedical applications. In general, biologically active molecules are restricted in therapeutic applications because of their poor solubility, limited bioavailability, and/or rapid elimination [33]. Furthermore, certain proteins, oligonucleotides, plasmid DNA, and interference RNA therapeutic agents are not successful as therapeutics, in part, because of their failure to reach the intracellular tissue targets. Furthermore, most of these biomacromolecules degrade in intra- and extracellular media, and their exposure to other cells frequently leads to uninvited side effects and cause toxicity [34].

The major concern in drug delivery is the ability to maintain an optimum therapeutic drug levels at the target sites for duration that is necessary for therapeutic outcome while minimizing the undesirable side effects. Therefore, suitable delivery systems are necessary that can protect the drug molecules and deliver them to the target site or the target cell population. It is expected that new nanotechnology-based drug delivery systems would be able to control the pharmacokinetics and pharmacodynamics of drugs to enhance their efficacy. To solve these problems, new materials are being developed including nanospheres, nanocapsules, micelles, liposome, and so on [35–43]. Although nanoscale size particles based on biodegradable polyesters and natural origin are well-known vehicles for low-molecular-weight drugs, oligonucleotides, peptides, and so on [44–50], the high cost of these materials, complexicity of their preparation methods, and low drug-loading capacity are some of the factors that limit their widespread applications in drug delivery. Therefore, there is a continuing interest to find new strategies and to design and develop novel drug delivery systems. Nanogels are of significant interest for drug delivery and other biomedical applications because of their slow degradation properties, biocompatibility, stimuli-reactive nature (e.g., pH- or temperature-sensitive), and the ability to develop targeted drug delivery system. In addition, nanogels have several advanced features that can resolve major challenges in drug delivery such as protection of drug from the body's defense mechanism, capacity to deliver required doses of drug to the targeted area, achieve prolonged blood circulation time and internalization efficiency by the target cells/tissue, and low cytotoxicity. Moreover, it is feasible to synthesize nanogels with a good control over their size; they are easy to bio-functionalize, and with internal cross-linking modulation the drug release can be controlled in response to a stimulus such as the change in the composition of physiological fluids under disease conditions (Figure 6.2).

6.2 CLASSIFICATION OF NANOGELS

Based on the type of linkages present in the network chains of gel structure, polymeric gels (including nanogel) are subdivided into two main categories: physical gels and

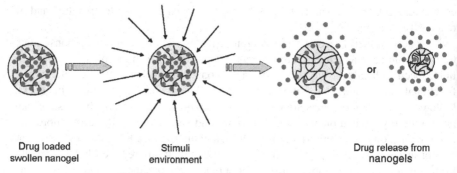

| Drug loaded | Stimuli | Drug release from |
| swollen nanogel | environment | nanogels |

Figure 6.2. Drug release model from nanogel.

chemical gels [2, 51]. Physical gels or pseudo gels are formed by weaker linkages through either (a) van der Waals forces, (b) hydrophobic, electrostatic interactions, or (c) hydrogen bonding. A few simple methods are available to obtain physical gels. For example, a warm polymer solution is cooled to obtain a gel or a polymer solution is cross-linked using physical methods (freeze–thaw or mechanical mixing), pH variation, and mixing with either polyanion, polycation, polyelectrolyte or multivalent ions. Since weaker linkages construct these physical gels, the changes in temperature, pH, or other parameters can disrupt the network structure of the gel. Physical gels can also be formed by the aggregation and/or self-assembly of polymeric chains. Chemically cross-linked gels are comprised of permanent chemical linkages (covalent bonds) throughout the gel networks and are difficult to either dissolve in solvents or perturb their network structure. Several methods have been developed for producing permanent gels—that is, via copolymerizing a monomer in the presence of cross-linker/multifunctional cross-linker and/or via conversion of hydrophilic/hydrophobic polymer into cross-linked gels, polymer cross-linking using radiation, and multifunctional reactive compounds. In this section we summarize different types of nanogels based on linkages in the gel networks.

6.2.1 Chemically Cross-Linked Nanogels

A number of distinguished chemically cross-linked network nanogel systems are documented in the literature. The properties of cross-linked gel system depend on the chemical linkages and functional groups present in the gel networks. Keeping this in view, nanogels have been synthesized using different strategies for chemical linking of polymeric chains.

Usually, hydrophilic polymers and hydrophilic–hydrophobic copolymers are obtained by the polymerization of vinyl monomers in the presence of multifunctional cross-linkers that are the launch cross-linking points within and between the polymeric chains. These cross-linking points allow modifying entire physicochemical properties of the gel systems. For this purpose, a few versatile cross-linking agents have been reported (Figure 6.3). Aliyar et al. [52] demonstrated a facile approach for nanogel (20–200 nm) preparation in which pendant thiol groups are incorporated into the polymeric chains

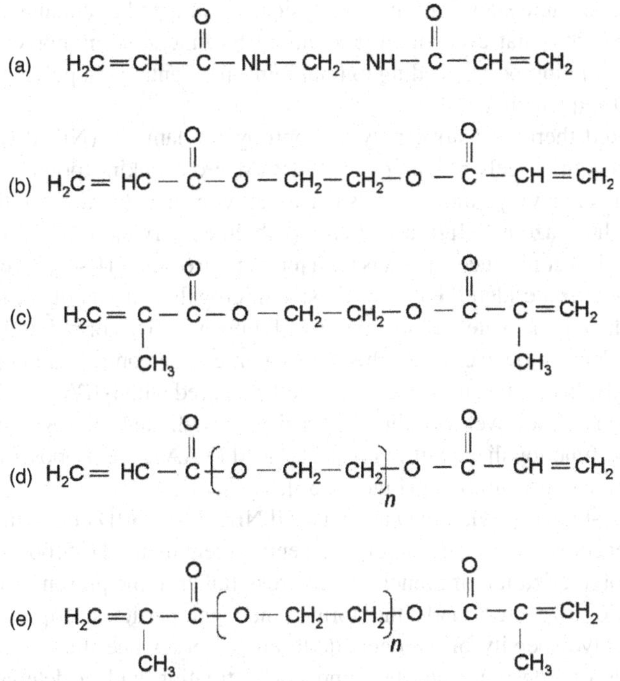

Figure 6.3. Chemical structures of cross-linking agents. (a) N,N'-methylenebisacrylamide, (b) ethyleneglycol diacrylate, (c) ethyleneglycol dimethacrylate, (d) poly(ethyleneglycol diacrylate), and (e) poly(ethyleneglycol dimethacrylate).

and their subsequent intramolecular disulfide cross-linking is achieved through "environmentally friendly chemistry" (green chemistry) as shown in Figure 6.4.

Doherty et al. [53,54] have reported poly(acrylamide)-based sparsely cross-linked nanogel matrixes as fluids—that is, mechanically stabilized networks alignment which were used for high-throughput microchannel DNA sequencing. In contrast to *ex situ* infinitely cross-linked gels, sparsely cross-linked discrete polymeric nanogels provide smaller colloidal dimensions with good fluidity. Sparsely cross-linked nanogel matrixes

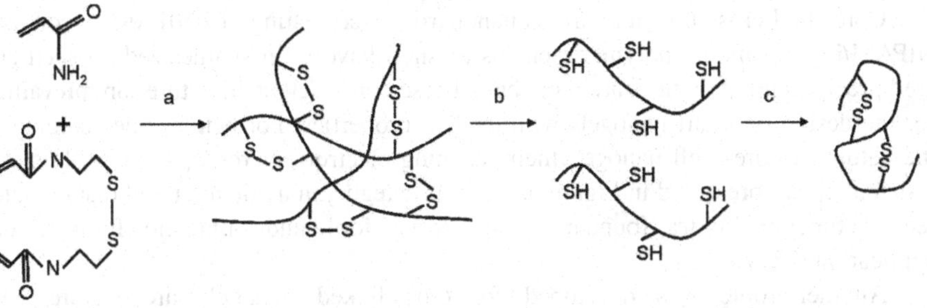

Figure 6.4. Cross-linked poly(acrylamide) nanogel formation. (a) copolymerization, (b) reduction, and (c) reduction to nanogel. (Modified from Ref. 52, Scheme 1.)

demonstrate reasonable stabilization of a physically entrapped or entangled sequencing matrix. It is possible that even small amounts of biofunctional monomer can provide chemical cross-linking points leading to final reinforced/entangled polymeric networks with improved sequencing.

Cross-linked thermosensitive poly(N-isopropylacrylamide) (NIPAM) is the most studied nanogels/microgels polymer [55]. NIPAM cross-linking polymerization with functional ionizable vinyl monomers such as acrylic acid (AAc), methacrylic acid (MAc), 1-vinylimidazole (VID), and 2-aminoethylmethacrylate (AEMA) in the presence of a cross-linker produce nanogels with temperature- and pH-sensitive architecture [56–61]. A novel route identified to synthesize microgels consists of well-defined NIPAM core with cationic water-soluble polymer [chitosan (CHS) or poly(ethyleneimine) (PEI)] shells. The responsiveness of this gel system can be controlled via pH and temperature [62,63]. In contrast to conventional gels prepared with NIPAM and acrylic acid or methacrylic acid, a novel swelling responsive cross-linked microgel system based on NIPAM and functionalized with vinylacetic acid (VAAc) are proposed to impart an ionization pattern over a narrow pH range [64].

MBA cross-linked acrylamide nanogels and NIPAM-co-VID copolymer have been prepared in liposome nanoscale reactor by an entrapment method [65,66]. Nanogels are obtained by polymerization of monomer and cross-linker in the presence of liposomes, where the lipid layer surrounds the formed nanogel particles. Liposomes control the size and polydispersity of nanogels quite effectively. Once the polymerization is over, the lipids of liposomes can be removed by treating with a detergent solution. More recently, Thienen et al. [67] followed the same path to produce biodegradable dextran-cl-2-hydroxyl ethyl methacrylate (Dex-cl-HEMA) nanogels with tunable degradation properties.

Vinogradov et al. [68–71] successfully developed a new class of highly dispersed and cross-linked poly(ethyleneimine) (ionic) and poly(ethylene oxide) (nonionic hydrophilic polymer) (PEI-cl-PEO) nanogels via condensation chemistry. These gels exhibit combined characteristics of polyelectrolytic and nonionic network and bind or encapsulate various types of negatively charged oligonucleotides (ODN) through ionic interactions. Though the charge in nanogels is completely neutralized by ODN chains, the stability of gel particles is achieved by PEO or PEG (Figure 6.5). The highly cross-linked poly(N,N-diethylamino ethylmethacrylate)/poly(ethylene dimethacrylate) with functionalized PEG shell nanogels offer conjugation sites for various ligands [72].

Core-shell cross-linked hydrogel nanoparticles consisting of PNIPAM as core and NIPAM/ionic/nonionic polymeric chains as shell have been synthesized by seed and feed precipitation polymerization method. These core-shell architectures are prevailing for the design of smart nanogels with tunable properties. For better understanding of the nature of core-shell nanogel, their scanning electron micrographs as well as size distribution are presented in Figure 6.6. In this category, anionic microgel that contains reactive functionality was found highly appropriate for ligand conjugation for microlens application [73–76].

Another route was developed for cross-linked nanogels architecture with temperature-responsive core and pH-sensitive arms—that is, via photo-cross-linking of poly(N-isorpopylacrylamide-g-dimethylmaleimide) (PNIPAM-g-DMIAAm) [77] (Figure 6.7). The formation of nanogel occurs above its lower critical solution temperature

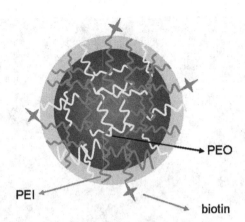

Figure 6.5. Poly(ethyleneimine) cross-linked poly(ethylene oxide) nanogel. See insert for color representation of this figure. (Reproduced with permission from Ref. 71, Figure 6.1. Copyright 2005, Elsevier Ltd.)

(LCST), where the entropy controls comparatively exothermic enthalpy due to hydrogen bonding, and therefore the total free energy becomes positive that leads to phase transition. At this stage, the existing chromophoric groups in the NIPAM graft copolymer undergo a cyclo-addition reaction under UV light that renders the formation of nanogel.

Polyampholytic nanogels contain both anionic and cationic polymeric networks in gel structure. These nanogels allow the network to swell with an increase in salt concentration when the negative and positive charges are well-balanced [18, 78–80]. Interpenetrated nanogels or combined nanogel networks are obtained by polymerizing monomer(s) in the presence of another type of network that forms an interpenetrated combination of different networks in the gel structure. Interpenetrated network nanogels composed of poly(methacrylic acid)/poly(ethylene glycol) (PMAc/PEG), poly(acrylamide) (PAAm) and poly(N-substituted acrylamide)/polyols, and epoxides/poly(alkylenepolyamines) [81–83] have been investigated, thus supporting the various condensation polymerization approaches [84–86].

6.2.2 Physically Cross-Linked Nanogels

6.2.2.1 Liposome Modified Nanogels.
A number of studies have been under investigation for specific delivery of drugs to a target site in order to improve their efficacy and reduce side effects [87]. It is rationally possible that liposomes can facilitate the drug release that can be triggered by physical and chemical stimuli, and hence many efforts have been made toward developing stimuli sensitive liposomes [88–95]. Among the liposomes, thermo-responsive liposomes (phospholipids with the liquid crystalline phase) have better medical applications since the efficacy of drug release can be enhanced above its phase transition temperature [89]. However, the decreased temperature sensitivity of lipids (do not release the contents effectively at the phase transition temperature of the membrane), their limited availability, and difficulty to release

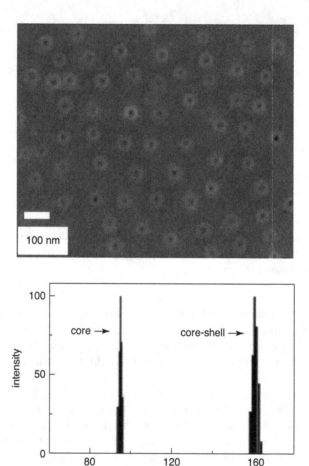

Figure 6.6. Scanning electron microscopy (SEM) of core-shell morphology and size distribution of nanogels. (Reproduced with permission from Ref. 73, Figures 6.1 and 6.6. Copyright 2004, American Chemical Society.)

large drug molecules from liposome as effectively as low-molecular-weight compounds promoted investigators to develop new formulations of liposomes which are modified with various polymers [87]. This approach is especially developed for the formulation of temperature- and pH-sensitive liposomes that are modified with synthetic membrane active polymers based on poly(ethyleneimine) (PEI), poly(2-ethylacrylic acid) (PEAAc), poly(propylacrylic acid) (PPAc), random copolymers of acrylic acid and ethyl acrylate (EA), and dioleoylphosphatidylethanolamine (DOPE). These liposomes have been investigated for their effectiveness for tumor targeting in conjunction with hyperthermia and for cytoplasmic delivery of membrane-impermeable molecules [96–100].

Kono et al. [101, 102] have disclosed liposomes bearing succinylated poly(glycidol); these liposomes undergo chain fusion below pH 5.5 that has been shown

Figure 6.7. Temperature-responsive core and pH-sensitive photo-cross-linked poly(*N*-isorpopylacrylamide-*g*-dimethylmaleimide) nanogels. (Reproduced with permission from Ref. 77, Scheme 3. Copyright 2002, American Chemical Society.)

to efficiently deliver calcein to the cytoplasm. Liposomes anchored by or modified with poly(*N*-isopropylacrylamide)-based copolymeric groups are suitable for thermo- and pH-responsive nanogels, which are being investigated for transdermal drug delivery [103–106]. Various PNIPAM-based copolymers have been used for this purpose, and their structures are depicted in Figure 6.8.

Schillemans et al. [107] synthesized novel bilayer-coated nanogels by a selective cross-linking polymerization method that permits monomers to polymerize inside liposomes but not exterior monomers. Somasundaran et al. [108] evaluated nanogels of

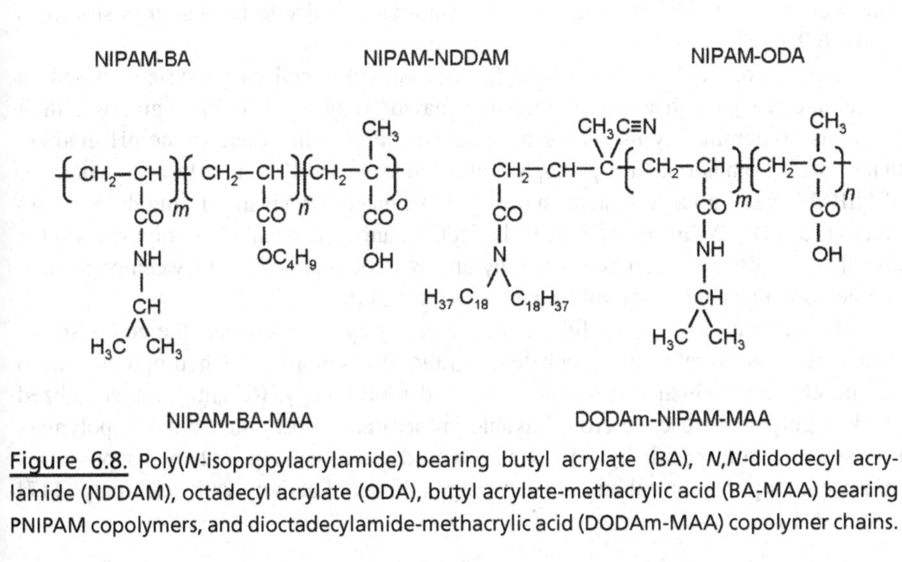

NIPAM-BA   NIPAM-NDDAM   NIPAM-ODA

NIPAM-BA-MAA   DODAm-NIPAM-MAA

Figure 6.8. Poly(*N*-isopropylacrylamide) bearing butyl acrylate (BA), *N,N*-didodecyl acrylamide (NDDAM), octadecyl acrylate (ODA), butyl acrylate-methacrylic acid (BA-MAA) bearing PNIPAM copolymers, and dioctadecylamide-methacrylic acid (DODAm-MAA) copolymer chains.

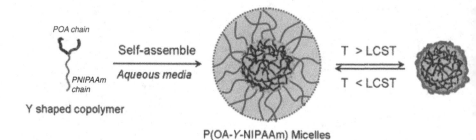

Figure 6.9. Y-shaped copolymer self-assembly to give micelle structures. See insert for color representation of this figure. (Reproduced with permission from Ref. 113, Figure 6.8. Copyright 2006, John Wiley & Sons, Inc.)

poly(acrylamide) (PAM)/poly(acrylic acid)(PAAc) and liposomes with a narrow size distribution for drug binding and release.

6.2.2.2 Micellar Nanogels. Polymer micellar nanogels can be obtained by the supramolecular self-assembly of amphiphilic block or graft copolymers in aqueous solutions [109–112]. They possess unique core-shell morphological structures, where a hydrophobic block segment in the form of a core is surrounded by hydrophilic polymer blocks as a shell (corona) that stabilizes the entire micelle. The core of micelles provides enough space for accommodating various drug or biomacromolecules by physical entrapment. Furthermore, the hydrophilic blocks may form hydrogen bonds with the aqueous media that lead to a perfect shell formation around the core of micelle. Therefore, the drug molecules in the hydrophobic core are protected from hydrolysis and enzymatic degradation. Li et al. [113] successfully developed highly versatile Y-shaped micelles of poly(oleic acid-Y-N-isopropylacrylamide) for drug delivery application. In this study, the delivery of prednisone acetate above its lower critical solution temperature (LCST) was demonstrated. A representation of micelle formation is shown in Figure 6.9.

The development of pH-sensitive (acid-sensitive) micellar nanosystems based on triggarable groups such as block copolymers having weak acidic or basic groups in their blocks is predominantly useful for drug delivery applications due to the pH gradient that exists in both normal and pathophysiological states [114]. Similarly, a large number of NIPAM-based micelle systems have been evaluated for advanced drug delivery devices [115–118]. Apart from this, novel micellar nanosystem called as polyion complex micelles or polyplexes can be formed by electrostatic interactions between oppositely charged polymers and macromolecular drugs [110, 119].

Micellar nanogels are useful for drug delivery systems because the composition, total molecular weight, and block length ratio (hydrophilic and hydrophobic ratio) can be changed, which control their size and morphology. Recently, functionalized block copolymers (reactive cross-linkable groups) and substitution block copolymers nanosystems with specific ligands have been used as a promising strategy that possesses improved temporal control and broader range of sites of activity with selectivity [117] (Figure 6.10).

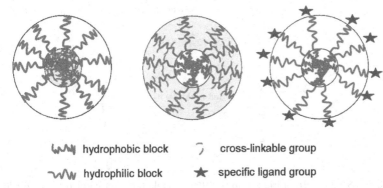

⎰⎱ hydrophobic block �socross-linkable group

⎰⎱ hydrophilic block ★ specific ligand group

Figure 6.10. Internal structural variation in micelle gels. See insert for color representation of this figure. (Reproduced with permission from Ref. 117, Figure 6.1. Copyright 2001, Elsevier, Ltd.)

Voets et al. [120] have prepared core and corona structured micelles from aqueous solutions of poly(2-(dimethylamino)-ethyl methacrylate)-*block*-poly(glyceryl methacrylate), PDMAEMA-*b*-PGMA, and poly(acrylic acid)-*b*-poly(acrylamide), PAAc-*b*-PAM, through complexation of oppositely charged polyelectrolyte blocks. The formed mixed micelles are monodispersed in nature with an average size of 15.2 nm.

6.2.2.3 Hybrid Nanogels. Hybrid nanogels are defined as a composite of nanogel particles dispersed in organic or inorganic matrices [121, 122]. Akiyoshi and co-workers [123–126] have demonstrated nanogel formation in an aqueous medium by self-assembly or aggregation of polymer amphiphiles, such as pullulan-PNIPAM, hydrophobized polysaccharides, and hydrophobized pullulan. This group has investigated cholesterol-bearing pullulan (CHP) nanogels. These nanogels have the ability to form complexes with various proteins, drugs, and DNA; and it is even possible to coat surfaces of liposomes, particles, and solid surfaces including cells [124, 127]. These hybrid nanogels are also capable of delivering insulin and anticancer drugs more effectively. CHP is composed of pullulan backbone and cholesterol branches. The CHP molecules self aggregate to form mono-dispersed stable nanogels through the association of hydrophobic groups that provide physical cross-linking points as shown in Figure 6.11 [128].

Morimoto et al. [129] have designed hybrid hydrogels with self-assembly of 2-methacryloyloxyethyl phosporylcholine (MPC) with methacryloyl-CHP (CHPMA). The nanogels made up of CHPMA domain dispersed in MPC polymer hydrogel matrix has been used as a promising material for protein reservoirs. Physical and chemical cross-linked structured nanogels have been synthesized using seed-nanogels with polymerizable groups as nanocarrier for carbonic anhydrase B (CAB) [130]. Self-assembled hydrogel nanoparticles have also been synthesized from carboxymethylated (CM)-curdlan, substituted with a sulfonylurea (SU) as a hydrophobic moiety for self-assembly [131].

Few tailor-made functional nanogels are being developed (nanogel engineering) by self-assembly associating building blocks of cholesteryl-bearing poly(amino acids),

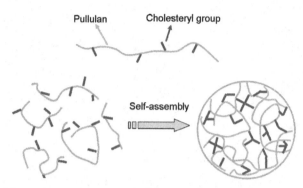

Figure 6.11. Schematic representation of CHP nanogel preparation by physical cross-linking (self-assembly). See insert for color representation of this figure. (Reproduced with permission from Ref. 128, Figure 6.1. Copyright 2004, Elsevier, Ltd.)

alky group modified PNIPAM-CHP, spiropyrane modified pullulan, deoxycholic acid modified chitosan, bile acid bearing dextran, and synthetic polyelectrolytes [123, 132–136]. Spiropyrane-bearing pullulan nanogels are self-assembly of functional associating polymeric chains as blocks. This system demonstrated excellent ability to control refolding of protein as an artificial molecular chaperone upon photo-stimulation [133]. Host–guest interactions could direct the inclusion complexes between hydrophobic molecules and the polar concavities of cyclodextrins described for nanogel self-assembly [137, 138]. Ichikawa and Fukomori [139] studies have revealed that composite of nanosized polymer gel dispersed in cellulose matrix demonstrates enhanced drug release at 50°C.

A different approach has been used by Shin et al. [137] to achieve a hybrid nanogel consisting of PNIPAM/silicate through a squeezing mechanism for sustainable positive thermosensitive drug release. Calcium phosphate-based hybrid nanoparticle systems are also being investigated as carriers for drugs [140, 141]. A novel nanohybrid of polymer-calcium phosphate was prepared from a dilute hydroxyapatite solution at moderate temperature and pH using a nanogel template mineralization method [142]. In this system, polymerizable nanogels (physically cross-linked) was employed as functional cross-linkers to fabricate hybrid hydrogels. Monodispersed CHP nanogel-quantum dot hybrids (~36 nm) have been obtained by simple mixing of the polymers. These nanogel complexes are more efficient in cellular uptake of quantum dots than conventional cationic liposomes [143].

6.3 SYNTHESIS OF NANOGELS

Recently, wide varieties of nanogels have been synthesized, and they are of great interest because these smart materials show native properties of both gels and colloids. Synthesis of these nanogels can be accomplished in several ways, but each method has its own advantages and drawbacks. Various approaches have been divided into three major techniques: (1) simultaneous cross-linking polymerization, (2) ideal intramolecular cross-linking, and (3) destruction of macrogel networks into nanoscale (conversion of macroscopic gels to nanogels).

6.3.1 Simultaneous Cross-Linking Polymerization

Simultaneous cross-linking polymerization allows one to prepare gel network in a single step in which polymerization as well as cross-linking processes occur simultaneously [144–148]. In these polymerizations, apart from monomer, a cross-linking agent (functionality ≥ 3) is used to allow three-dimensional network formation. In such polymerization process, the free radicals formed in the initiation step of the reaction tend to react with monomer and cross-linker molecules to obtain polymeric chains along with some cross-linking points. Intermolecular cross-linking leads to the formation of branched and highly network structure that ultimately is responsible for macroscopic gel product. The intramolecular cross-linking process is responsible for the formation internally cross-linked polymeric chain networks. Figure 6.12 illustrates the difference between inter- and intra-cross-linking behaviors in the simultaneous cross-linking polymerization process.

Micro- and nanogels can be successfully prepared by the cross-linking polymerizations employing (a) bulk polymerization, (b) solution polymerization, (c) emulsion polymerization, (d) suspension polymerization, and (e) nonconventional polymerization methods.

6.3.1.1 Bulk Polymerization. Bulk polymerization is rarely used preparative method for nanogels, but it is highly suitable to generate macroscopic gels [2–7]. Because of the auto-acceleration effect in the polymer reaction, the viscosity of the reaction mixture increases and it may not be possible to control the formation of macrogelation product [149, 150]. Most of the simultaneous cross-linking polymerizations are not entirely homogeneous systems. Even before macrogelation, the system persists with different inhomogeneous density of cross-links, which can be separated at the molecular level. According to the classical gelation theory, homogeneous growth of linear polymeric chains and its cross-linking should not lead to individual microgels in the intermediate stage. However, recent investigations revealed that free-radical simultaneous bulk polymerization is effective in producing nanoscale gel particles [151, 152]. Tetraethoxylated

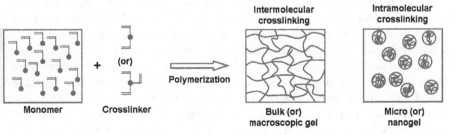

Figure 6.12. Inter- and Intramolecular cross-linking modes in the simultaneous cross-linking polymerization.

bisphenol A dimethacrylate with styrene or divinyl benzene cross-linking polymerizations showed gel particles of 10–40 nm at different stages of the cross-linking process [151,152]. Simulation results of various multifunctional monomers demonstrated structural inhomogeneity in the microgel network formation [153–155]. Utilization of this type of polymerization still needs to be further explored for the synthesis of nanogels.

6.3.1.2 Solution Polymerization. Free-radical cross-linking polymerization is not adequate to produce nanogels, since it results in a broader size distribution than any other conventional polymerization techniques. However, this technique is versatile for understanding the gelation kinetics and there is no need of additional surfactants. The difficult part of the polymerization system is to control the propagation of linear polymeric chains and inter- and intramolecular cross-linking balance that may lead to macrogelation. Free-radical cross-linking polymerization mechanisms are documented in a few critical reviews [156,157]. According to these mechanisms, even much diluted systems can also result in the formation of macrogel. In order to avoid macrogelation, the concentration of polyfunctional precursors (co-monomers and multifunctional cross-linkers) is usually kept below its critical gelation concentration, and the solvent for each polymer system is critically selected to avoid macrogelation [84,158]. To obtain nanogels by solution cross-linking polymerization, one needs to balance the presence of dead chains, polymeric chains (active or propagating chains), and chain loops that act as good steric stabilizers. The polymerization reaction is conducted in a solvent in dilute condition, and the solvent solubility parameter (δ) value is kept similar to that of the corresponding polymer. There is also a possibility of controlling the macrogelation in solution polymerization by using chain transfer agents (cobalt porphyrin) [159,160].

Maitra's group [161,162] reported NIPAM-based copolymer gel (~50 nm) using solution polymerization employing ammonium persulfate/ferrous ammonium sulfate (APS/FAS) as the initiating system at room temperature for delivery of nonsteroidal anti-inflammatory drugs. We have recently developed poly(*N*-isopropylacrylamide-*co*-vinyl pyrrolidone) [poly(NIPAM-*co*-VP)] nanogels with an average size of 40 nm for the delivery of anticancer drugs (Figure 6.13). Other research groups have explored novel methods to form smart core-shell polymeric gels of PNIPAM/PEI and PNIPAM/PEI/chitosan using the same polymerization technique [62,63]. Similarly, well-defined amphiphilic core-shell polymer nanospheres are obtained (60–160 nm) via graft copolymerization of methyl methacrylate (MMA) from water-soluble polymer chains containing amino groups and casein [39,40]. Furthermore, the temperature-sensitive liposomes bearing or modified with poly(*N*-isopropylacrylamide) are possible to obtain by solution polymerization in dioxane or organic solvent using 4,4[1]-azobisisobutyronitrile (AIBN) initiator [87,106]. Molecularly imprinted, soluble, highly cross-linked acrylamide–arginine–tyrosine nanogels have been prepared in dimethylsulfoxide (DMSO) [163]. Highly cross-linked microgels based on ethylene dimethacrylate-methyl methacrylate-trimethylolpropane trimethacrylate (EDMA-MMA-TRIM) have been successfully prepared in different solvent systems using free-radical solution polymerization [85]. A facile and direct homogeneous solution polymerization approach for hydroxyethyl methacrylate and methacrylic acid nanogels was developed in a single step without the use of oleophilic surfactants [164]. It is possible to produce more complex

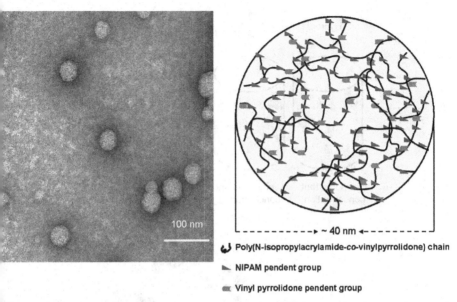

Poly(N-isopropylacrylamide-co-vinylpyrrolidone) chain

NIPAM pendent group

Vinyl pyrrolidone pendent group

Figure 6.13. Transmission electron microscope (TEM) image of nanogel and nanogel internal cross-linking network structure.

architectural nanogels using anionic polymerization in solution and as small as 3–30 nm in size [165–167].

6.3.1.3 Emulsion Polymerization. Emulsion polymerization is widely used method to synthesize both micro- and nanogels. To prepare nanogels using this method, the droplets of monomer solution are added to a stable emulsion (Figure 6.14) in which the system contains initiator molecules that generate free radicals in the liquid phase (usually water) to start the polymerization reaction. In this manner, the polymer chains grow in the surfactant protected layers and then the surrounding solvent is removed by evaporation, extraction, or dialysis. Using this method of polymerization, the size of gel or nanoparticles can be controlled effectively by maintaining the size of the droplets in the water–oil emulsions [168, 169]. In this method of polymerization, monofunctional monomers can give polymer nanoparticles where multifunctional monomers form internally cross-linked network polymers or gels. In the case of a monofunctional monomer system, polymerization can be readily terminated but multifunctional system restricts this behavior and the propagation and cross-linking steps of polymerization process leads to the formation of gel networks. The prepared nanogels/microgels employing this method are generally incompatible with biological macromolecules. To obtain better compatibility with biomacromolecules, water–water emulsions could be employed for the preparation of nanoparticles or gels, but controlling the size of gels to lower nanometers is difficult [170, 171].

Microemulsion, inverse emulsion, and surfactant-free emulsion polymerization techniques are modified methods of emulsion polymerization from which nearly

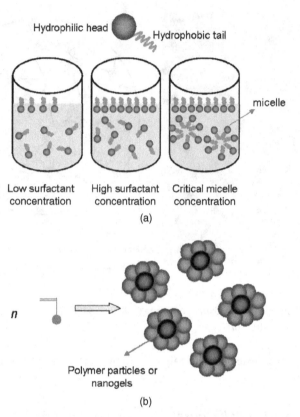

Figure 6.14. Classical emulsion polymerization technique for nanogel preparation. See insert for color representation of this figure.

monodispersed microgels can be obtained [172–179]. These methods differ in the selection of particular conditions, but the main principle is the same as emulsion polymerization. The major variations in these methods are as follows:

Microemulsion. Constructed with a critical concentration of emulsifier where all the monomer molecules are present in micelles and not in the form of monomer droplets, and the polymerization is initiated under such condition.

Inverse Emulsion. This technique is just opposite to the emulsion polymerization, that is, polymerization of hydrophilic monomers is carried out in organic hydrophobic phase instead of in an aqueous phase.

Surfactant-Free Emulsion Polymerization. This method of polymerization is carried out without any surfactant but monomer and the formed polymer chains themselves act as emulsifier.

Various authors have studied extensively the kinetics, mechanism, and size distribution of gel particle formation using these methods.

Pelton and co-workers [16, 58, 180–183] have systematically developed PNIPAM latexes (gel particles) in water. These gel particles were prepared using NIPAM with potassium persulfate as an initiator, sodium dodecyl sulfate (SDS) as a surfactant, and N,N'-methlylenebisacrylamide (MBA) as a cross-linker in aqueous media. The authors reported that under these conditions, the gel particles are formed only above the critical temperature (i.e., 55°C). Using similar procedure, Murray et al. [15, 80] have tailored various colloidal microgels from NIPAM, NIPAM with acrylic acid, and/or other vinyl monomers. Antonietti et al. [184] disclosed the critical values of microgel cross-link density for gel transition in microemulsion polymerization, where the gels are stabilized with polymeric chains itself.

Recently, nanogels composed of NIPAM, AAc, and N-vinylimidazole have been reported [15, 60]. Ito et al. [60] adopted an oil-free (surfactant-free) redox emulsion polymerization for NIPAM-based anionic and cationic polyelectrolyte nanogels in the presence of sodium dodecylbenzenesulfonate (NaDBS). Ishii et al. [185] employed surfactants other than NaDBS to obtain nanogels with longer self-life. Ramanan et al. [186] described in their studies in-depth the information about how the particle size of polyNIPAM gel changes with change in concentration of SDS. Lyon's group [75, 76] prepared core-shell cross-linked gels with PNIPAM as core and NIPAM/acrylic acid polymeric arms using seed and feed precipitation emulsion polymerization.

6.3.1.4 Suspension Polymerization. This method is less suitable for producing nanogels or nanoparticles [149, 156, 187]. The nanogel preparation using this method follows different steps. Usually, suspension polymerization proceeds with a suitable initiator with a dilute system of monomers; the cross-linker is dispersed by mechanical agitation in the above liquid phase to obtain macromolecular chains. However, as these chains reach a critical length, they collapse to form precursor gel particles. These precursor particles continue to increase in their size due to aggregation of macromolecular chains. The particles formed in this way possess good colloidal stability; however, the main deficiency of this method is the difficulty in controlling the particle size because the colloidal stability is dependent on monomer to cross-linker composition, initiator, and temperature.

6.3.1.5 Nonconventional Polymerization Methods. It is widely speculated that the conventional simultaneous cross-linking polymerization methods followed by initiation with thermally decomposed initiators to give fragments (free radicals) are responsible for activation of polymerization and subsequent cross-linking reactions to acquire cross-linked network for nanogels [149, 150]. These techniques produce end polymeric gels that contains unreacted species including cross-linker, initiator, reactive intermediates, and surfactants. The unreacted or active species are physically entrapped in their polymeric networks and are chemically bound. The presence of unreacted species in the final product can cause an adverse effect and hence is a concern. Furthermore, the existence of cross-linker can lead to structural nonhomogeneity that can harm the phase transition of gels [58, 188–190].

Alternative cross-linking polymerization paths such as UV polymerization, radiation polymerization, and ultrasound/microwave polymerizations are other convenient

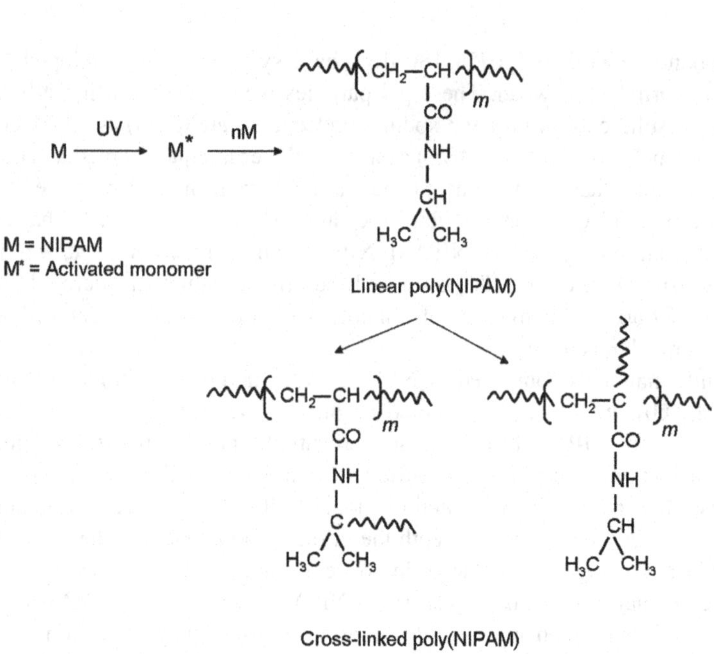

Figure 6.15. Formation of cross-linked PNIPAM nanogels under UV irradiation. (Reproduced with permission from Ref. 198, Figure 6.5. Copyright 2006, Elsevier, Ltd.)

methods to prepare nanogels [191–195]. To achieve highly homogeneous cross-linked gels, the UV-curing polymerization is considered quite appropriate. The benefit of this method is that monomers employed are multifunctional in which photosensitive initiators are introduced to obtain highly cross-linked polymers [196]. Gao and Frisken [197] reported a convenient way to prepare cross-linker free poly(N-isopropylacrylamide) nanospheres. Initially, linear poly(N-isopropylacrylamide) is prepared using a potassium persulfate/N,N'-tetramethylethylenediamine (KPS/TMEDA) initiating system, which is then exposed to He–Ne laser beam irradiation at high temperature (70°C) to form intramolecular cross-linked PNIPAM gel networks. Nanogels can also be prepared using this method with monomers encapsulated/entrapped in liposomes [65]. Figure 6.15 demonstrates the mechanism of formation of nanogels from monomer/polymeric chains through photopolymerization [198]. The yield was higher at pH 2 media, where the polymerization reaction accelerates with H^+ ions. Another convenient UV-polymerization method, is based on double layer. In this method, the upper layer contains droplets and UV is exposed to the lower layer that gives internally cross-linked PEG nanogels [199].

6.3.2 Ideal Intramolecular Cross-Linking

The gels prepared by chemical cross-linking polymerization must be extra-pure to fulfill the required specifications of the Food and Drug Administration (FDA) [200]. Generally, vinyl monomers and cross-linkers/initiators are highly toxic, carcinogenic, or teratogenic in nature. The practical applications of cross-linked hydrogels/micro- and

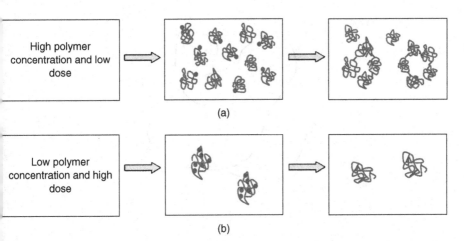

Figure 6.16. Radiation mechanism for (a) bulk/micro gel and (b) nanogel formation. See insert for color representation of this figure. (Reproduced with permission from Ref. 203, Figure 6.2. Copyright 2003, American Chemical Society.)

nanogels in drug delivery are still limited due to the purity issue. A few additive-free methods have been developed to prepare gels using radiation/UV irradiation [201, 202]. In this method, unlike other techniques, no problems arise with regard to removing the potentially harmful substances such as unreacted monomer, initiator, cross-linker, stabilizers, and so on, from the final product. This technique provides nanogel with a variety of compositions, and it is easy to control their size. Nanogel formation (polymer cross-linkability) depends on the function of polymer chemical structure, polymer concentration, type of irradiation (pulsed versus continuous), and the rate of radiation dose. A high-dose pulse irradiation of a very dilute polymer solution leads to cross-link formation within single polymer chains and highly favors the formation of nanogels. The low-dose pulse irradiation, however, forms highly cross-linked gel nanoparticles [203] (Figure 6.16).

Rosiak and co-workers [203, 204] have employed the radiation induction method to produce nanogels based on neutral hydrophilic polymers such as poly(vinyl alcohol) (PVA), poly(vinyl pyrrolidone) (PVP), and poly(acrylic acid) (PAAc) through intramolecular cross-linking of the linear polymers using high-energy dose electron pulse. Nurkeeva et al. [205] extended the PAAc nanogels interpolymer complex studies with various nonionic polymers in aqueous solutions using different physicochemical methods. The study explains the structural variations between linear polymer chains and intramolecularly cross-linked nanogels (Figure 6.17). The intramolecular cross-linking of macromolecules for nanogel formation was achieved by applying higher radiation doses on polymers, which is responsible for the decrease in their radius of gyration and molecular weight [205].

Furusawa et al. [206] have reported the preparation of nanosized gelatin particles by means of gamma-ray irradiation; the obtained nanoparticles consisted of highly and randomly packed gelatin macromolecular chains with stable conformation against

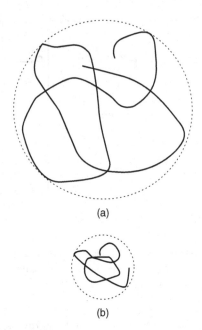

Figure 6.17. (a) Micro and (b) nano poly(acrylic acid) gel structure. (Reproduced with permission from Ref. 205, Figure 6.1. Copyright 2006, Elsevier, Ltd.)

temperature and with a small radius of gyration. So far, this radiation additive-free method has been explored for few hydrophilic polymers (PVA, PVP, and PAAc) and hence an extensive research is needed to obtain nanogel materials using other synthetic polymers. It is possible to obtain intramolecular chemically cross-linking of single chains of water-soluble polymers by reacting them in the presence of a suitable cross-linking agent in very dilute solutions. The cross-linkers are capable of reacting with functional groups of the polymer. In this category, poly(vinyl alcohol) is internally cross-linked with glutaraldehyde [207]. The cross-linking reaction takes place between –OH groups of PVA chains and aldehyde group of glutaradehyde. Similarly, intramolecular cross-linked poly(allylamine), hydroxypropylcellulose, and hyaluronic acid-based nanogels were obtained [208–212]. A diluted polymer solution is used in the case of chemical cross-linking to avoid macrogelation. Nanogels are also synthesized by photo-cross-linking of various polymers and copolymers of N-isopropylacrylamide (NIPAM) by UV-irradiation of polymer solutions in water. The size of nanogels is controlled by the concentration of chromophore in the photopolymer and surfactant concentrations [213–216].

6.3.3 Conversion of Macroscopic Gels to Nanogels

Several synthetic methodologies are identified to prepare macroscopic gel networks (bulk gel networks or wall-to-wall cross-links) and are easy to prepare, because it is not necessary to control the synthetic parameters as are required in nanogel or microgel synthesis to control the size [2–9]. The macroscopic gel networks are generally prepared by bulk polymerization, which produce a solid and the network structure with macroporous blocks. These blocks are then crushed, grounded, and sieved to obtain gels of desired particle size. However, this is a time- and energy-consuming process and results in significant loss of material. Nevertheless, micro- and nanogels obtained from this method have particles of different shape and sizes.

6.4 PROPERTIES OF NANOGELS

A characteristic of nanogels is their swelling property in aqueous media. The extent of swelling capacity and driving forces of nanogels/microgels are the same as their bulk or macrogels, but the most beneficial feature is their rapid swelling/de-swelling characteristics. Like other colloidal particles, poly(NIPAM) and its copolymeric nanogels/microgels have surface-active property. The various properties of microgels/nanogels such as particle size, swellability, electromobility, rheology, and colloidal stability are well-described in the literature [17, 58, 80, 180, 183, 217–222]. Such properties of nanogels/microgels are significantly influenced by temperature, presence of hydrophilic/hydrophobic groups in the polymeric networks, the cross-linking density of the gels, surfactant concentration, and type of cross-links present in the polymer networks.

Various types of cross-linkers are being used for making gels, and each cross-linker can form its integral structural network in the gel [223–226]. N,N'-methlylenebisacrylamide (MBA) is the most widely used cross-linker because it is a more quickly used reactive cross-linker due to its greater reactivity as compared to other monomers. Thus, the resulting gel formed with this cross-liner has a certain extent of high cross-link density zones in each of particle. Since the MBA networks are highly hydrophilic in nature, the dense MBA network structures are present in the inner portion of gel and less cross-linked networks are outside at the water interface. The cross-linker concentration results in a various range of products with different cross-linked dense networks of various nanometer size and distribution [224, 227, 228]. McPhee et al. [182] investigated the extent of swelling behavior for NIPAM microgels with a different amount of MBA present during polymerization. The network structures determine the swelling characteristic of gel particles [224]. Woodward et al. [227] investigated in detail the influence of varying cross-linker concentration of MBA (0.25–30.0 wt%) on the properties of gel system using turbidimetric, light-scattering, and differential scanning calorimetric (DSC) analyses. Cross-link density of the gel network determines the micro- or macrogelation transition [184]. As the cross-linked density increases in a microgel system, their viscosity decreases. The major contribution of varied cross-link densities leads to formation of soft to very hard gels systems which exhibit different particle interactions and swelling behaviors [15, 17, 229]. Figures 6.18 and 6.19 provide an insight of cross-linking effect on the size of nanogel and swelling characteristics [230].

The global properties of nanogel particles depend on the nature of monomeric units present in the polymeric chains of gel networks. These functional groups have a tremendous effect on drug-carrying and drug-releasing properties, and some functional

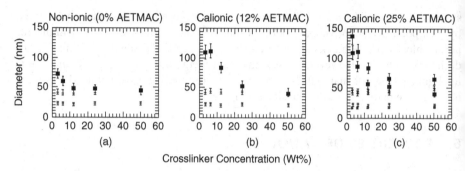

Figure 6.18. Hydrodynamic diameter of micelles and nanogels versus cross-linker concentrations. (Reproduced with permission from Ref. 230, Figure 6.2. Copyright 2002, American Chemical Society.)

groups have the potential to conjugate with drugs/antibodies for targeting applications. These pendent functional groups of polymeric chains contribute toward establishing hydrogen bonding or van der Waals forces of interactions within the gel network and thus facilitate the drug-carrying efficiency. Moreover, the presence of functional groups at interface with drug/protein molecules is also responsible for higher loading.

Nano- and microgel networks composed of repeating units such as N-isopropylacrylamide, N-isopropylmethacrylamide (NIPMAM), N-diethylacrylamide (NDEAM), N-vinylisobutaramide (NVIBAM), N-acryloylpyrrolidone (NAP), and N-vinylcaprolactam (NVCL) are well-known synthetic aqueous thermosensitive systems [14,17,183]. The most widely studied thermosensitive nanogel/microgel system is based on poly(N-isopropylacrylamide) because it has volume phase transition temperature (VPTT) or LCST very close to the body temperature, whereas the other thermosensitive systems exhibit VPTT either below or above the body temperature. Figure 6.20

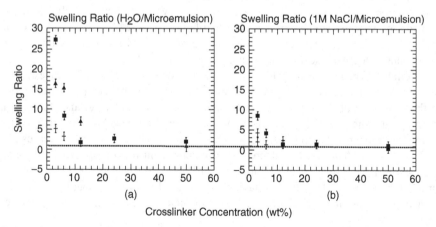

Figure 6.19. (A) Swelling ratio of nanogels in water versus heptane micro emulsion. (B) Swelling ratio of nanogels in 1 M versus heptane micro emulsion. (Reproduced with permission from Ref. 230, Figure 6.3. Copyright 2002, American Chemical Society.)

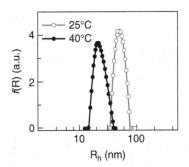

Figure 6.20. PNIPAM gel particles size distribution below and above the LCST. (Reproduced with permission from Ref. 231, Figure 6.1. Copyright 2005, John Wiley & Sons, Inc.)

demonstrates the distribution of hydrodynamic radius of PNIPAM-thermosensitive nanogels below and above the LCST [231]. The hydrodynamic radius of nanogel particles is considerably higher below the LCST of gels. The same gel particles have lower particle size above its LCST due to shrinking of the gel networks. The particle size and LCST or swelling pattern of the nanogels can be controlled by introducing other repeating units in the gel networks. For example, hydrophilic monomeric units improve LCST and hydrophobic units decrease their LCST position.

The properties of nanogels depend upon the pH. The micro- or nanogels containing pH-dependent hydrophobic/hydrophilic repeating units or networks are considered to be better drug delivery carriers. For example, the nanogels can absorb positively charged drugs at alkaline pH through electrostatic attractions and can release them in acidic pH. Similarly, nanogels can absorb molecules with low polarity in acidic pH when the core of nanogels is relatively hydrophobic and can release them in neutral pH where nanogels become more hydrophilic. These nanogels changes their swelling or particles size with respect to pH [232–236].

Core-shell nanogels have both temperature- and pH-responsive polymeric chains within a gel network structure [73–76]. In these cases, there are two distinguished spherical structural layers in which every individual core is surrounded with a shell. Thus, the combined temperature- and pH-responsive characteristics can be imparted to these nanogels.

As mentioned above, the majority of successful methods to prepare nanogels are based on emulsion polymerization in which the use of surfactant is necessary. Strong interactions have been identified between gel particles and ionic surfactants, but there are no interactions with nonionic surfactant [15, 17, 175, 178, 237–239]. The interactions of aqueous surfactant vary with chain length. Anionic surfactants have a larger effect on microgel properties than do cationic surfactants. In the presence of SDS (anionic surfactant), the size of PNIPAM microgel particles increases due to SDS accumulating within the gel network, causing internal electrostatic repulsion and therefore expanding the polymer network structure. This effect also results in the shift to higher conformational transition temperature (LCST). The cationic surfactants such as dodecylpyridine bromide and dodecyltrimethylammonium bromide are not effective in raising the LCST of gel particles

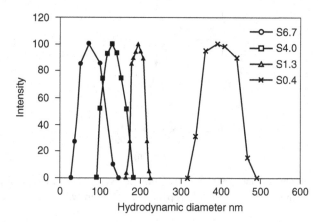

Figure 6.21. The intensity-weighed size distributions of the microgels in aqueous suspensions at 20°C (DLS). S0.4 = 0.4 mM; S1.3 = 1.3 mM; S4.0 = 4.0 mM; and S6.7 = 6.7 mM of SDS concentration, respectively, in the formulations. (Reproduced with permission from Ref. 244, Figure 6.1. Copyright 2006, John Wiley & Sons, Inc.)

in the same way as anionic surfactants. The surfactant binding to the nanogels/microgels is influenced by swelling/size of the nanogel [239–244]. The surfactant binding to polymeric chains networks leads to improved size, swelling, and VPTT properties of gels (Figure 6.21).

As nanogels have potential applications in drug delivery, the gel particles should be stable in aqueous media—that is, should have high dispersibility over a period. In addition, they must not aggregate under physiological conditions. Specifically, electrostatic and steric interactions between the swollen gel particles are liable for the dispersion stability. The Hamaker constant (a measuring constant to dispersion stability or dispersion) for nanogels/microgels at equilibrium stage is negligible, since a large volume fraction of solvent present in the swollen stage; that is, the dispersion of gels at equilibrium stage is more stable than in the dry state. McAllister et al. [230] have verified the stability of nanogels after 1 month of storage at ambient temperature and 6 months stored at 4°C. Recently, we confirmed that the PNIPAM-based nanogels do not cause any change in nanogel size even after 4 months of storage at 4°C (Figure 6.22).

6.5 DRUG DELIVERY APPLICATIONS OF NANOGELS

Despite availability of several potent agents against various diseases, their effective delivery remains a major challenge. Several issues such as solubility, stability, specificity, sustained availability, penetrability across anatomical and physiological barriers, nonspecific toxicity, and so on, limit the successful transformation of these agents into drugs. Recent advances in nanotechnology offer many strategies to address the various problems associated with drug delivery. Nanostructures, due to their smaller size, can effectively shuttle therapeutics across different physiological and anatomical barriers.

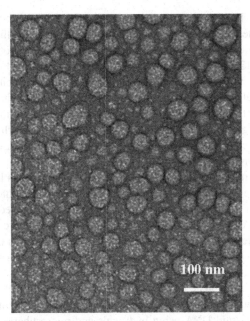

Figure 6.22. Transmission electron microscopic image of NIPAM-based nanogel after 3 months' storage.

The strategies that are being pursued are active targeting by modifying and functionalizing the surface of nanoparticle with tissue-specific antibodies or ligands. In passive targeting, the advantage of leaky vasculature associated with inflammatory milieu or in tumor is explored. However, the challenge in targeting is that nanoparticles attract the natural body defense system and hence are easily identified and taken up by the reticuloendothelial system. Therefore, successful nanostructure design requires assembly of appropriate surface characteristics as well as targeting ligands to achieve long systemic circulation and subsequent interactions with the receptors in the target tissue [245]. Nanogel formulations provide a better platform for designing of nanocarriers because of their three-dimensional cross-linkage structure with hydrophilic surface properties that prevent their clearance by the reticuloendothelial system. In addition, nanogels demonstrate unique properties, such as the ability to trap biomolecules inside the gel structure and rapidly respond to an external stimulus [246]. Due to these inherent qualities, nanogels have been investigated for protein and gene delivery applications [15, 17, 68–70, 230]. Recent studies using novel photosensitive nanogels have demonstrated increase molecular chaperone-like activity upon photostimulation [133]. Similarly, the other group has shown improved thermal stability of the enzyme associated with artificial chaperons by using nanogels [247]. Since many drugs act as protagonists or antagonists to different chemicals in the body, a delivery system that can respond to the concentrations of certain molecules in the body is invaluable [248]. Yamagat et al. [249] demonstrated that pH-sensitive hydrogels protect insulin from gastric and intestinal enzymes.

Recently, there has been significant interest in developing nonviral vectors for synthetic vaccines designed to prime the adaptive immune system that are sought for a broad range of infectious diseases and for the treatment of cancer in both prophylactic and therapeutic settings [250–252]. However, many obstacles associated with the successful delivery of synthetic vaccines remain such as antigen loading capacity, maintaining the integrity of encapsulated proteins, and minimizing the nonspecific antigen–antibody reaction. Therefore, the delivery system that can effectively carry the protein antigens to antigen-presenting cells is most desirable. The feasibility of achieving these goals has been demonstrated using submicron-sized hydrophilic particles, which were loaded with high doses of protein [253].

One other important area of research where these nanogels can be effectively used is in the regenerative medicine. The major challenge is how to achieve significant expression of genes in highly compromised injured tissue surrounded by inflammatory environment [254]. Although delivery of genes encoding therapeutic protein is an alternative for outright delivery of protein, stability of plasmid DNA is not much different compared to proteins. The successful therapeutic gene delivery for tissue engineering requires a system that can provide a control over the DNA release, facilitate cellular uptake of DNA, maintain gene expression, and provide support for homing and successful differentiation of infiltrated stem cells [255]. Recently, several studies explored controlled-release plasmid DNA using hydrogels and microspheres [256, 257]. These studies have reported that complexation with gelatin can reduce the degradation of DNA from nucleases and may facilitate cellular entry through interaction of positively charged complexes with negatively charged cell membranes [258]. Furthermore, a novel hydrogel composite of oligo(poly(ethylene glycol) fumarate) (OPF) and cationized gelatin microspheres (CGMS) have been demonstrated to have potential for controlled gene delivery in tissue engineering applications [255].

Most recent studies have demonstrated applications of nanogels composed of amphiphilic polymers and cationic polyethylenimine for delivery of cytotoxic nucleoside analogs $5'$-triphosphates (NTPs) into cancer cells [259]. Similarly, other studies have demonstrated that nanogels composed of N-isopropylacrylamide (NIPAM) and N-vinylpyrrolidone (VP) cross-linked with N,N'-methylenebisacrylamide (MBA) coated with polysorbate 80 can be used to encapsulate N-hexylcarbamoyl-5-fluorouracil (HCFU) for targeting to the brain [260]. Thus, the nanogel-based delivery systems have opened a new avenue for drug delivery applications.

6.6 SILENT FEATURES OF NANOGELS

- Nanogels are able to solubilize hydrophobic drugs and diagnostic agents in their core or networks of gel.
- They are easy to load with different types of therapeutics, usually involving mixing of the formed nanogels with drugs of interest.
- Nanogels can be modified or functionalized with targeting ligands to alter biodistribution of the loaded therapeutics.

- Nanogels typically range in size of 20–200 nm in diameter and hence are effective in avoiding the rapid renal exclusion but are small enough to avoid the uptake by the reticuloendothelial system.

- Nanogels or polymeric micellar nanogel systems have better stability over the surfactant micelles and exhibit lower critical micelle concentrations, slower rates of dissociation, and longer retention of loaded drugs.

- Nanogels exhibit sharp stimuli/physiological-responsive properties that can be used for selective delivery of therapeutic drug molecules.

- Nanogels could be prepared without employing energy or harsh conditions such as sonication or homogenization, which is critical for encapsulating biomacromolecules.

6.7 CONCLUSIONS

The drug delivery research is expected to address a series of problems associated with old and modern therapeutics such as unfavorable pharmacokinetics and pharmacodynamics causing nonspecific effects, poor stability in biological environment, poor drug solubility affecting drug absorption, transient drug retention in the target site, and so on. Nanogels are poised to play an important role in the field of drug delivery to address some of the problems. One will be able to control the release of the loaded drug in response to an external stimulus such as pH and temperature. It is also conceivable to develop bioresponsive nanogel where the specific changes in the environment under pathological conditions could trigger the release of the loaded therapeutics. Furthermore, their small size, ability to load different types of therapeutics, and functionalize surface with targeting moiety offer unique opportunities to use nanogels effectively to address the problems associated with most modern therapeutics.

ACKNOWLEDGMENTS

Grant support from the National Institutes of Health (1R01 EB 003975, 1R01 EB005822, 1R21 CA121751-01), American Heart Association, Heartland Affiliate (Award #0515489Z), and Nebraska Research Initiative is gratefully acknowledged.

REFERENCES

1. Wichterle O, Lim D. Hydrophilic gels for biological use. *Nature* 1960;185:117–118.
2. Hoffman AS. Hydrogels for biomedical applications. *Adv Drug Deliv Rev* 2002;54:3–12.
3. Peppas NA. Hydrogels in Medicine and Pharmacy, *Vol. 1*. Fundamentals. Boca Raton, FL: CRC Press; 1986.
4. Rosiak JM. In Clough RL, Shalaby SW, editors. *Radiation Effect on Polymers*. Washington, DC: ACS Series; 1991, Chapter 17, pp. 271–285.

5. Buchholz FL. Polyacrylamides and polyacrylic acids. Elvers B, Hawkins A, Schulz S, editors. *Ullmann's Encyclopedia of Industrial Chemistry.* Weinheim; Germany: VCH; 1992, pp. 143–146.

6. Buchholz FL. Preparation and structure of polyacrylates. In Peppas BL, Harland RS, editors. *Absorbent Polymer Technology.* New York: Elsevier Science Publishing Company; 1990, Chapter 2, pp. 23–44.

7. Brannon-Peppas L. Preparation and characterization in crosslinked hydrophilic networks. In Peppas BL, Harland RS, editors. *Absorbent Polymer Technology.* New York: Elsevier Science Publishing Company; 1990, Chapter 2, pp. 45–66.

8. Schild HG. Poly(*N*-isopropylacrylamide): Experiment, theory and application. *Prog Polym Sci* 1992;17:163–249.

9. Tanaka Y, Gong JP, Osada Y. Novel hydrogels with excellent mechanical performance. *Prog Polym Sci* 2005;30:1–9.

10. Gil ES, Hudson SM. Stimuli-reponsive polymers and their bioconjugates. *Prog Polym Sci* 2004;29:1173–1222.

11. Osada Y, Gong JP, Tanaka T. Polymer gels. *J Macromol Sci Part C: Polym Rev* 2004;44:87–112.

12. Murali Mohan Y, Keshava Murthy PS, Mohana Raju K. Synthesis, characterization and effect of reaction parameters on swelling properties of acrylamide–sodium methacrylate superabsorbent copolymers. *React Funct Polym* 2005;63:11–26.

13. Hoffman AS. Intelligent polymers. In Park K, editor. *Controlled Drug Delivery.* Washington, DC: American Chemical Society; 1997, pp. 485–498.

14. Staudinger H, Husemann E. Highly polymerized compounds. CXVI. The limiting swelling capacity of polystyrene. *Ber Dtsch Chem Ges* 1935;68:1618–1634.

15. Murray MJ, Snowden MJ. The preparation, characterisation and applications of colloidal microgels. *Adv Colloid Interface. Sci* 1995;54:73–91.

16. Shashoua VE, Beaman RG. Microgel: An idealized polymer molecule. *J Polym Sci* 1958;33:101–117.

17. Pelton R. Temperature-sensitive aqueous microgels. *Adv Colloid Interface Sci* 2000;85:1–33.

18. Ogawa K, Nakayama A, Kokufuta E. Preparation and characterization of thermosensitive polyampholyte nanogels. *Langmuir* 2003;19:3178–3184.

19. Duan Vo C, Kuckling D, Adler H-JP, Schönhoff M. Preparation of thermosensitive nanogels by photo-cross-linking. *Colloid Polym Sci* 2002;280:400–409.

20. Chilkoti A, Dreher MR, Meyer DE, Raucher, D. Targeted drug delivery by thermally responsive polymers. *Adv Drug Deliv Rev* 2002;54:613–630.

21. Biffis A. Functionalized microgels: novel stabilizers for catalytically active metal colloids. *J Mol Catal A: Chem* 2001;165:303–307.

22. Morris GE, Vincent B, Snowden MJ. The interaction of thermosensitive, anionic microgels with metal ion solution species. *Prog Colloid Polym Sci* 1997;105:16–22.

23. Kiser PF, Wilson G, Needham D. Lipid-coated microgels for the triggered release of doxorubicin. *J Control Rel* 2000;68:9–22.

24. Peppas NA, Bures P, Leobandung W, Ichikawa H. Hydrogels in pharmaceutical formulations. *Eur J Pharm Biopharm* 2000;50:27–46.

25. Reese CE, Mikhonin AV, Kamenjicki M, Tikhonov A, Asher SA. Nanogel nanosecond photonic crystal optical switching. *J Am Chem Soc* 2004;126:1493–1496.

26. Sakai T, Yoshida R. Self-oscillating nanogel particles. *Langmuir* 2004;20:1036–1038.

27. Sakai T, Hara Y, Yoshida R. Phase transition behaviors of self-oscillating polymer and nano-gel particles. *Macromol Rapid Commun* 2005;26:1140–1144.

28. Ito Y, Nogawa M, Yoshida R. Temperature control of the Belousov–Zhabotinsky reaction using a thermoresponsive polymer. *Langmuir* 2003;19:9577–9579.

29. Yoshida R, Sakai, T, Ito S, Yamaguchi T. Self-oscillation of polymer chains with rhythmical soluble-insoluble changes. *J Am Chem Soc* 2002;124:8095–8098.

30. Crook CJ, Smith A, Jones RAL, Ryan AJ. Chemically induced oscillations in a pH-responsive hydrogel. *Phys Chem Chem Phys* 2002;4:1367–1369.

31. Ryan AJ, Crook CJ, Howse JR, Topham P, Jones AL, Geoghegan M, Parnell AJ, Ruiz-Perez L, Martin SJ, Cadby A, Menelle A, Webster JR, Gleeson AJ, Bras W. Responsive brushes and gels as components of soft nanotechnology. *Faraday Discuss* 2005;128:55–74.

32. Varga I, Szalai I, Mészaros R, Gilnyi T. Pulsating pH-responsive nanogels. *J Phys Chem B* 2006;110:20297–20301.

33. Singla AK, Garg A, Aggarwal D. Paclitaxel and its formulations. *Int J Pharm* 2002;235:179–192.

34. Benjamin RS. Pharmacokinetics of adriamycin (NSC-123127) in patients with sarcomas. *Cancer Chemother Rep* 1974;58:271–273.

35. Kayser O, Lemke A, Hernández-Trejo N. The impact of nanobiotechnology on the development of new drug delivery systems. *Curr Pharm Biotech* 2005;6:3–5.

36. Vasir JK, Reddy MK, Labhasetwar V. Nanosystems in drug targeting: Opportunities and challenges. *Currt Nanosci* 2005;1:47–64.

37. Wilkinson JM. Nanotechnology applications in medicine. *Med Device Technol* 2003;14:29–31.

38. Roco MC. Nanotechnology: Convergence with modern biology and medicine. *Curr Opin Biotechnol* 2003;14:337–346.

39. Zhu J, Li P. Synthesis and characterization of poly(methyl methacrylate)/casein nanoparticles with a well-defined core-shell structure. *J Polym Sci Part A: Polym Chem* 2003;41:3346–3353.

40. Li P, Zhu J, Sunintaboon P, Harris FW. New route to amphiphilic core-shell polymer nanospheres: Graft copolymerization of methyl methacrylate from water-soluble polymer chains containing amino groups. *Langmuir* 2002;18:8641–8646.

41. Sahoo SK, Labhasetwar V. Nanotech approaches to drug delivery and imaging. *Drug Discov Today* 2003;8:1112–1120.

42. Gao X, Cui Y, Levenson RM, Chung LWK, Nie S. *In vivo* cancer targeting and imaging with semiconductor quantum dot. *Nat Biotechnol* 2004;22:969–976.

43. Derfus AM, Chan WCW, Bhatia SN. Intracellular delivery of quantum dots for live cell labeling and organelle tracking. *Adv Mater* 2004;16:961–966.

44. Panyam J, Labhasetwar V. Biodegradable nanoparticles for drug and gene delivery to cells and tissue. *Adv Drug Deliv Rev* 2003;55:329–347.

45. Lamprecht A, Ubrich N, Yamamoto H, Schäfer U, Takeuchi H, Maincent P, Kawashima Y, Lehr C-M. Biodegradable nanoparticles for targeted drug delivery in treatment of inflammatory bowel disease. *J Pharmacol Exp Ther* 2001;299:775–781.

46. Brigger I, Dubernet C, Couvreur P. Nanoparticles in cancer therapy and diagnosis. *Adv Drug Deliv Rev* 2002;54:631–651.

47. Lambert G, Fattal E, Couvreur P. Nanoparticulate systems for the delivery of antisense oligonucleotides. *Adv Drug Deliv Rev* 2001;47:99–112.

48. Monsky WL, Fukumura D, Gohongi T, Ancukiewcz M, Weich HA, Torchilin VP, Yuan F, Jain RK. Augmentation of transvascular transport of macromolecules and nanoparticles in tumors using vascular endothelial growth factor. *Cancer Res* 1999;59:4129–4135.

49. Winet H, Hollinger JO, Stevanovic M. Incorporation of polylactide-polyglycolide in a cortical defect: Neoangiogenesis and blood supply in a bone chamber. *J Orthop Res* 1995;13:679–689.

50. Guzman LA, Labhasetwar V, Song C, Jang Y, Lincoff AM, Levy R, Topol EJ. Local intra-luminal infusion of biodegradable polymeric nanoparticles. A novel approach for prolonged drug delivery after balloon angioplasty. *Circulation* 1996;94:1441–1448.

51. Sahiner N, Godbey WT, McPherson GL, John VT. Microgel, nanogel and hydrogel–hydrogel semi-IPN composites for biomedical applications: Synthesis and characterization. *Colloid Polym Sci* 2006;284:1121–1129.

52. Aliyer HA, Hamilton PD, Remsen EE, Ravi N. Synthesis of polyacrylamide nanogels by intramolecular disulfide cross-linking. *J Bioact Compat Polym* 2005;20:169–181.

53. Doherty EAS, Kan CW, Barron AE. Sparsely cross-linked "nanogels" for microchannel DNA sequencing. *Electrophoresis* 2003;24:4170–4180.

54. Doherty EAS, Kan CW, Paegel BM, Yeung SHI, Cao S, Mathies RA, Barron AE. Sparsely cross-linked "nanogel" matrixes as fluid, mechanically stabilized polymer networks for high-throughput microchannel DNA sequencing. *Anal Chem* 2004;76:5249–5256.

55. Bromberg LE, Ron ES. Temperature-responsive gels and thermogelling polymer matrices for protein and peptide delivery. *Adv Drug Deliv Rev* 1998;31:197–221.

56. Dong LC, Hoffman AS. A novel approach for preparation of pH-sensitive hydrogels for enteric drug delivery. *J Control Rel* 1991;15:141–152.

57. Kratz K, Hellweg T, Eimer W. Influence of charge density on the swelling of colloidal poly(N-isopropylacrylamide-co-acrylic acid) microgels. *Colloids Surf A* 2000;170:137–149.

58. Wu X, Pelton RH, Hamielec AE, Woods DR, McPhee W. The kinetics of poly(N-isopropylacrylamide) microgel latex formation. *Colloid Polym Sci* 1994;272:467–477.

59. Meunier F, Elaïssari A, Pichot C. Preparation and characterization of cationic poly(n-isopropylacrylamide) copolymer latexes. *Polym Adv Tech* 1995;6:489–496.

60. Ito S, Ogawa K, Suzuki H, Wang B, Yoshida R, Kokufuta E. Preparation of thermosensitive submicrometer gel particles with anionic and cationic charges. *Langmuir* 1999;15:4289–4294.

61. Habeck M. Temperature-sensitive gels: From tissue engineering to drug delivery. *Drug Discov Today* 2001;6:553–554.

62. Leung MF, Zhu J, Harris FW, Li P. New route to smart core-shell polymeric microgels: Synthesis and properties. *Macromol Rapid Commun* 2004;25:1819–1823.

63. Leung MF, Zhu J, Harris FW, Li P. Novel synthesis and properties of smart core-shell microgels. *Macromol Symp* 2005;226:177–185.

64. Hoare T, Pelton R. Highly pH and temperature responsive microgels functionalized with vinylacetic acid. *Macromolecules* 2004;37:2544–2550.

65. Kazakov S, Kaholek M, Teraoka I, Levon K. UV-induced gelation on nanometer scale using liposome reactor. *Macromolecules* 2002;35:1911–1920.

66. Kazakov S, Kaholek M, Kudasheva D, Teraoka I, Cowman MK, Levon K. Poly(*N*-isopropylacrylamide-*co*-1-vinylimidazole) hydrogel nanoparticles prepared and hydrophobically modified in liposome reactors: Atomic force microscopy and dynamic light scattering study. *Langmuir* 2003;19:8086–8093.

67. Thienen TGV, Lucas B, Flesch FM, Nostrum CFV, Demeester J, Smedt SCD. On the synthesis and characterization of biodegradable dextran nanogels with tunable degradation properties. *Macromolecules* 2005;38:8503–8511.

68. Vinogradov SV, Batrakova EV, Kabanov AV. Poly(ethylene glycol)-polyethyleneimine nanogel particles: Novel drug delivery systems for antisense oligonucleotides. *Colloids Surf, B Biointerfaces* 1999;16:291–304.

69. Lemieux P, Vinogradov SV, Gebhart CL, Guerin N, Paradis G, Nguyen HK, Ochietti B, Suzdaltseva YG, Bartakova EV, Bronich TK, St. Pierre Y, Alakhov VY, Kabanov AV. Block and graft copolymers and nanogel copolymer networks for DNA delivery into cell. *J Drug Target* 2000;8:91–105.

70. Vinogradov SV, Bronich TK, Kabanov AV. Nanosized cationic hydrogels for drug delivery: Preparation, properties and interactions with cells. *Adv. Drug Delivery Rev* 2002;54:135–147.

71. Vinogradov SV, Zeman AD, Batrakova EV, Kabanov AV. Polyplex nanogel formulations for drug delivery of cytotoxic nucleoside analogs. *J Control Rel* 2005;107:143–157.

72. Hayashi H, Iijima M, Kataoka K, Nagasaki Y. pH-sensitive nanogels possessing reactive PEG tethered chains on the surface. *Macromolecules* 2004;37:5389–5396.

73. Li X, Zuo J, Guo Y, Yuan X. Preparation and characterization of narrowly distributed nanogels with temperature-responsive core and pH-responsive shell. *Macromolecules* 2004;37:10042–10046.

74. Hermanson GT. *Bioconjugate Techniques*. San Diego: Academic Press; 1996.

75. Nayak S, Lyon LA. Ligand-functionalized core/shell microgels with permselective shells. *Angew Chem Int Ed* 2004;43:6706–6709.

76. Kim J, Singh N, Lyon LA. Label-free biosensing with hydrogel microlenses. *Angew Chem Int Ed* 2006; 45:1446–1449.

77. Kuckling D, Duan Vo C, Wohlrab SE. Preparation of nanogels with temperature-responsive core and pH-responsive arms by photo-cross-linking. *Langmuir* 2002;18:4263–4269.

78. Ogawa K, Nakayama A, Kokufuta E. Electrophoretic behavior of ampholytic polymers and nanogels. *J Phys Chem B* 2003;107:8223–8227.

79. Braun O, Selb J, Candau F. Synthesis in microemulsion and characterization of stimuli-responsive polyelectrolytes and polyampholytes based on *N*-isopropylacrylamide. *Polymer* 2001;42:8499–8510.

80. Murray M, Rana F, Haq I, Cook J, Chowdhry BZ, Snowden MJ. Colloidal microgel systems: Phase transition properties in aqueous solution of poly(*N*-isopropylacrylamide). *J Chem Soc Chem Commun* 1994; 1803–1804.

81. Ichikawa H, Peppas NA. pH dependent swelling of nanosized poly(methacrylic acid–g-ethylene glycol) gels. *Polym Prepr* 1999;40:363–364.

82. Gregor HP, Samuelson E, Dalven PI, Gregor CD. Homopolymers and copolymers of acrylamide, N-substituted acrylamide or N-substituted methacrylamide solid state crosslinked with polyamines or polyols. US Patent. 5,280,078; 1994.

83. Ward EL. Reaction product of an epoxide and a polyalkylenepolyamine in bead form and a method for preparing same. US Pat. 4,189,539; 1980.

84. Graham NB, Cameron A. Nanogels and microgels: The new polymeric materials playground. *Pure Appl Chem* 1998;70:1271–1275.

85. Biffis A, Graham NB, Siedlaczek G, Stalberg S, Wulff G. The synthesis, characterization and molecular recognition properties of imprinted microgels. *Macromol Chem Phys* 2001;202:163–171.

86. Wulff G, Chong B-O, Kolb U. Soluble single-molecule nanogels of controlled structure as a matrix for efficient artificial enzymes. *Angew Chem Int Ed* 2006;45:2955–2958.

87. Kono K. Thermosensitive polymer-modified liposomes. *Adv Drug Delivery Rev* 2001;53:307–319.

88. Simões S, Moreira JN, Fonseca C, Düzgüne N, Pedroso de Lima MC. On the formulation of pH-sensitive liposomes with long circulation times. *Adv Drug Deliv Rev* 2004;56:947–965.

89. Yatvin MB, Weinstein JN, Dennis WH, Blumenthal R. Design of liposomes for enhanced local release of drugs by hyperthermia. *Science* 1978;202:1290–1293.

90. Zou Y, Yamagishi M, Horikoshi I, Ueno M, Gu X, Perez-Soler R. Enhanced therapeutic effect against liver W256 carcinosarcoma with temperature-sensitive liposomal adriamycin administered into the hepatic artery. *Cancer Res* 1993;53:3046–3051.

91. Iga K, Ogawa Y, Toguchi H. Heat-induced drug release rate and maximal targeting index of thermosensitive liposome in tumor-bearing mice. *Pharm Res* 1992;9:658–662.

92. Yatvin MB, Kreutz W, Horwitz, BA, Shinitzky M. pH-sensitive liposomes: Possible clinical implications. *Science* 1980;210:1253–1255.

93. Maeda M, Kumano A, Tirrell, DA. H+-induced release of contents of phosphatidylcholine vesicles bearing surface-bound polyelectrolyte chains. *J Am Chem Soc* 1988;110:7455–7459.

94. Kusumi A, Nakahama S, Yamaguchi K. Liposomes that can be disintegrated by photo-irradiation. *Chem Lett* 1989; 433–436.

95. Anderson VC, Thompson DH. Triggered release of hydrophilic agents from plasmalogen liposomes using visible light or acid. *Biochim Biophys Acta Biomembr* 1992;1109:33–42.

96. Godbey WT, Wu KK, Mikos AG. Poly(ethyleneimine) and its role in gene delivery. *J Control Rel* 1999;60:149–160.

97. Thomas JL, Tirrell DA. Polyelectrolyte-sensitized phospholipid vesicles. *Acc Chem Res* 1992;25:336–342.

98. Chung JC, Gross DJ, Thomas JL, Rirrell DA, Opsahlong L. pH-sensitive, cation-selective channels formed by a simple synthetic polyelectrolyte in artificial bilayer membranes. *Macromolecules* 1996;29:4636–4641.

99. Murthy N, Robichaud JR, Tirrell DA, Stayton PS, Hoffman AS. The design and synthesis of polymers for eukaryotic membrane disruption. *J Control Rel* 1999;61:137–143.

100. Drummond DC, Zignani M, Leroux J-C. Current status of pH-sensitive liposomes in drug delivery. *Prog Lipid Res* 2000;39:409–460.

101. Kono K, Igawa T, Takagishi T. Cytoplasmic delivery of calcein mediated by liposomes modified with a pH-sensitive poly(ethylene glycol) derivative. *Biochim Biophys Acta* 1997;1325:143–154.

102. Kono K, Zenitani KI, Takagishi T. Novel pH-sensitive liposomes: Liposomes bearing a poly(ethylene glycol) derivative with carboxyl groups. *Biochim Biophys Acta* 1994;1193:1–9.

103. Roux E, Stomp R, Giasson S, Pézolet M, Moreau P, Leroux J-C. Steric stabilization of liposomes by pH-responsive N-isopropylacrylamide copolymer. *J Pharm Sci* 2002;91:1795–1802.

104. Hayashi H, Kono K, Takagishi T. Temperature-controlled release property of phospholipid vesicles bearing a thermo-sensitive polymer. *Biochim Biophys Acta (BBA)-Biomembranes* 1996;1280:127–134.

105. Kono K, Henmi A, Yamashita H, Hayashi H, Takagishi T. Improvement of temperature-sensitivity of poly(N-isopropylacrylamide)-modified liposomes. *J Control Rel* 1999;59:63–75.

106. Kono K, Hayashi H, Takagishi T. Temperature-sensitive liposomes: Liposomes bearing poly(N-isopropylacrylamide). *J Control Rel* 1994;30:69–75.

107. Schillemans JP, Flesch FM, Hennink WE, Nostrum CFV. Synthesis of bilayer-coated nanogels by selective cross-linking of monomers inside liposomes. *Macromolecules* 2006; 39:5885–5890.

108. Somasundaran P, Liu F, Chakraborty S, Gryte CC, Deo N, Somasundaran T. Novel nanogels for drug binding and delivery. *Polymeric Drug Delivery II.* ACS Symposium Series, Vol. 924. Washington, DC: American Chemical Society; 2006, pp. 69–87.

109. Kataoka K, Harada A, Nagasaki Y. Block copolymer micelles for drug delivery: Design, characterization and biological significance. *Adv Drug Deliv Rev* 2001;47:113–131.

110. Kakizawa Y, Kataoka K. Block copolymer micelles for delivery of gene and related compounds. *Adv Drug Deliv Rev* 2002;54:203–222.

111. Lavasanifar A, Samuel J, Kwon GS. Poly(ethylene oxide)-*block*-poly(L-amino acid) micelles for drug delivery. *Adv Drug Deliv Rev* 2002;54:169–190.

112. Kabanov AV, Batrakova EV, Alakhov VY. Pluronic® block copolymers as novel polymer therapeutics for drug and gene delivery. *J Control Rel* 2002;82:189–212.

113. Li Y-Y, Zhang X-Z, Kim G-C, Cheng H, Cheng S-X, Zhuo R-X. Thermosensitive Y-shaped micelles of poly(oleic acid-Y-N-isopropylacrylamide) for drug delivery. *Small* 2006;2:917–923.

114. Gilles ER, Frechet, JMJ. Development of acid-sensitive copolymer micelles for drug delivery. *Pure Appl Chem* 2004;76:1295–1307.

115. Cammas S, Suzuki K, Sone C, Sakurai Y, Kataoka K, Okano T. Thermo-responsive polymer nanoparticles with a core-shell micelle structure as site-specific drug carriers. *J Control Rel* 1997;48:157–164.

116. Chung JE, Yokoyama M, Okano T. Inner core segment design for drug delivery control of thermo-responsive polymeric micelles. *J Control Rel* 2000;65:93–103.

117. Rösler A, Vadermeulen GWM, Klok H-A. Advanced drug delivery devices via self-assembly of amphiphilic block copolymers. *Adv Drug Deliv Rev* 2001;53:95–108.

118. Chung JE, Yokoyama M, Yamato M, Aoyagi T, Sakurai Y, Okano T. Thermo-responsive drug delivery from polymer micelles constructed using block copolymers of poly(N-isopropylacrylamide) and poly(butyl methacrylate). *J Control Rel* 1999;62:115–127.

119. Harada-Shiba M, Yamauchi K, Harada A, Takamisawa I, Shimokado K, Kataoka K. Polyion complex micelles as vectors in gene therapy-pharmacokinetics and *in vivo* gene transfer. *Gene Ther* 2002;9:407–414.

120. Voets IK, Keizer AD, Stuart MAC. Core and corona structure of mixed polymeric micelles. *Macromolecules* 2006;39:5952–5955.

121. Parak WJ, Gerion D, Pellegrino T, Zanchet D, Micheel C, Williams SC, Boudreau R, Le Gros MA, Larabell CA, Alivisatos AP. Biological applications of colloidal nanocrystals. *Nanotechnology* 2003;14:R15–R17.

122. Katz E, Willner I. Integrated nanoparticle-biomolecule hybrid systems: Synthesis, properties, and applications. *Angew Chem Int Ed* 2004;43:6042–6108.

123. Akiyoshi K, Kang E-C, Kuromada S, Sumamoto J, Principi T, Winnik FM. Controlled association of amphiphilic polymers in water: Thermosensitive nanoparticles formed by self-assembly of hydrophobically modified pullulans and poly(*N*-isopropylacrylamides). *Macromolecules* 2000;33:3244–3249.

124. Nishikawa T, Akiyoshi K, Sunamoto J. Macromolecular complexation between bovine serum albumin and the self-assembled hydrogel nanoparticle of hydrophobized polysaccharides. *J Am Chem Soc* 1996;118:6110–6115.

125. Akiyoshi K, Sasaki Y, Sunamoto J. Molecular chaperone-like activity of hydrogel nanoparticles of hydrophobized pullulan: Thermal stabilization with refolding of carbonic anhydrase B. *Bioconjug Chem* 1999;10:321–324.

126. Kuroda K, Fujimoto K, Sunamoto J, Akiyoshi K. Hierarchical self-assembly of hydrophobically modified pullulan in water: Gelation by networks of nanoparticles. *Langmuir* 2002;18:3780–3786.

127. Nishikawa T, Akiyoshi K, Sunamoto J. Supramolecular assembly between nanoparticles of hydrophobized polysaccharide and soluble-protein complexation between the self-aggregate of cholesterol-bearing pullulan and α-chymotrypsin. *Macromolecules* 1994;27:7654–7659.

128. Lee I, Akiyoshi K. Single molecular mechanics of a cholesterol-bearing pullulan nanogel at the hydrophobic interfaces. *Biomaterials* 2004;25:2911–2918.

129. Morimoto N, Endo T, Iwasaki Y, Akiyoshi K. Design of hybrid hydrogels with self-assembled nanogels as cross-linkers: Interaction with proteins and chaperone-like activity. *Biomacromolecules* 2005;6:1829–1834.

130. Morimoto N, Endo T, Ohtomi M, Iwasaki Y, Akiyoshi K. Hybrid nanogels with physical and chemical cross-linking structures as nanocarriers. *Macromol Biosci* 2005;5:710–716.

131. Na K, Park K-H, Kim SW, Bae YH. Self-assembled hydrogel nanoparticles from curdlan derivatives: characterization, anti-cancer drug release and interaction with a hepatoma cell line (HepG2). *J Control Rel* 2000;69:225–236.

132. Akiyoshi K, Ueminami A, Kurumada S, Nomura Y. Self-association of cholesteryl-bearing poly(L-lysine) in water and control of its secondary structure by host-guest interaction with cyclodextrin. *Macromolecules* 2000;33:6752–6756.

133. Hirakura T, Nomura Y, Aoyama Y, Akiyoshi K. Photoresponsive nanogels formed by the self-assembly of spiropyrane-bearing pullulan that act as artificial molecular chaperones. *Biomacromolecules* 2004;5:1804–1809.

134. Lee KY, Jo WH, Kwon IC, Kim Y-H, Jeong SY. Physicochemical characteristics of self-aggregates of hydrophobically modified chitosans. *Langmuir* 1998;14:2329–2332.

135. Nichifor M, Lopes A, Carpov A, Melo E. Aggregation in water of dextran hydrophobically modified with bile acids. *Macromolecules* 1999;32:7078–7085.

136. Yusa S-I, Kamachi M, Morishima Y. Hydrophobic self-association of cholesterol moieties covalently linked to polyelectrolytes: Effect of spacer bond. *Langmuir* 1998;14:6059–6067.

137. Shin Y, Chang JH, Liu J, Williford R, Shin Y-K, Exarhos GJ. Hybrid nanogels for sustainable positive thermosensitive drug delivery release. *J Control Rel* 2001;73:1–6.
138. Gref R, Amiel C, Molinard K, Daoud-Mahammed S, Sebille B, Gillet B, Beloeil J-C, Ringard C, Rosilio V, Poupaert J, Couvreur P. New self-assembled nanogels Based on host–guest interactions: Characterization and drug loading. *J Control Rel* 2006; 111:316–324.
139. Ichigawa H, Fukumori Y. A novel positively thermosensitive controlled-release microcapsule with membrane of nano-sized poly(*N*-isopropylacrylamide) gel dispersed in ethyl cellulose matrix. *J Control Rel* 2000;63:107–119.
140. Perkin KK, Turner JL, Wooley KL, Mann S. Fabrication of hybrid nanocapsules by calcium phosphate mineralization of shell cross-linked polymer micelles and nanocages. *Nano Lett* 2005;5:1457–1461.
141. Welzel T, Mayer-Zaika W, Epple M. Continuous preparation of functionalised calcium phosphate nanoparticles with adjustable crystallinity. *Chem Commun* 2004; 1204–1205.
142. Sugawara A, Yamane S, Akiyoshi K. Nanogel-templated mineralization: Polymer–calcium phosphate hybrid nanomaterials. *Macromol Rapid Commun* 2006; 27:441–446.
143. Hasegawa U, Nomura SM, Kaul SC, Hirano T, Akiyoshi K. Nanogel–quantum dot hybrid nanoparticles for live cell imaging. *Biochem Biophys Res Commun* 2005;33:917–921.
144. Okay O. Macroporous copolymer networks. *Prog Polym Sci* 2000;25:711–779.
145. Liu Y, Velada JL, Huglin MB. Thermoreversible swelling behaviour of hydrogels based on *N*-isopropylacrylamide with sodium acrylate and sodium methacrylate. *Polymer* 1999;40:4299–4306.
146. Tanaka T. Gels. *Sci Am* 1981;244:110–123.
147. Murali Mohan Y, Murthy PSK, Raju KM. Preparation and swelling behavior of macroporous poly(acrylamide-*co*-sodium methacrylate) superabsorbent hydrogels. *J Appl Polym Sci* 2006; 101:3202–3214.
148. Murali Mohan Y, Murthy PSK, Sreedhar B, Raju KM. Swelling and thermal characteristics of pH sensitive crosslinked poly(acrylamide-co-calcium methacrylate) superabsorbent copolymers. *J Appl Polym Sci* 2006;102:1–12.
149. Odian G. *Principles of Polymerization*, 3rd edition. New York: Wiley; 1991.
150. Flory PJ. *Principles of Polymer Chemistry*. Ithaca: University Press; 1953.
151. Rey L, Galy J, Sautereau H. Reaction kinetics and morphological changes during isothermal cure of vinyl/dimethacrylate networks. *Macromolecules* 2000;33:6780–6786.
152. Rey L, Duchet J, Galy J, Sautereau H, Vouagner D, Carrion L. Structural heterogeneities and mechanical properties of vinyl/dimethacrylate networks synthesized by thermal free radical polymerization. *Polymer* 2002;43:4375–4384.
153. Ghiass M, Rey AD, Dabir B. Microstructure evolution and simulation of copolymerization reaction using a percolation kinetic gelation model. *Polymer* 2002;43:989–995.
154. Hutchison JB, Anseth KS. Off-lattice approach to simulate radical chain polymerizations of tetrafunctional monomers. *Macromol Theory Simul* 2001;10:600–607.
155. Ghiass M, Rey AD, Dabir B. Simulation of non-linear free-radical polymerization using a percolation kinetic gelation model. *Macromol Theory Simul* 2001;10:657–667.
156. Funke W, Okay O, Joos-Müller B. Microgels: Intramolecularly crosslinked macromolecules with a globular structure. *Adv Polym Sci* 1998;136:139–234.

157. Okay O. Phase separation in free-radical crosslinking copolymerization: Formation of heterogeneous polymer networks. *Polymer* 1999;40:4117–4129.

158. Frank RS, Downey JS, Yu K, Stöver HDH. Poly(divinylbenzene-*alt*-maleic anhydride) microgels: Intermediates to microspheres and macrogels in cross-linking copolymerization. *Macromolecules* 2002;35:2728–2735.

159. Costello PA, Martin IK, Slark AT, Sherrington DC, Titterton A. Branched methacrylate copolymers from multifunctional monomers: Chemical composition and physical architecture distributions. *Polymer* 2002;43:245–254.

160. O'Brien N, McKee A, Sherrington DC, Slark AT, Titteron. A. Facile, versatile and cost effective route to branched vinyl polymers. *Polymer* 2000;41:6027–6031.

161. Sahoo SK, De TK, Ghosh PK, Maitra A. pH- and thermo-sensitive hydrogel nanoparticles. *J Colloid Polym Sci* 1998;206:361–368.

162. Gupta AK, Madan S, Majumdar DK, Maitra A. Ketorolac entrapped in polymeric micelles: Preparation, characterization and ocular anti-inflammatory studies. *Int J Pharm* 2000;209:1–14.

163. Maddock SC, Pasetto P, Resmini M. Novel imprinted soluble microgels with hydrolytic catalytic activity. *Chem Commun* 2004; 536–537.

164. Leon JW, Bennett JR, Che W, Mourey TH, Qiao T. Biocompatible nanogels via homogeneous polymerization in water. *PMSE Preprints* 2006; 95:341–342.

165. Balsara NP, Tirrell M, Lodge TP. Micelle formation of BAB triblock copolymers in solvents that preferentially dissolve the A block. *Macromolecules* 1991;24:1975–1986.

166. Saito R, Akiyama Y, Ishizu K. Flower shaped microgel 3: Synthesis of poly(2-hydroxyethylmethacrylate) core-polystyrene shell type microgels. *Polymer* 1999;40:655–660.

167. Okay O, Funke W. Conditions of microgel formation in homogeneous anionic polymerization of 1,4-divinylbenzene. *Die Makromol Chemie* 1990;191:1565–1573.

168. Munshi N, De TK, Maitra A. Size modulation of polymeric nanoparticles under controlled dynamics of microemulsion droplets. *J Colloid Interface Sci* 1997;190:387–391.

169. Kriwet B, Walter E, Kissel T. Synthesis of bioadhesive poly(acrylic acid) nano- and microparticles using an inverse emulsion polymerization method for the entrapment of hydrophilic drug candidates. *J Control Rel* 1998;56:149–158.

170. Stenekes RJH, Franssen O, Van Bommel EM, Crommelin DJA, Hennink WE. The preparation of dextran microspheres in an all-aqueous system: Effect of the formulation parameters on particle characteristics. *Pharm Res* 1998;15:557–561.

171. Franssen O, Stenekes RJH, Hennink WE. Controlled release of a model protein from enzymatically degrading dextran microspheres. *J Control Rel* 1999;59:219–228.

172. Tobita H, Yamamoto K. Network formation in emulsion cross-linking copolymerization. *Macromolecules* 1994;27:3389–3396.

173. Tobita H, Kumagai M, Aoyagi N. Microgel formation in emulsion polymerization. *Polymer* 2000;41:481–487.

174. Jabbari E. Monte Carlo simulation of tri-functional branching and tetra-functional crosslinking in emulsion polymerization of butadiene. *Polymer* 2001;42:4873–4884.

175. Landfester K. Synthesis of colloidal particles in miniemulsions. *Ann Rev Mater Res* 2006;36:231–279.

176. Gröhn F, Antonietti M. Intermolecular structure of spherical polyelectrolyte microgels in

salt-free solution. 1. Quantification of the attraction between equally charged polyelectrolytes. *Macromolecules* 2000;33:5938–5949.

177. Antonietti M, Briel A, Gröhn F. Spherical polyelectrolyte microgels in salt-free solution. 2. Combined analysis of static structure and viscosity and quantitative testing of the mode-mode coupling approximation. *Macromolecules* 2000;33:5950–5953.

178. Lovell PA, El Asser MS. *Emulsion Polymerization and Emulsion Polymers*. Wiley Chichester, UK: Wiley; 1997.

179. Piirma I. *Emulsion Polymerization*. New York: Academic Press; 1982.

180. Pelton RH, Chibante P. Preparation of aqueous latices with N-isopropylacrylamide. *Colloids Surf* 1986;20:247–256.

181. Pelton RH, Pelton HM, Morphesis A, Rowell RL. Particle sizes and electrophoretic mobilities of poly(N-isopropylacrylamide) latex. *Langmuir* 1989;5:816–818.

182. McPhee W, Tam KC, Pelton RH. Poly(N-isopropylacrylamide) latexes prepared with sodium dodecyl sulfate. *J Colloid Interface Sci* 1993;156:24–30.

183. Pelton R, Wu X, McPhee W, Tam KC. The preparation and characterization of polyNIPAAm latexes. Goodwin JW, Buscall R, editors. *Colloidal Polymer Particles*. London: Academic; 1995, pp. 81–99.

184. Antonietti M, Bremser W, Pakula T. Synthesis of critically cross-linked microgels by dispersion polymerization. *Acta Polym* 1995;46:37–44.

185. Ishii K. Synthesis of microgels and their application to coatings. *Colloids Surface A: Physicochem Eng Aspects* 1999;153:591–595.

186. Ramanan RMK, Chellamuthu P, Tang L, Nguyen KT. Development of a temperature-sensitive composite hydrogel for drug delivery applications. *Biotechnol Prog* 2006; 22:118–125.

187. Saunders BR, Vincent B. Microgel particles as model colloids: Theory, properties and applications. *Adv Colloid Interface Sci* 1999;80:1–25.

188. Ohta H, Ando I, Fujishige S, Kubota K. Molecular motion and ^{1}H NMR relaxation of aqueous poly(N-isopropylacrylamide) solution uder high pressure. *J Polym Sci Part B: Polym Phys* 1991;29:963–968.

189. Tokuhiro T, Amiya T, Mamada A, Tanaka T. NMR study of poly(N-isopropylacrylamide) gels near phase transition. *Macromolecules* 1991;24:1936–2943.

190. Griffiths PC, Chowdhry BZ, Snowden MJ. pGSE-NMR studies of solvent diffusion in poly(N-isopropylacrylamide) colloidal microgels. *J Colloid Polym Sci* 1995;273:405–411.

191. Carenza M, Veronese FM. Entrapment of biomolecules into hydrogels obtained by radiation-induced polymerization. *J Control Rel* 1994;29:187–193.

192. Ding Z-L, Yoshida M, Asano M, Ma Z-T, Omichi H, Katakai R. Thermo-responsive behavior of a methacryloyl-DL-alanine methyl ester polymer gel prepared by radiation-induced polymerization. *Radiat Phys Chem* 1994;44:263–272.

193. Woods RJ, Pikaev AK. *Applied Radiation Chemistry: Radiating Processing*. New York: Wiley-Interscience; 1993.

194. Mason TJ, Lorimer JP. *Sonochemistry, Theory and Uses of Ultrasound in Chemistry*. New York: Wiley-Interscience; 1989.

195. Price GJ. Applications of high intensity ultrasound in polymer chemistry. In Eldik RV, Hubbard CD, editors. *Chemistry Under Extreme or Non-classical Conditions*. New York: Wiley; p. 381.

196. Peppas SP. *UV—Curing Science and Technology*, Vols. 1–3. Stamford CT: Technology Marketing Corporation; 1985.

197. Gao J, Frisken BJ. Cross-linker-free *N*-isopropylacrylamide gel nanospheres. *Langmuir* 2003;19:5212–5216.

198. Qiao X, Zhang Z, Yao S, Preparation of initiator and cross-linker-free poly(*N*-isopropylacrylamide) nanogels by photopolymerization. *J Photochem Photobiol A: Chem* 2006; 177:191–196.

199. Russell RJ, Axel AC, Shields KL, Pishko MV. Mass transfer in rapidly photopolymerized poly(ethylene glycol) hydrogels used for chemical sensing. *Polymer* 2001;42:4893–4901.

200. Gehrke SH. Synthesis, equilibrium swelling, kinetics, permeability and applications of environmentally responsive gels. *Adv Polym Sci* 1992;110:81–144.

201. Rosiak JM, Ulanski P. Synthesis of hydrogels by irradiation of polymers in aqueous solution. *Radiat Phys Chem* 1999;55:139–151.

202. Rosiak JM, Janik I, Kadlubowski S, Kozicki M, Kujawa P, Stasica P, Ulanski P. Nano-, micro- and macroscopic hydrogels synthesized by radiation technique. *Nucl Instr Meth Phys Res B.* 2003;208:325–330.

203. Kadlubowski S, Grobelny J, Olejniczak W, Cichomski M, Ulanski P. Pulses of fast electrons as a tool to synthesize poly(acrylic acid) nanogels. Intramolecular cross-linking of linear polymer chains in additive-free aqueous solution. *Macromolecules* 2003;36:2484–2492.

204. Ulanski P, Rosiak JM. Radiation formation of polymeric nanogels. *Radiat Phys Chem* 1998;52:289–294.

205. Nurkeeva ZS, Khutoryanskiy VV, Mun GA, Bitekenova AB, Kadlubowski S, Shilina YA, Ulanski P, Rosiak JM. Interpolymer complexes of poly(acrylic acid) nanogels with some non-ionic polymers in aqueous solutions. *Colloids Surf A: Physicochem Eng Asp* 2004;236:141–146.

206. Furusawa K, Terao K, Nagasawa N, Yoshii F, Kubota K, Dobashi T. Nanometer-sized gelatin particles prepared by means of gamma-ray irradiation. *Colloid Polym Sci* 2004;283:229–233.

207. Brasch U, Burchard W. Preparation and solution properties of microhydrogels from poly(vinyl alcohol). *Macromol Chem Phys* 1996;197:223–235.

208. Inagaki HP. *Cellulose Utilization*. London: Elsevier Applied Science; 1989.

209. Gebben B, Van Den Berg HWA, Bargeman D, Smolders CA. Intramolecular crosslinking of poly(vinyl alcohol). *Polymer* 1985;26:1737–1740.

210. Arbogast W, Horvath A, Vollmert B. Darstellung und [η]-M̄-beziehung von mikrogelen aus polyvinylalkohol. *Die Macromol Chem* 1980;181:1513–1524.

211. Lu X, Hu Z, Gao J. Synthesis and light scattering study of hydroxypropyl cellulose microgels. *Macromolecules* 2000;33:8698–8702.

212. Al-Assaf S, Phillips GO, Deeble DJ, Parson B, Starnes H, Von Sonntag C. The enhanced stability of the cross-linked hylan structure to hydroxyl (OH) radicals compared with the uncross-linked hyaluronan. *Radiat Phys Chem* 1995;46:207–217.

213. Kuckling D, Vo CD, Adler H-JP, Völkel A, Cölfen H. Preparation and characterization of photo-cross-linked thermosensitive PNIPAAm nanogels. *Macromolecules* 2006; 39:1585–1591.

214. Kuckling D, Harmon ME, Frank CW. Photo-cross-linkable PNIPAAm copolymers. 1. Synthesis and characterization of constrained temperature-responsive hydrogel layers. *Macromolecules* 2002;35:6377–6383.

215. Harmon, ME, Kuckling D, Frank CW. Photo-cross-linkable PNIPAAm copolymers. 2. Effects of constraint on temperature and pH-responsive hydrogel layers. *Macromolecules* 2003;36:162–172.

216. Harmon ME, Kuckling D, Frank CW. Photo-cross-linkable PNIPAAm copolymers. 5. Mechanical properties of hydrogel layers. *Langmuir* 2003;19:10660–10665.

217. Snowden MJ, Chowdhry BZ, Vincent B, Morris GE. Colloidal copolymer microgels of N-isopropylacrylamide and acrylic acid: pH, ionic strength and temperature effects. *J Chem Soc Faraday Trans* 1996;92:5013–5016.

218. Duracher D, Elaïssari A, Pichot C. Preparation of poly(N-isopropylmethacrylamide) latexes kinetic studies and characterization. *J Polym Sci, Part A: Polym Chem* 1999;37:1823–1837.

219. Senff H, Richtering W. Influence of cross-link density on rheological properties of temperature-sensitive microgel suspensions. *Colloid Polym Sci* 2000;278:830–840.

220. Senff H, Richtering W. Temperature sensitive microgel suspensions: Colloidal phase behavior and rheology of soft spheres. *J Chem Phys* 1999;111:1705–1711.

221. Duracher D, Elaissari A, Pichot C. Characterization of cross-linked poly(N-isopropylmethacrylamide) microgel latexes. *Colloid Polym Sci* 1999;277:905–913.

222. Jones CD, Lyon LA. Synthesis and characterization of multiresponsive core-shell microgels. *Macromolecules* 2000;33:8301–8306.

223. Tan BH, Tam KC, Lam YC, Tan CB. Microstructure and rheology of stimuli-responsive microgel systems—Effect of cross-linked density. *Adv Colloid Interface Sci* 2005;113:111–120.

224. Ma X, Cui Y, Zhao X, Zheng S, Tang X. Different deswelling behavior of temperature-sensitive microgels of poly(N-isopropylacrylamide) crosslinked by polyethyleneglycol dimethacrylates. *J Colloid Interface Sci* 2004;276:53–59.

225. Nolan CM, Reyes CD, Debord JD, Garcia, AJ, Lyon LA. Phase transition behavior, protein adsorption, and cell adhesion resistance of poly(ethylene glycol) crosslinked microgel particles. *Biomacromolecules* 2005;6:2032–2039.

226. Kaneda I, Sogabe A. Rheological properties of water swellable microgel polymerized in a confined space. *Colloids Surf A: Physicochem Eng Aspects* 2005; 270–271: 163–170.

227. Woodward NC, Chowdhry BZ, Snowden MJ, Leharne SA, Griffiths PC, Winnington AL. Calorimetric investigation of the influence of cross-linker concentration on the volume phase transition of poly(N-isopropylacrylamide) colloidal microgels. *Langmuir* 2003;19:3202–3211.

228. Hellweg T, Eimer W, Pouget S, Kratz K. Neutron spin echo study of the dynamics in BIS cross-linked poly(N-isopropyl acrylamide) microgels: Dependence on the cross-linker concentration. *Lecture Notes in Physics Neutron Spin Echo Spectroscopy*, 2003;601:291–301.

229. Wolfe MS. Dispersion and solution rheology control with swellable microgels. *Prog Org Coat* 1992;20:487–500.

230. McAllister K, Sazani P, Adam M, Cho MJ, Rubinstein M, Samulski RJ, DeSimone JM. Polymeric nanogels produced via inverse microemulsion polymerization as potential gene and antisense delivery agents. *J Am Chem Soc* 2002;124:15198–15207.

231. Arleth L, Xia X, Hjelm RP, Wu J, Hu Z. Volume transition and internal structures of small poly(N-isopropylacrylamide) microgels. *J Polym Sci, Part B: Polym Phys* 2005;43:849–860.

232. Pich, A, Tessier A, Boyko V, Lu Y, Adler H-JP. Synthesis and characterization of poly(vinylcaprolactam)-based microgels exhibiting temperature and pH-sensitive properties. *Macromolecules* 2006; 39:7701–7707.

233. Dupin D, Fujii S, Armes SP, Reeve P, Baxter SM. Efficient synthesis of sterically stabilized pH-responsive microgels of controllable particle diameter by emulsion polymerization. *Langmuir* 2006; 22:3381–3387.

234. Das M, Mardyani S, Chan WCW, Kumacheva E. Biofunctionalized pH-responsive microgels for cancer cell targeting: Rational design. *Adv Mater* 2006; 18:80–83.

235. Tan BH, Tam KC, Lam YC, Tan CB. Microstructure and rheological properties of pH-responsive core-shell particles. *Polymer* 2005;46:10066–10076.

236. Amalvy JI, Wanless EJ, Li Y, Michailidou V, Armes SP, Duccini Y. Synthesis and characterization of novel pH-responsive microgels based on tertiary amine methacrylates. *Langmuir* 2004;20:8992–8999.

237. Zhu PW, Napper DH. Studies of aggregation kinetics of polystyrene latices sterically stabilized by poly(N-isopropylacrylamide). *Phys Rev E* 1994;50:1360–1366.

238. Wu C, Zhou S, Au-Yeung SCF, Jiang S. Volume phase transition of spherical microgel particles. *Die Ang Makro Chemie* 1996;240:123–136.

239. Wu C, Zhou S. Effects of surfactants on the phase transition of poly(N-isopropylacrylamide) in water. *J Polym Sci, Part B: Polym Phys* 1996;34:1597–1604.

240. Tam KC, Ragaram S, Pelton RH. Interaction of surfactants with poly(N-isopropylacrylamide) microgel latexes. *Langmuir* 1994;10:418–422.

241. Mears SJ, Deng Y, Cosgrove T, Pelton R. Structure of sodium dodecyl sulfate bound to a poly(NIPAM) microgel particle. *Langmuir* 1997;13:1901–1906.

242. Wang G, Pelton R, Zhang J. Sodium dodecyl sulfate binding to poly(N-isopropylacrylamide) microgel latex studied by isothermal titration calorimetry. *J Colloid Surf A: Physiochem Eng Aspects* 1999;153:335–340.

243. Woodward NC, Chowdhry BZ, Leharne SA, Snowden MJ. The interaction of sodium dodecyl sulphate with colloidal microgel particles. *Eur Polym J* 2000;36:1355–1364.

244. Andersson M, Maunu SL. Structural studies of poly(N-isopropylacrylamide) microgels: Effect of SDS surfactant concentration in the microgel synthesis. *J Polym Sci, Part B: Polym Phys* 2006;44:3305–3314.

245. Emerich DF, Thanos CG. The pinpoint promise of nanoparticle-based drug delivery and molecular diagnosis. *Biomol Eng* 2006;23:171–184.

246. Hu Z, Lu X, Gao J, Wang C. Polymer gel nanoparticle networks. *Adv Mater* 2000;12:1173–1176.

247. Nomura Y, Sasaki Y, Takagi M, Narita T, Aoyama Y, Akiyoshi K. Thermoresponsive controlled association of protein with a dynamic nanogel of hydrophobized polysaccharide and cyclodextrin: Heat shock protein-like activity of artificial molecular chaperone. *Biomacromolecule* 2005;6:447–452.

248. Peppas NA, Kavimandan NJ. Nanoscale analysis of protein and peptide absorption: Insulin absorption using complexation and pH-sensitive hydrogels as delivery vehicles. *Eur J Pharm Sci* 2006;29:183–197.

249. Yamagata T, Morishita M, Kavimandan NJ, Peppas NA, Takayama K. Characterization of insulin protection properties of complexation hydrogels in gastric and intestinal enzyme fluids. *J Control Rel* 2006;112:343–349.

250. Ada G. Vaccines and vaccination. *N Engl J Med* 2001;345:1042–1053.

251. Pardoll DM. Tumor reactive T cells get a boost. *Nat Biotechnol* 2002;20:1207–1208.

252. Raychaudhuri S, Rock KL. Fully mobilizing host defense: Building better vaccines. *Nat Biotechnol* 1998;16:1025–1031.

253. Jain S, Yap WT, Irvine DJ. Synthesis of protein-loaded hydrogel particles in an aqueous two-phase system for coincident antigen and CpG oligonucleotide delivery to antigen-presenting cells. *Biomacromolecules* 2005;6:2590–2600.

254. Fu K, Klibanov AM, Langer R. Protein stability in controlled-release systems. *Nat Biotechnol* 2000;18:24–25.

255. Kurtis Kasper F, Simon Y, Kazuhiro T, Michael AB, Yasuhiko T, John AJ, Antonios GM. Characterization of DNA release from composites of oligo(poly(ethylene glycol) fumarate) and cationized gelatin microspheres in vitro. *J Biomed Mater Res, Part A* 2006; 78:823–835.

256. Fukunaka Y, Iwanaga K, Morimoto K, Kakemi M, Tabata Y. Controlled release of plasmid DNA from cationized gelatin hydrogels based on hydrogel degradation. *J Control Rel* 2002;80:333–343.

257. Kushibiki T, Tomoshige R, Fukunaka Y, Kakemi M, Tabata Y. *In vivo* release and gene expression of plasmid DNA by hydrogels of gelatin with different cationization extents. *J Control Rel* 2003;90:207–216.

258. Kushibiki T, Tabata Y. A new gene delivery system based on controlled release technology. *Curr Drug Deliv* 2004;1:153–163.

259. Vinogradov SV, Kohli E, Zeman AD. Comparison of nanogel drug carriers and their formulations with nucleoside 5′-triphosphates. *Pharm Res* 2006; 23:920–930.

260. Soni S, Babbar AK, Sharma RK, Maitra A. Delivery of hydrophobised 5-fluorouracil derivative to brain tissue through intravenous route using surface modified nanogels. *J Drug Target* 2006; 14:87–95.

7

TARGETED GOLD NANOPARTICLES FOR IMAGING AND THERAPY

Raghuraman Kannan and Kattesh V. Katti

7.1 INTRODUCTION

Gold nanoparticles (AuNPs) have been known for over 2500 years; however, their potential applications in medicine have gained burgeoning attention in the twenty-first century [1–10]. Bulk gold is well known for being inert however, the nanoparticulate sizes of gold display astronomically high chemical reactivity [11–16]. The high reactivity of AuNPs combined with a plethora of chemical and biochemical vectors continue to produce a wide spectrum of targeted AuNPs that have sizes within the cellular regime yet retaining most/all of the properties specific to nanosizes [4, 17–22]. Gold nanoparticles have found numerous biological applications due to their biocompatibility, dimensions, and ease of characterization. Their rich surface chemistry allows surface modification reactions to better suit the needs of biomedical applications [1–6, 11–22]. Targeted AuNPs are produced by the interaction of highly reactive nascent AuNPs with chemical functionalities present on simple chemical molecules or on specific molecules of biological interest (for example, biomolecules including peptides and proteins) [11, 17, 18, 21, 23–25]. The size similarity of targeted AuNPs with cellular species within the biological domain has motivated researchers toward applications of nanotechnology

Biomedical Applications of Nanotechnology. Edited by Vinod Labhasetwar and Diandra L. Leslie-Pelecky
Copyright © 2007 John Wiley & Sons, Inc.

for site/cell-specific labeling of targeted nanoparticles for the detection and therapy of different diseases including cancer [26–35]. The multifunctionality and the unique reactivities of AuNPs would allow building targeted nanoparticles carrying imaging and therapeutic probes to target cancer cells [26–33, 36].

Nanoparticles of gold continue to play pivotal roles in the design and development of nanoscale devices and tumor-specific nanosensors [37–42]. The ubiquitous place of gold in nanoscience stems from its unique chemical property of serving in the unoxidized state at the nano level, whereas most of the surface of less noble metals gets oxidized to a depth of several nanometers or more, often obliterating the nano-scale properties. The high reactivity of AuNPs juxtaposed with their biocompatibility has spawned a major interest in the utility of AuNPs for *in vivo* imaging agents. Most of the recent work is centered on the development of hybrid AuNPs starting from nascent metal nanoparticles. Tumor-targeting nanoparticles are formed by coating AuNPs with tumor-cell-specific biomolecules including monoclonal antibodies, aptamers, peptides, and various receptor-specific substrates [43–54]. Receptor-specific hybrid nanoparticles are used mainly for targeting three different markers that are overexpressed on cancer cells. They include matrix metalloproteases (MMPs), epidermal growth factor receptor (EGFR), and oncoproteins that are associated with human papillomavirus (HPV) infection [55].

In this chapter, we summarize the recent advances in synthesis of targeted gold nanoparticles and their applications in imaging and therapy (Figure 7.1). Specifically, we discuss the use of gold nanoparticles as site-specific contrast agents for various imaging modalities. The utilization of gold nanoparticles as therapeutic agents for tumor is discussed.

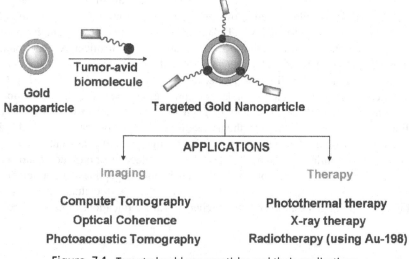

Figure 7.1. Targeted gold nanoparticles and their applications.

7.2 SYNTHESIS OF GOLD NANOPARTICLES

7.2.1 Spherical Gold Nanoparticles

Gold nanoparticles of different sizes have been synthesized from $AuCl_4^-$, predominantly, using either citrate or sodium borohydride as reducing agents (Scheme 7.1) [56]. Citrate molecules serve dual roles of reduction of the metal and capping of nanoparticles for subsequent stabilization. By varying the concentrations of the citrate, different sizes of AuNPs can be generated. Citrate-capped AuNPs have been widely used as precursors for conjugating with biomolecules [57–60].

In the sodium borohydride method, thiols serve as capping agents [61–63]. Reduction reactions of Au^{3+} using $NaBH_4$ have been performed in the presence of a large excess of thiols. The high reactivity of thiols with AuNPs provides instant coating of nanoparticle surfaces. Reactions utilizing thiol-containing carboxylic acids in the place of thiols have produced nanoparticles with –COOH groups on the surface. These –COOH groups containing AuNPs have been used as precursors for conjugating with target- specific biomolecules or proteins [61–63].

7.2.2 Gold Nanoshells

Gold nanoshells are prepared by using aminated silica particles with spherical gold particles (1–3 nm as diameter) as precursors (Scheme 7.2) [64,65]. The interactions of these precursors have resulted in the surface adsorption of gold nanoparticles over silica core. Suspending these particles in $HAuCl_4$ with formaldehyde as reducing agents have resulted in layers of gold encapsulated with silica core. Gold nanoshells possess attractive photophysical properties, and thus they can serve as excellent contrast agents for optoacoustic tomography. Targeted gold nanoshells have been synthesized by attaching target-specific biomolecules to the surface [64,65].

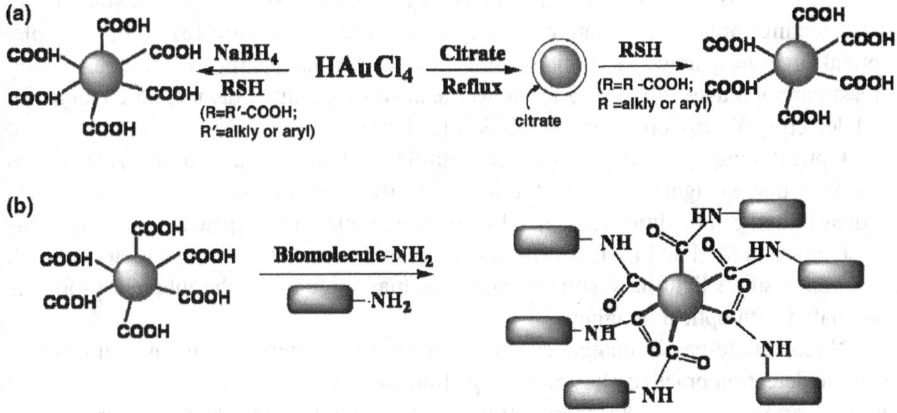

Scheme 7.1. (a) Synthesis of gold nanoparticles. (b) Synthesis of targeted gold nanoparticles.

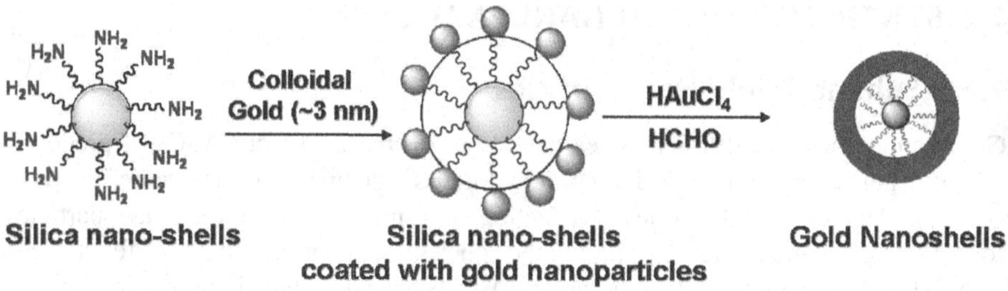

Silica nano-shells   **Silica nano-shells**   **Gold Nanoshells**
                                coated with gold nanoparticles

Scheme 7.2. Synthesis of gold nanoshells.

7.2.3 Gold Nanocages

Nanocages of gold are synthesized using galvanic replacement reaction between nanocages of silver and Au^{3+} ions. Concentrations of Au^{3+} and silver nanocages have been varied to synthesize gold nanocages of various sizes. A surface plasmon resonance peak at around 800 nm has been found to be characteristic for gold nanocages of sizes ~40 nm. These nanocages, when functionalized using tumor-specific biomolecules, can serve as excellent contrast agents for their potential applications in optical coherence tomography [36, 66].

7.3 GOLD NANOPARTICLES AS IMAGING AGENTS

The unique photophysical properties of hybrid nanoparticles (coated with cancer-cell-specific peptides/proteins) allows for tuning of optical absorption from the near-UV to the mid-IR region of the optical spectra [67, 68]. These attractive photophysical properties of hybrid AuNPs have rejuvenated interest in the development of optical imaging methods using inexpensive fiber-optic confocal imaging systems capable of being fed through ducts and capillaries [69]. Recent studies in animals have demonstrated that hybrid AuNPs coated with cancer markers can be used with a dielectric silicon core for detecting molecular signatures of cancer via optical imaging [67]. For example, optical imaging studies in animals using gold/silicon nanoparticles conjugated with breast-carcinoma-specific Her-2 biomolecule have successfully demonstrated targeting and detecting microscopic tumors in breast [67].

Optical imaging of tumors still faces significant challenges due to tissue absorption and reflection of light. However, the inherent advantage of low cost associated with optical imaging modalities, compared to more complex and expensive imaging techniques such as MRI and PET, is expected to create further advances in nanotechnology to produce smart light-harvesting vectors that may circumvent the inherent problems associated with optical imaging.

Nanoparticle-based imaging can combine complementary imaging modalities into a single detection practice, thus providing clinicians with the means of detecting tumor margins prior to or even during resection. Such dual imaging modalities would be particularly useful for the successful treatment of brain malignancies [70]. Combination of MRI and optical imaging would provide new opportunities to understand preoperative tumor localization and intraoperative imaging of tumor margins using optical fluorescence [71]. Recent studies have demonstrated the feasibility of the multimodality imaging approach by using a nanoparticle contrast agent consisting of an optically detectable NIRF fluorochrome conjugated to an MRI-detectable iron oxide core in surgical resection of brain tumors [71]. These studies, for the first time, provide the clinical utility of multimodal nanoparticle probes for the preoperative visualization of brain tumors by serving as an MRI contrast agent and afford an intraoperative discrimination of tumors from brain tissue because of its near-IR fluorescence. Such approaches of dual imaging have the realistic potential of increasing the precision of surgical resection and improving the outlook for brain cancer patients [71].

7.3.1 X-Ray Contrast Agent

Clinically approved and extensively used CT imaging agents, Omnipaque® or Barium sulfate, contain high-atomic-weight elements iodine or barium. Contrasts induced by these agents provide clear CT images of organs or tissues. However, a major disadvantage in using iodine or barium-based contrast agents is associated with nonspecificity that impedes selective molecular imaging of cells or organs. Therefore, there is considerable interest in developing target-specific molecular imaging CT agents [8, 72]. Among the elements with higher atomic number, gold is relatively nontoxic and possesses excellent X-ray absorption properties (atomic number 79; absorption coefficient (100 keV) 5.16 cm^2g^{-1}; K-edge 80.7 keV), which are pivotal in developing site-specific contrast agents. Additionally, gold can be converted into nanoparticulate forms by readily available methods under clinical settings. Nanosized gold provides a inherent advantage toward increasing sensitivity. Each AuNP contains hundreds of inner and surface atoms. Surface atoms can be easily conjugated with target-specific biomolecules, such as tumor-avid peptide or protein, for site-specific delivery of contrast agents. Furthermore, gold nanoparticles, due to their inherent ability to accumulate in leaky tumor vasculature, provide attractive opportunities to serve as site-specific contrast agents.

Hainfeld et al. [8, 72] have demonstrated the feasibility of using 1.9-nm AuNPs to detect mammary tumors in mice models. In these experiments, AuNPs were targeted toward tumor vasculature. Indeed, injected AuNPs accumulated in tumor in one thigh of mice and produced better CT images of tumor (Figure 7.2). Pharmacokinetic data of AuNPs obtained using graphite flame atomic absorption spectroscopy showed that tumor-to-muscle gold ratio was 3.4 at 15 min post-injection and increased to 9.6 after 24 hours [8, 72]. From pharamockinetic and toxicity data, these authors concluded that injected AuNPs cleared through renal pathway and are harmless (LD_{50} = 3.2 g Au kg^{-1}). These experiments show considerable promise for gold nanoparticles for use in site-specific molecular imaging via CT [8, 72].

7.3.2 Optical Coherence Tomography

Optical coherence tomography (OCT) is a novel nonioinizing imaging method that uses differential scattering and absorption of light waves for different tissue types. OCT images of biological tissues are generated with 10- to 15-μm resolution. The quality of

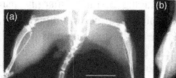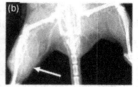

<u>Figure 7.2.</u> Radiographs of mouse hind legs before and after gold nanoparticle (2.7 g Au/Kg) injection. (a) Before injection; (b) 2 minutes after injection. Significant contrast is seen in the leg with the tumor (indicated by arrow) compared with the normal contralateral leg. (Thanks are due to authors and publishing agency for granting permission to reprint the figure from: *Phys Med Biol* 2004; 49: N309–315. Copyright © 2004 Institute of Physics Publishing).

OCT images depends directly on the inherent optical properties of the biological tissues. The sensitivity of OCT can be increased by using optical contrast agent. The unique, size- and shape-dependent photophysical properties of gold nanoparticles allow for tuning of optical absorption characteristics from the near-UV to the mid-IR region of the optical spectra. These attractive photophysical properties of AuNPs have sparked interest in the development of gold-based optical contrast agents for OCT. Among different sizes and shapes of gold nanoparticles, gold nanocages and nanoshells have ideal properties for use as optical contrast agents via OCT modality [36, 65, 73–76].

Younan Xia, Xingde Li, and co-workers have recently reported preliminary results on using gold nanocages of 40 nm in dimension as OCT contrast agents [36]. These authors have compared the OCT images of phantom samples with and without nanocages (Figure 7.3). From their measurements, the scattering cross section ($\sim$8.1 $\times$ 10^{-16} m^2) and absorption cross section (7.26 $\times$ 10^{-15} m^2) of nanocages were obtained. The absorption cross section of gold nanocages has been found to be five orders of magnitude greater than the conventional dye, indocyanine green (2.90 $\times$ 10^{-20} m^2) at 800 nm. These authors have reported a successful synthetic strategy for the bioconjugation of gold nanocages with biomolecules. These results demonstrate that gold nanocages are attractive candidates for the development of site-specific OCT contrast agents. Drezek and co-workers [73–76] have demonstrated that gold nanoshells can serve as contrast agents for OCT. These authors have found that nanoshells, with silica core radius of 100 nm and a gold shell thickness of 20 nm, exhibit stronger backscattering signals than polystyrene microspheres. *in vivo* studies have been performed in mice and hamsters. Nanoshells of certain concentrations were injected into tail the vein of mice and hamsters. These authors have compared the OCT images of dorsal skin of mice and hamsters both prior to and after the injection and demonstrated that nanoshells exhibit significant brightness in the dermis. These studies establish that targeted nanoshells could be a potential contrast agent for OCT imaging modality [65].

7.3.3 Photoacoustic Tomography (PAT)

Photoacoustic tomography, or optoacoustic tomography (OAT), is a state-of-the-art imaging modality that utilizes both light and ultrasound to image tumors. This method is

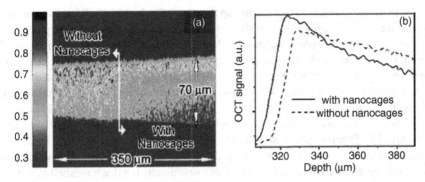

Figure 7.3. (a) OCT image of a gelatin phantom embedded with TiO2, and the concentration of TiO2 was controlled at 1 mg/mL to nimic the background scattering of soft tissues. The right portion of the phantom contained 1 nM of gold nanocages while the left portion did not contain any gold nanocages. (b) Plots of the OCT signals on a log scale as a function of depth. Note that the OCT signal recorded from the portion of phantom with gold nanocages decays faster than the portion without nanocages. (Reprinted with permission from *Nano Lett* 2005; 5(3): 473–477. Copyright © 2006 American Chemical Society.) See insert for color representation of this figure.

based on the inherent light absorption contrast characteristics of biological tissues. Absorption of light waves by tissues leads to thermal expansion with concomitant creation of ultrasound waves. Acoustic waves are detected by sensitive electronics to develop images of biological tissues. Normal and tumor tissues show excellent absorption contrast; for example, the contrast between normal breast and tumor breast tissues is around 300% for 1064-nm light [77]. Despite the unique advantages of this imaging modality, there are still problems associated with lack of optimum sensitivity in detecting tumor at early stages. This problem is presumably due to the low endogeneous absorption contrast between early stage tumor and normal tissue. However, this situation can be circumvented by using exogeneous contrast agents such as gold nanoparticles (AuNPs). Gold nanospheres and nanoshells have ideal optical properties for optoacoustic imaging [78]. In particular, gold nanoshells encapsulated with silica cores offer the best potential for optoacoustic imaging applications. The optical resonance of these shells can be varied to shift the absorption to NIR region (prefered region for imaging biological tissues) by changing the core size. Optical extinction coefficients of nanospheres and nanoshells are around 1000 times higher in magnitude than the commonly used dyes such as indocyanine green. Nanoshells can be targeted to tumor sites usng the leaky properties of tumor vasculature.

Copland et al. [79] have used 40-nm-diameter gold nanoparticles conjugated to herceptin, a monoclonal antibody that binds HER2/neu, to selectively target human SK-BR-2 breast cancer cells. These authors have imaged breast cancer cells by OAT, in a gelatin phantom, using AuNPs as exogeneous optical contrast agents. These experiments show that even as few as 9 or 10 AuNPs were detectable by OAT at a depth of 6 cm. These studies have demonstrated that even the smallest tumor can be detected using AuNPs based OAT [28, 79, 80].

Wang et al. [78] have reported the application of gold nanoshells as intravascular contrast agent for imaging brain of live rat using photoacoustic tomography (PAT). In these studies, gold nanoshells made of 125-nm silica core were coated with 10- to 12-nm-thick gold shell with an optical absorption maximum at 800 nm were employed. Three successive doses of $\sim 0.8 \times 10^9$ nanoshells/g of body weight were injected through the tail vein of rats. Dynamic distribution of nanoshells in rat brain was imaged using NIR light at a wavelength of 800 nm. The optical absorption of blood vessels increases by $\sim 100\%$ after administering gold nanoshells. Photoacoustic angiographs are shown in Figure 7.4. Figures 7.4a and 7.4b are preinjection and post-injection (~ 20 min) photoacoustic images of the brain [78]. The vasculature of rat brain is seen with greater clarity in Figure 7.4b. These studies establish a new dimension that nanoshell-based PAT imaging has potential to emerge as an important molecular imaging modality [78]. In fact, gold nanoshells could potentially be conjugated to target-specific biomolecules for potential applications as molecular imaging probes for specific tumors [78].

7.4 GOLD NANOPARTICLES AS THERAPEUTIC AGENTS

Metallic nanoparticles, because of their size and their chemical and physical properties, are particularly attractive as therapeutic probes in treating cancer (Figure 7.5). Central to any clinical advances in nanoparticulate-based therapy will be to produce hybrid nanoparticles that can be targeted to vascular, extracellular, or cell surface receptors. Development of hybrid nanoparticles that specifically target cancer vasculature has received considerable attention [29,81]. Most cancers have leaky vasculature; in addition, the defective vascular architecture (created due to the rapid vascularization necessary to serve fast growing cancers), in combination with poor lymphatic drainage, allows increased permeation and retention effects. The leaky vasculature, because of higher porosity and permeability, serves as a natural high-affinity target to metallic nanoparticles [29, 81]. Overexpression of specific antigens by cancerous cells serves as the most effective pathway to target nanoparticle-based therapeutic agents. Hybrid nanoparticles produced via conjugation of peptides (and various other biomolecules) with proven high affinity toward receptors on cancer cells will provide an effective means of targeting therapeutic nanoparticles on cancerous tissue/cells (Figure 7.5).

7.4.1 Photothermal Therapy

Application of heat to control/eradicate specific tumors is an attractive medical modality in the treatment of cancer. The noninvasive use of heat for tumor ablation is referred to as hyperthermia and involves the application of microwaves, ultrasound, or laser light energy sources to deliver quantum of lethal dose of heat at specific tumor sites [82–87]. Hyperthermia produces temperatures around 40°C at confined tumor volumes to induce lipid transitions and conformational changes in RNA and DNA. It has been shown that thermally induced cellular injury/death occurs due to extensive protein denaturation ensuing at 40°C and above [82–87]. However, care must be

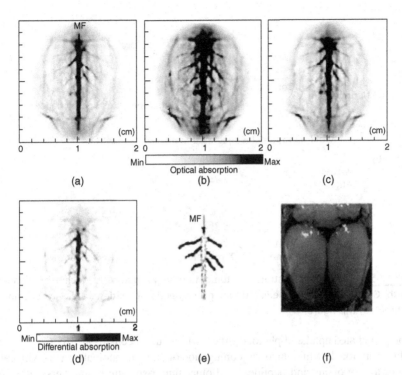

Figure 7.4. Noninvasive PAT of a rat brain *in vivo* employing the nanoshell contrast agent and NIR light at a wavelength of 800 nm. (a) Photoacoustic image acquired before the administrations of nanoshells. *MF*; median fissure. (b) Photoacoustic image obtained ~20 min after the third administration of nanoshells. (c) Photoacoustic image acquired ~350 min after the third administration of nanoshells. (d) Differential image that was obtained by subtracting the pre-injection image from the post-injection image (Image d Image b - Image a). (e) Large blood vessels in the cerebral cortex segmented from the PAT images (shown with solid regions). (f) Open-skull photograph of the rat brain cortex obtained after the data acqusition for PAT. (Reprinted with permission from: *Nano Lett* 2004; 4(9): 1689–1692. Copyright © 2006 American Chemical Society.)

taken to ensure that thermal treatement is limited to tumor regions with minimal/no injury to surrounding tissue. Photothermal therapy, wherein lasers are used for thermal treatment of tumors, is gaining prominence in recent years. Pigment- and water-mediated efficient absorption of energy in healthy tissue have been the cause of major concern and have impeded the effectiveness of heat uptake within tumor tissue. In order to achieve increased effectiveness, current efforts are directed toward reduction of nonspecific heat uptake and consequent maximization of heat absorption within tumor cells. Various approaches being developed toward the selective absorption of heat energy on tumor sites include (a) site-specific delivery of photoabsorbing dyes for efficient absorption of laser light energy with consequent conversion into heat and (b)

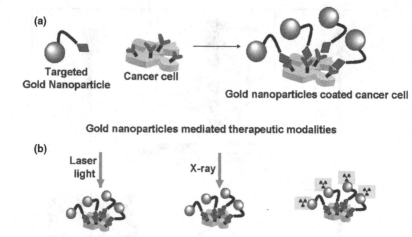

Figure 7.5. (a) Schematic illustration of tumor targeting and surface attachment of nanoparticles. (b) Gold nanoparticles mediated tumor therapeutic modalities. See insert for color representation of this figure.

receptor-mediated uptake of photoabsorbers within tumor cells/tissue [65,67,74,88,89]. The latter approach, which involves conjugation of efficient photoabsorbers with cancer receptor avid proteins and peptides (and other tumor-specific biomarkers), appears to be a highly attractive approach to achieve higher therapeutic heat load on tumor sites [65,67,74,88,89]. Therefore, conjugation approaches that embed laser light absorbers within cancer-target-specific biomolecules would provide the most attractive modality toward maximization of heat dose on tumor sites. Recent findings have demonstrated that strongly enhanced absorption of hybrid gold nanoparticles and the ability of functionalized gold nanoparticles to convert absorbed light into localized heat could provide effective avenues for photothermal therapy [65,67,74,88–90]. Indeed, El-Sayeed et al. [90] have pioneered the utility of gold nanoparticulates for photoabsorption and consequent heat transfer to destroy malignant cells. Their studies have unequiovocally demonstrated that immunotargeted gold nanoparticles conjugated to anti-EGFR antibodies selectively lowered laser energy required to destroy oral cancel cells in vitro upon irradiation with a 514-nm argon ion laser as compared the energy needed to kill nonmalignant cells. These results have provided a solid foundation for further research in the area of nanoparticle-mediated photothermal therapy.

Near-infrared (NIR) radiation is capable of inducing heat in tumor tissue and within cancer cells. Indeed, irreversible heat damage in the carcinoma cells has been demonstrated by exposing tumor tissue—localized hybrid AuNPs—to NIR radiation for 4 min. Further experiments in tumor-bearing mice has provided the first proof of principle that gold-coated nanoshells can be used to achieve dual optical imaging and NIR-mediated optical therapy. Tumor-bearing mice that underwent a combination of optical detection and NIR treatment experienced a marked increase in survival compared to animals treated with NIR without prior administration of gold-coated nanoshells [88,89].

7.4.2 X-Ray Therapy

Seminal discovery by Hainfeld et al. [8] demonstrated that AuNPs can help kill cancerous cells very effectively in mice. This work involved injecting cancer cells into mice, followed by the administration of AuNPs and (a few minutes later) irradiation of animals with 250-kV X-rays. The results from this work are startling because administration of AuNPs followed by X-ray treatment reduced the size of the tumors, and in some instances tumors were completely eradicated. This study also found that AuNPs alone had no therapeutic effect and that tumors continued to grow in animals that had received only the X-ray treatment. One-year survival rate in animals that had received combined gold nanoparticle–X-ray treatment was 85% versus 20% for X-ray therapy alone and dropped to zero for gold nanoparticle treatment without X-rays. Leaky vasculature of tumors presumably maximizes localization of AuNPs in tumors. This selective accumulation of AuNPs in tumors results in strong absorption of X-rays as compared to normal tissue. These results represent the 'tip of the iceberg' on the potential applications of metallic nanoparticle-based therapeutic approaches for combating cancer [8]. Because AuNPs can be seen through CT and planar X-rays, AuNPs (if systematically targeted) could serve as dual imaging and therapeutic probes in the detection and therapy of cancer.

7.4.3 Radiotherapy

Another attractive approach toward the application of nanotechnology to nanomedicine is the utility of nanoparticles that display inherent therapeutic properties. For example, radioactive AuNPs present attractive prospects in therapy of cancer [29, 81]. The radioactive properties of Au-198 ($\beta_{max} = 0.96$, MeV; $t_{1/2} = 2.7$ days) and Au-199 ($\beta_{max} = 0.46$ MeV; $t_{1/2} = 3.14$ days) make them ideal candidates for use in radiotherapeutic applications. In addition, they both have imagable gamma emissions for dosimetry and pharmacokinetic studies. Gold nanoparticles are of interest for treatment of diseases because they can be delivered directly into cells and cellular components with a higher concentration of radioactivity (e.g., higher dose of radioactivity) to cancerous tumor cells. Therapeutic agents derived from radioactive AuNPs provide higher therapeutic payload on tumor sites because each gold nanoparticle contains multiple atoms of gold. This unique advantage of achieving a substantial increase in therapeutic dose to the tumor site, coupled with the feasibility of tagging nanoparticles of Au-198 with oligonucleotides and peptides that are selective for receptors overexpressed by diseased tissue, presents unprecedented opportunities in the design and development of tumor-specific nanotherapeutic agents for the treatment of cancer. Another approach to using radioisotopes of gold in tumor therapy involves encapsulation of radioisotopes within a nanocomposite device (NCD) of a defined size and surface properties. NCDs act as vehicles to transport radioactive AuNPs to tumor sites [29, 81]. Therefore, the radioactivity delivered to a tumor site can be increased by controlling the particles size of the NCD or by increasing the number of gold atoms without destroying the targeting ability of the nanocomposite device. Gold NCDs are synthesized as monodisperse hybrid nanoparticles composed of radioactive guests immobilized by dendritic polymer hosts. Commercially available polymers, including poly(amidoamine) PAMAM dendrimers, and tecto dendrimers, are used to fabricate nanocomposites. Typically, Au-198 encapuslated within PAMAM dendrimers are prepared by mixing dilute solutions of PAMAM dendrimer with an aqueous solution of $HAuCl_4$. Salt formation between the tetrachlroaurate anions and the dendrimer nitrogens would ensure effective encapsulation of gold within the dendrimer matrix. Activation of elemental gold into Au-198 within the dendrimer matrix is carried out both in solid and in solution forms by direct neutron irradiation. Biodistribution studies using 5-nm-sized tritium-labeled PAMAM dendrimers in tumor models, including mouse B16 melanoma, human prostate DU 145, and human KB squamous cell carcinoma mouse xenograft models, have been performed. Uptake of tritium activity in tumor tissue and also retention of activity for several weeks provides a proof of principle that radioactive isotope encapsulated nano devices (such as PAMAMs with β-emitting Au-198) could be very useful in tumor therapy [29, 81].

7.5 SUMMARY

Synthesis of functionalized hybrid nanoparticles will dictate a plethora of future advances toward their applications in biomedicine. From the aforementioned discussions, it is clear that nanotechnology will have a major impact on the future of new imaging and therapeutic modalities for combating deadly human diseases including cancer. Although the penetration of nanotechnology within the biomedical front (for drug diagnostic and therapeutic design) is limited to date, the synergistic advantages of nanoparticle-mediated imaging and therapy to the field of oncology is opening up new horizons in clinical modalities related to treating cancer. The size relationship of nanoparticles with biological cells, coupled with the inherent diagnostic/therapeutic properties of specific nanoparticles and the availability of sophisticated targeting vectors to target nanoparticles onto tumor cells/tumor tissue, are all poised to dramatically increase our ability to detect and treat cancer in its earliest stages.

REFERENCES

1. Daniel M-C, Astruc D. Gold nanoparticles: Assembly, supramolecular chemistry, quantum-size-related properties, and applications toward biology, catalysis, and nanotechnology. *Chem Rev* 2004;104(1):293–346.

2. Perez-Luna VH, Aslan K, Betala P. Colloidal gold. *Encycl Nanosci Nanotechnol* 2004;2:27–49.

3. McMillan RA. Biomolecular templates: nanoparticles align. *Nat Mater* 2003;2(4):214–215.

4. Csaki A, Moller R, Fritzsche W. Gold nanoparticles as novel label for DNA diagnostics. *Expert Rev Mol Diagn* 2002;2(2):187–193.

5. Whitesides GM. The "right" size in nanobiotechnology. *Nat Biotechnol* 2003;21(10): 1161–1165. Available from <Go to ISI>://000185647200026.

6. Lowe CR. Nanobiotechnology: The fabrication and applications of chemical and biological nanostructures. *Curr Opin Struct Biol* 2000;10(4):428–434. Available from <Go to ISI>://000088685300007.

7. Hainfeld JF, Powell RD, Hacker GW. Nanoparticle molecular labels. *Nanobiotechnology* 2004:353–86.

8. Hainfeld JF, Slatkin D, Smilowitz HM. The use of gold nanoparticles to enhance radiotherapy in mice. *Phys Med Biol* 2004;49(18):N309–N315.

9. Hainfeld JF, Powell RD. New frontiers in gold labeling. *J Histochem and Cytochem* 2000;48(4):471–480.

10. Hainfeld JF, Robinson JM. New frontiers in gold labeling: Symposium overview. *J Histochem Cytochem* 2000;48(4):459–460.

11. Xue C, Arumugam G, Palaniappan K, Hackney SA, Liu H, Liu J. Construction of conjugated molecular structures on gold nanoparticles via the Sonogashira coupling reactions. *Chem Commun* 2005(8):1055–1057.

12. Cui H, Zhang Z-F, Shi M-J. Chemiluminescent Reactions Induced by Gold Nanoparticles. *J Phy Chem B* 2005;109(8):3099–3103.

13. Woehrle GH, Brown LO, Hutchison JE. Thiol-functionalized, 1. 5-nm gold nanoparticles through ligand exchange reactions: Scope and mechanism of ligand exchange. *J Am Chem Soc* 2005;127(7):2172–2183.

14. Pengo P, Polizzi S, Pasquato L, Scrimin P. Carboxylate-imidazole cooperativity in dipeptide-functionalized gold nanoparticles with esterase-like activity. *J Am Chem Soc* 2005;127(6):1616–1617.

15. Potapenko DV, Horn JM, Beuhler RJ, Song Z, White MG. Reactivity studies with gold-supported molybdenum nanoparticles. *Surf Sci* 2005;574(2–3):244–258.

16. Kell AJ, Donkers RL, Workentin MS. Core size effects on the reactivity of organic substrates as monolayers on gold nanoparticles. *Langmuir* 2005;21(2):735–742.

17. Letsinger RL, Mirkin CA, Elghanian R, Mucic RC, Storhoff JJ. Chemistry of oligonucleotide–gold nanoparticle conjugates. *Phosphorus, Sulfur and Silicon and the Related Elements* 1999;144–146:359–362.

18. Park J-E, Atobe M, Fuchigami T. Sonochemical synthesis of inorganic–organic hybrid nanocomposite based on gold nanoparticles and polypyrrole. *Chem Letts* 2005;34(1):96–97.

19. Goodrich GP, Helfrich MR, Overberg JJ, Keating CD. Effect of macromolecular crowding on DNA:Au nanoparticle bioconjugate assembly. *Langmuir* 2004;20(23):10246–10251.

20. Hong R, Emrick T, Rotello VM. Monolayer-controlled substrate selectivity using noncovalent enzyme–nanoparticle conjugates. *J Am Chem Soc* 2004;126(42):13572–13573.

21. Lee K-B, Kim E-Y, Mirkin CA, Wolinsky SM. The use of nanoarrays for highly sensitive and selective detection of human immunodeficiency virus type 1 in plasma. *Nano Lett* 2004;4(10):1869–1872.

22. Dubertret B, Calame M, Libchaber AJ. Single-mismatch detection using gold-quenched fluorescent oligonucleotides. *Nat Biotechnol* 2001;19(4):365–370.

23. Shaffer AW, Worden JG, Huo Q. Comparison study of the solution phase versus solid phase place exchange reactions in the controlled functionalization of gold nanoparticles. *Langmuir* 2004;20(19):8343–8351.

24. Lioubashevski O, Chegel VI, Patolsky F, Katz E, Willner I. Enzyme-catalyzed bio-pumping of electrons into Au-nanoparticles: A surface plasmon resonance and electrochemical study. *J Am Chem Soc* 2004;126(22):7133–7143.

25. Nam J-M, Stoeva SI, Mirkin CA. Bio-bar-code-based DNA detection with PCR-like sensitivity. *J Am Chem Soc* 2004;126(19):5932–5933.

26. Paciotti GF, Myer L, Weinreich D, Goia D, Pavel N, McLaughlin RE, et al. Colloidal Gold: A novel nanoparticle vector for tumor directed drug delivery. *Drug Deliv* 2004;11(3):169–183.

27. Paciotti Giulio F, Myer L, Weinreich D, Goia D, Pavel N, McLaughlin Richard E, et al. Colloidal gold: A novel nanoparticle vector for tumor directed drug delivery. *Drug Deliv* 2004;11(3):169–183.

28. Eghtedari MA, Copland JA, Popov VL, Kotov NA, Motamedi M, Oraevsky AA. Bioconjugated gold nanoparticles as a contrast agent for detection of small tumors. *Proc SPIEi—Intl Soc Opt Eng* 2003;4960(Biomedical Optoacoustics IV):76–85.

29. Balogh LP, Nigavekar SS, Cook AC, Minc L, Khan MK. Development of dendrimer–gold radioactive nanocomposites to treat cancer microvasculature. *PharmaChem* 2003;2(4):94–99.

30. Sokolov K, Follen M, Aaron J, Pavlova I, Malpica A, Lotan R, et al. Real-time vital optical imaging of precancer using anti-epidermal growth factor receptor antibodies conjugated to gold nanoparticles. *Cancer Res* 2003;63(9):1999–2004.

31. Sokolov K, Robinson C, Collier T, Richards-Kortum R, Follen M, Lotan R. Metal nanoparticles as biospecific contrast agents for cancer imaging. *Trends Opt Phot* 2002;71(OSA Biomedical Topical Meetings, 2002):376–378.

32. Ma Z, Wu J, Zhou T, Chén Z, Dong Y, Tang J, et al. Detection of human lung carcinoma cell using quartz crystal mićrobalance amplified by enlarging Au nanoparticles *in vitro*. *N J Chem* 2002;26(12):1795–1798.

33. Ma Z, Liang R, Jiang W, Zhou T, Chen Z, Duan M, et al. Colorimetric detection of human lung carcinoma cell by enlarging Au-nanoparticles *in situ*. *Chemistry Lett* 2002(6):570–571.

34. Robertson JFR, Come SE, Jones SE, Beex L, Kaufmann M, Makris A, et al. Endocrine treatment options for advanced breast cancer—the role of fulvestrant. *Eur J Cancer* 2005;41(3):346–356.

35. Abrams MJ, Murrer BA. Metal compounds in therapy and diagnosis. *Science* 1993;261(5122):725–730.

36. Chen J, Saeki F, Wiley BJ, Cang H, Cobb MJ, Li Z-Y, et al. Gold Nanocages: Bioconjugation and their potential use as optical imaging contrast agents. *Nano Lett* 2005;5(3):473–477.

37. Nath N, Chilkoti A. Immobilized gold nanoparticle sensor for label-free optical detection of biomolecular interactions. *Proc SPIE—Int Soc Opt Eng* 2002;4626 (Biomedical Nanotechnology Architectures and Applications):441–448.

38. Tokareva I, Minko S, Fendler JH, Hutter E. Nanosensors based on responsive polymer brushes and gold nanoparticle enhanced transmission surface plasmon resonance spectroscopy. *J Am Chem Soc* 2004;126(49):15950–15951.

39. Khlebtsov NG. Optical models for conjugates of gold and silver nanoparticles with biomacromolecules. *J Quant Spectrosc Radiat Transfer* 2004;89(1–4):143–153.

40. Aslan K, Lakowicz JR, Geddes CD. Tunable plasmonic glucose sensing based on the dissociation of Con A-aggregated dextran-coated gold colloids. *Anal Chim Acta* 2004;517(1–2):139–144.

41. Haes AJ, Zou S, Schatz GC, Van Duyne RP. A nanoscale optical biosensor: The long range distance dependence of the localized surface plasmon resonance of noble metal nanoparticles. *J Phys Chem B* 2004;108(1):109–116.

42. Tokareva I, Minko S, Fendler Janos H, Hutter E. Nanosensors based on responsive polymer brushes and gold nanoparticle enhanced transmission surface plasmon resonance spectroscopy. *J Am Chem Soc* 2004;126(49):15950–15951.

43. Oh E, Hong M-Y, Lee D, Nam S-H, Yoon HC, Kim H-S. Inhibition assay of biomolecules based on fluorescence resonance energy transfer (FRET) between quantum dots and gold nanoparticles. *J Am Chem Soc* 2005;127(10):3270–3271.

44. Jimenez OA, Chikneyan S, Baca AJ, Wang J, Zhou F. Sensitive detection of sulfhydryl groups in surface-confined metallothioneins and related species via ferrocene-capped gold nanoparticle/streptavidin conjugates. *Environ Sci Technol* 2005;39(5):1209–1213.

45. Raschke G, Franzl T, Kowarik S, Soennichsen C, Klar TA, Feldmann J, et al. Biomolecular sensor based on optical spectroscopy of single gold nanoparticles. *Trends Opt Phot* 2004;96/B(Conference on Lasers and Electro-Optics, 2004):CThI2/1–CThI2/2.

46. Schwartzberg AM, Grant CD, Wolcott A, Talley CE, Huser TR, Bogomolni R, et al. Unique gold nanoparticle aggregates as a highly active surface-enhanced Raman scattering substrate. *J Phys Chem B* 2004;108(50):19191–19197.

47. Jorgenson RC, Basiji DA, Ortyn WE. Imaging platform for nanoparticle detection applied to SPR biomolecular interaction analysis Application: US (Amnis Corporation, USA). 2004:20 pp, Cont-in-part of US Ser No 628662.

48. Tang DP, Yuan R, Chai YQ, Zhong X, Liu Y, Dai JY, et al. Novel potentiometric immunosensor for hepatitis B surface antigen using a gold nanoparticle-based biomolecular immobilization method. *Anal Biochem* 2004;333(2):345–350.

49. Aslan K, Luhrs CC, Perez-Luna VH. Controlled and reversible aggregation of biotinylated gold nanoparticles with streptavidin. *J Phys Chem B* 2004;108(40):15631–39.

50. Stevens MM, Flynn NT, Wang C, Tirrell DA, Langer R. Coiled-coil peptide-based assembly of gold nanoparticles. *Adv Mater (Weinheim, Germany)* 2004;16(11):915–18.

51. Baca AJ, Zhou F, Wang J, Hu J, Li J, Wang J, et al. Attachment of ferrocene-capped gold nanoparticle/streptavidin conjugates onto electrode surfaces covered with biotinylated biomolecules for enhanced voltammetric analysis. *Electroanalysis* 2004;16(1–2):73–80.

52. Schroedter A, Weller H. Ligand design and bioconjugation of colloidal gold nanoparticles. *Angew Chem, Int Ed* 2002;41(17):3218–3221.

53. Maxwell DJ, Taylor JR, Nie S. Self-assembled nanoparticle probes for recognition and detection of biomolecules. *J Am Chem Soc* 2002;124(32):9606–9612.

54. Tang DP, Yuan R, Chai YQ, Zhong X, Liu Y, Dai JY, et al. Novel potentiometric immunosensor for hepatitis B surface antigen using a gold nanoparticle-based biomolecular immobilization method. *Anal Biochem* 2004;333(2):345–350.

55. Sokolov K, Aaron J, Hsu B, Nida D, Gillenwater A, Follen M, et al. Optical systems for *in vivo* molecular imaging of cancer. *Technol Cancer Res Treatment* 2003;2(6):491–504.

56. Turkevich J, Stevenson PC, Hillier J. The nucleation and growth processes in the synthesis of colloidal gold. *Discussions Faraday Soc* 1951;11:55–75.

57. Wang Z, Levy R, Fernig DG, Brust M. The peptide route to multifunctional gold nanoparticles. *Bioconjug Chem* 2005;16(3):497–500.

58. Zhu H, Tao C, Zheng S, Li J. One step synthesis and phase transition of phospholipid-modified Au particles into toluene. *Colloids Surf, A: Physicochem Eng Aspects* 2005;257–258:411–414.

59. Pei L, Mori K, Adachi M. Formation process of two-dimensional networked gold nanowires by citrate reduction of $AuCl_4^-$ and the shape stabilization. *Langmuir* 2004;20(18):7837–7843.

60. Zhu T, Vasilev K, Kreiter M, Mittler S, Knoll W. Surface modification of citrate-reduced colloidal gold nanoparticles with 2-mercaptosuccinic acid. *Langmuir* 2003;19(22):9518–9525.

61. Brust M, Walker M, Bethell D, Schiffrin DJ, Whyman R. Synthesis of thiol-derivatized gold nanoparticles in a two-phase liquid–liquid system. *J Chem Soc, Chem Commun* 1994(7):801–802.

62. Green M, O'Brien P. A simple one phase preparation of organically capped gold nanocrystals. *Chem Commun* 2000(3):183–184.

63. Isaacs Steven R, Cutler Erin C, Park J-S, Lee TR, Shon Y-S. Synthesis of tetraoctylammonium-protected gold nanoparticles with improved stability. *Langmuir: ACS J Surf Ccolloids* 2005;21(13):5689–5692.

64. Pham T, Jackson JB, Halas NJ, Lee TR. Preparation and characterization of gold nanoshells coated with self-assembled monolayers. *Langmuir* 2002;18(12):4915–4920.

65. Loo C, Lin A, Hirsch L, Lee M-H, Barton J, Halas N, et al. Nanoshell-enabled photonics-based imaging and therapy of cancer. *Technol Cancer Res Treatment* 2004;3(1):33–40.

66. Chen J, Saeki F, Wiley Benjamin J, Cang H, Cobb Michael J, Li Z-Y, et al. Gold nanocages: bioconjugation and their potential use as optical imaging contrast agents. *Nano Lett* 2005;5(3):473–477.

67. Hirsch LR, Stafford RJ, Bankson JA, Sershen SR, Rivera B, Price RE, et al. Nanoshell-mediated near-infrared thermal therapy of tumors under magnetic resonance guidance. *Proc Natal Acad Scie USA* 2003;100(23):13549–13554.

68. O'Neal DP, Hirsch Leon R, Halas Naomi J, Payne JD, West Jennifer L. Photo-thermal tumor ablation in mice using near infrared-absorbing nanoparticles. *Cancer Lett* 2004;209(2):171–176.

69. Sung KB, Liang C, Descour M, Collier T, Follen M, Malpica A, et al. Near real time *in vivo* fibre optic confocal microscopy: Sub-cellular structure resolved. *J Microscopy* 2002;207(Pt 2):137–145.

70. Black P. Management of malignant glioma: Role of surgery in relation to multimodal therapy. *J Neurovirol* 1988;4:227–236.

71. Kircher MF MU, King RS, Weissleder R, Josephson L. A multimodal nanoparticle for pre-operative magnetic resonance imaging and intraoperative optical brain tumor delineation. *Cancer Res* 2003;63:8122–8125.

72. Hainfeld JF, Slatkin DN, Focella TM, Smilowitz HM. Gold nanoparticles: A new X-ray contrast agent. *Br J Radiol* 2006;79(939):248–253. Available from <Go to ISI>://000236114200012.

73. Loo C, Hirsch L, Lee M-H, Chang E, West J, Halas N, et al. Gold nanoshell bioconjugates for molecular imaging in living cells. *Opt Lett* 2005;30(9):1012–1014.

74. Loo C, Lowery A, Halas N, West J, Drezek R. Immunotargeted nanoshells for integrated cancer imaging and therapy. *Nano Lett* 2005;5(4):709–711.

75. Hirsch LR, Gobin AM, Lowery AR, Tam F, Drezek RA, Halas NJ., et al. Metal nanoshells. *Ann Biomed Eng* 2006;34(1):15–22.

76. Lin AWH, Lewinski NA, West JL, Halas NJ, Drezek RA. Optically tunable nanoparticle contrast agents for early cancer detection: Model-based analysis of gold nanoshells. *J Biomed Opt* 2005;10(6):064035.

77. Wang X, Xu Y, Xu M, Yokoo S, Fry ES, Wang LV. Photoacoustic tomography of biological tissues with high cross-section resolution: Reconstruction and experiment. *Med Phys* 2002;29(12):2799–2805.

78. Wang Y, Xie X, Wang X, Ku G, Gill KL, O'Neal DP, et al. Photoacoustic tomography of a nanoshell contrast agent in the *in vivo* rat brain. *Nano Lett* 2004;4(9):1689–1692.

79. Copland John A, Eghtedari M, Popov Vsevolod L, Kotov N, Mamedova N, Motamedi M, et al. Bioconjugated gold nanoparticles as a molecular based contrast agent: Implications for imaging of deep tumors using optoacoustic tomography. *Mol Iimaging Biol (MIB: The official publication of the Academy of Molecular Imaging)* 2004;6(5):341–349.

80. Li P-C, Wei C-W, Liao C-K, Chen C-D, Pao K-C, Wang C-RC, et al. Multiple targeting in photoacoustic imaging using bioconjugated gold nanorods. *Proc SPIE—Int Soc Opt Eng* 2006;6086 (Photons Plus Ultrasound: Imaging and Sensing):60860M/1–60860M/10.

81. Bielinska A, Eichman JD, Lee I, Baker JR, Jr, Balogh L. Imaging {Au0-PAMAM} Gold-dendrimer Nanocomposites in Cells. *J Nanoparticle Res* 2002;4(5):395–403.

82. Mirza AN, Fornage BD, Sneige N, Kuerer HM, Newman LA, Ames FC, et al. Radiofrequency ablation of solid tumors. *Cancer J (Sudbury, MA)* 2001;7(2):95–102.

83. Overgaard J. The current and potential role of hyperthermia in radiotherapy. *Int J Radiat Oncol, Biol, Phys* 1989;16(3):535–549.

84. Seki T, Wakabayashi M, Nakagawa T, Imamura M, Tamai T, Nishimura A, et al. Percutaneous microwave coagulation therapy for patients with small hepatocellular carcinoma: comparison with percutaneous ethanol injection therapy. *Cancer* 1999;85(8):1694–1702.

85. Jolesz Ferenc A, Hynynen K. Magnetic resonance image-guided focused ultrasound surgery. *Cancer J (Sudbury, MA)* 2002;8 (Suppl 1):S100–S112.

86. Philipp CM, Rohde E, Berlien HP. Nd:YAG laser procedures in tumor treatment. *Semin Surg Oncol* 1995;11(4):290–298.

87. Prudhomme M, Tang J, Rouy S, Delacretaz G, Salathe RP, Godlewski G. Interstitial diode laser hyperthermia in the treatment of subcutaneous tumor. *Lasers Surg Med* 1996;19(4):445–450.

88. Loo CH, Lee M-H, Hirsch LR, West JL, Halas NJ, Drezek RA. Nanoshell bioconjugates for integrated imaging and therapy of cancer. *Proc SPIE—Int Soc Opt Eng* 2004;5327 (Plasmonics in Biology and Medicine):1–4.

89. O'Neal DP, Hirsch LR, Halas NJ, Payne JD, West JL. Photo-thermal tumor ablation in mice using near infrared-absorbing nanoparticles. *Cancer Lett (Amsterdam, Netherlands)* 2004;209(2):171–176.

90. El-Sayed IH, Huang X, El-Sayed MA. Selective laser photo-thermal therapy of epithelial carcinoma using anti-EGFR antibody conjugated gold nanoparticles. *Cancer Lett* 2006;239(1):129–135.

8

BUILDING BLOCKS OF NUCLEIC ACID NANOSTRUCTURES: UNFOLDING THERMODYNAMICS OF INTRAMOLECULAR DNA COMPLEXES

Luis A. Marky, Souvik Maiti, Chris Olsen, Ronald Shikiya, Sarah E. Johnson, Mahima Kaushik and Irine Khutsishvili

8.1 INTRODUCTION

The function of nucleic acids is carried out via interactions with other molecules. In the readout of genetic information and in the control of gene regulation, the presence of proteins and other ligands is essential. Distinctive conformations are associated with the function of DNA. Sites of interaction of DNA with gene regulatory proteins have been found to exhibit characteristic conformational properties [1]. For instance, origins of replication in both prokaryotes and eukaryotes appear to possess characteristic structural features. To understand how nucleic acids carry out their biological roles, it is essential to have a complete physical description of these interacting systems. For these reasons, there is now unprecedented interest in the conformational fine structure of nucleic acids and the forces that determine their structure. In principle, the folding of nucleic acids is controlled by the local base sequence in a precise and potentially predictable way. However, knowledge of the structure of a particular duplex alone cannot provide an understanding of the forces responsible for maintaining the distinct structures in nucleic acids. The overall physical properties of a nucleic acid molecule depend on its

Biomedical Applications of Nanotechnology. Edited by Vinod Labhasetwar and Diandra L. Leslie-Pelecky
Copyright © 2007 John Wiley & Sons, Inc.

chemical architecture, but globally on contributions from base pairing, base stacking, polyelectrolyte behavior (ion binding), and hydration (water binding).

In recent years, the discovery of chromosomal sequences has led scientists to postulate the formation of unusual structures, including the Holliday junction, H-DNA, Z-DNA, telomeres, i-motifs, and triplexes. In the particular case of nucleic acid triplexes, the formation of a triplex between two strands of poly (rU) with one strand of poly (rA) was first reported by Felsenfeld et al. in 1957 [2]. It was not until 25 years later that interest in these triplexes was stimulated by the discovery of single-strand-specific S1 endonuclease-hypersensitive sites in the upstream regions of several eukaryotic genes [3] and later with homopurine-homopyrimidine tracts undergoing a superhelix-induced structural transition into a novel DNA conformation termed H-DNA [4–8]. The most common triplex formation involves binding a homopyrimidine strand in the major groove of a duplex containing complementary homopurine–homopyrimidine sequences, with the third strand in a parallel orientation with respect to the homopurine strand. In this structure, sequence-specificity is achieved by Hoogsteen pairing of thymine with dA·dT and of protonated cytosine with dG·dC Watson–Crick base pairs [9–12]. The binding of a third strand to double-helical DNA via triple helix formation has led to potential applications of triplexes as sequence-specific artificial nucleases [9]. Thymines and cytosines in the third strand can be arranged to recognize the dA·dT and dG·dC base pairs of many gene sequences [9, 13–16]. Furthermore, oligonucleotide-directed triplex formation has been implicated as a possible means of controlling cellular processes, both by an endogenous or exogenous mechanism [17–22], and as intermediates in homologous recombination [23, 24]. The inhibition of gene expression via triple helix formation is feasible due to the natural abundance of homopurine-homopyrimidine tracts in genes and their 5′ flanking regions [3, 25–28], providing evidence for the biological relevance of triplexes. It is also reasonable to think that their presence in biological systems may be exerting a variety of effects, justifying their complete physical characterization in terms of stability as a function of sequence and solution conditions.

Furthermore, it has been demonstrated that palindromic sequences imbedded in plasmids developed cruciform structures in response to a topological stress [29, 30]. Specific endonucleases will then cleave supercoiled plasmid in the middle of these palindromic sequences [31, 32]. These findings suggest that the presence of hairpin loops in DNA may play an important role in biological processes. Nowadays, there is considerable interest in both the structure and overall physical properties for the folding (and unfolding) of nucleic acid hairpin loops. The formation of an oligonucleotide intramolecular DNA complex takes place with a lower entropy penalty and their unfolding resemble the pseudo-monomolecular melting of a DNA polymer with convenient transition temperatures. One latest application of the use of long single-stranded DNA molecules complexes is in their formation of nanoscale two-dimensional shapes (DNA Origami) [33].

The major focus of this chapter is to report the relative thermodynamic contributions of base pairing, base stacking, and ionic effects to the stability and melting behavior of a variety of intramolecular DNA complexes. Specifically, we will report complete thermodynamic profiles for several oligonucleotide model systems, including hairpin loops, double hairpins, three-way junctions, pseudoknots, triplexes, and G- and

C-quadruplexes. This information should improve our picture of how sequence, loops, and ion binding control the stability and melting behavior of a nucleic acid molecule. The resulting data should supplement existing nearest-neighbor thermodynamic parameters to help in the prediction of secondary structure from a given sequence, as demonstrated earlier [34–37].

8.2 MATERIALS AND METHODS

8.2.1 Materials

All oligonucleotides were synthesized in the Eppley Institute Molecular Biology Core Facility at UNMC, reverse-phase HPLC purified, desalted on a G-10 Sephadex column, and lyophilized to dryness prior to experiments. The concentration of each oligomer solution was determined from absorbance measurements at 260 nm and 80°C using the molar absorptivities in $mM^{-1}cm^{-1}$ of strands of Table 8.1. These values were calculated

Table 8.1. Sequences of Deoxyoligonucleotides Investigated

Deoxyoligonucleotide (5′ to 3′ Sequence)	Designation	ϵ (mM^{-1} cm^{-1})
Initial Hairpins		
$GA_3T_2C_5A_2T_3C$	*Hairpin*	164
$GA_3T_2GCGCT_5GCGCGTGCT_5GCACA_2T_3C$	*Hammer*	344
$CGCGCGT_4GA_3T_2CGCGCGT_4A_2T_3C$	*Pseudoknot*	309
Triplexes		
$A_7C_5T_7C_2$	*7ATDu*	197
$A_4C_5T_4C_5T_4$	*4TAT*	195
$A_5C_5T_5C_5T_5$	*5TAT*	220
$A_6C_5T_6C_5T_6$	*6TAT*	246
$A_7C_5T_7C_5T_7$	*7TAT*	273
$A_8C_5T_8C_5T_8$	*8TAT*	307
$A_9C_5T_9C_5T_9$	*9TAT*	336
$A_{11}C_5T_{11}C_5T_{11}$	*11TAT*	394
Triplex–Duplex Complexes		
$A_6GCGCT_5GCGCT_6C_5T_6$	*Complex-1*	317
$A_6GCGCT_5GCGCT_6C_5T_6C$	*Complex-2*	324
$AGA_4GCGCT_5GCGCT_4CTC_5TCT_4$	*Complex-3*	319
$AGAGA_2GCGCT_5GCGCT_2CTCTC_5TCTCT_2$	*Complex-4*	316
$GCGCT_5GCGC$	*GCGChp*	108
$AGCGCT_5GCGCT$	*AGCGChp*	131
G–Quadruplexes		
$G_2T_2G_2TGTG_2T_2G_2$	*G2*	146
$G_3T_2G_3TGTG_3T_2G_3$	*G3*	191
i-Motifs		
$C_3TA_2C_3TA_2C_3TA_2C_3TA_2$	*C3TA2*	220

by extrapolation of the tabulated values of the dimers and monomer bases from 25°C to high temperatures, using procedures reported earlier [38,39]. Buffer solutions consisted of 10 mM sodium phosphate buffer adjusted to different sodium concentrations with NaCl or KCl. All chemicals used in this study were reagent grade. We also use other buffers, such as a 10 mM sodium acetate buffer or 100 mM NaCl at pH 4.2 or 4.8.

8.2.2 Overall Experimental Protocol

Initially, we use temperature-dependent UV spectroscopy to measure the UV unfolding of each molecule. Analysis of the resulting UV melting curves yields transition temperatures that indicate the thermal stability of the molecule. We then test if the folding of each DNA complex takes place intramolecularly, by following the dependence of the transition temperature on strand concentration; if the transition temperature remains constant, complex formation is intramolecular. The next step is to use circular dichroism spectroscopy to (a) check for the particular conformation that each complex adopts and (b) confirm the formation of specific complexes using the spectral characteristics or physical signatures of each type of complex. Then, we use differential scanning calorimetry to determine complete thermodynamic profiles for the temperature unfolding of each DNA complex. Finally, the measurement of the differential binding of counterions and/or protons between the helical and random coil states is done from the dependence of the transition temperatures on salt and pH, respectively, obtained in UV melting curves, and the calorimetric melting curves, as described in the following sections.

8.2.3 Temperature-Dependent UV Spectroscopy

Absorbance versus temperature profiles (UV melting curves) as a function of strand, salt concentration, and/or pH were measured at 260 nm with a thermoelectrically controlled Aviv 14 DS UV–vis spectrophotometer (Lakewood, New Jersey). The absorbance was scanned with a temperature ramp of $\sim$0.4°C/min. The analysis of the shape of the melting curves yielded transition temperatures, T_M, which correspond to the inflection point of the order–disorder transitions. To determine the molecularity of the transition(s) of each DNA complex, we investigated the dependence of T_M over at least a 10-fold range of total strand concentration, 4–80 μM. If the T_M remains constant in this range of strand concentration, this indicates a monomolecular or intramolecular transition; otherwise the transition takes place with higher molecularity.

8.2.4 Circular Dichroism Spectroscopy (CD)

We use a thermoelectrically controlled Aviv circular dichroism spectrometer Model 202SF (Lakewood, New Jersey) to measure the CD spectrum of each oligonucleotide. The analysis of these spectra yielded the conformation adopted by the helical state of each oligonucleotide. Typically, we prepared an oligonucleotide solution with an absorbance of 1 (total strand concentration of $\sim$4 μM) in appropriate buffered solutions, and the CD spectrum is measured from 320 to 200 nm every 1 nm, using a strained free quartz cuvette with a pathlength of 1 cm, and at temperatures that the particular oligonucleotide is 100% in the helical state. The reported spectra correspond to an average of at least two scans.

8.2.5 Differential Scanning Calorimetry (DSC)

Heat capacity functions of the helix-coil transition of each oligonucleotide were measured with a Microcal VP-DSC (Northampton, MA) instrument. Two cells, the sample cell containing 0.5 ml of a DNA solution ($\sim$0.15 mM in total strands) and the reference cell filled with the same volume of buffer solution, were heated from 0°C to 90°C at a heating rate of 0.75°C/min. Analysis of the resulting thermographs yielded T_{MS} and standard thermodynamic unfolding profiles: ΔH_{cal}, ΔS_{cal}, and $\Delta G^{\circ}_{(T)}$ [40]. These parameters are measured from DSC experiments using the following relationships: $\Delta H_{cal} = \int \Delta C_p^a \, dT$ and $\Delta S_{cal} = \int (\Delta C_p^a)/T \, dT$, where ΔC_p^a represents the anomalous heat capacity during the unfolding process [40]. The free energy, $\Delta G^{\circ}_{(T)}$, is obtained at any temperature with the Gibbs relationship: $\Delta G^{\circ}_{(T)} = \Delta H_{cal} - T \Delta S_{cal}$. Alternatively, $\Delta G^{\circ}_{(T)}$ can be obtained using the following equation: $\Delta G^{\circ}_{(T)} = \Delta H_{cal} (1 - T/T_M)$, which can be applied rigorously for intramolecular transitions. For multiphasic unfolding curves, the heat associated with each transition are obtained using a non-two-state transition deconvolution procedure of the Origin software, provided by the VP-DSC instrument.

8.2.6 Differential Binding of Counterions and Protons

The helical and coil states of an oligonucleotide are associated with a different number of bound ions and protons; therefore, their Helix $\rightarrow$ Coil transition is accompanied by a differential release (or uptake) of counterions and protons. The differential release (or uptake) of each of these species can be measured experimentally using the following reaction and Na^+ ions as an example:

$$\text{Helix}(a\text{Na}^+, b\text{H}^+) \rightarrow \text{Coil}(x\text{Na}^+, y\text{H}^+) + \Delta n_{Na^+} \text{Na}^+ + \Delta n_{H^+} \text{H}^+,$$

where $\Delta n_{Na^+} = x - a$ and $\Delta n_{H^+} = y - b$; each of these terms corresponds to the differential binding of counterions and protons, respectively, and is written on the right-hand side of the reaction to indicate "releases"; however, if there is an "uptake" of any of these species, they should be written on the left-hand side.

The corresponding reaction constant can be written as

$$K = \{(\text{Coil})/(\text{Helix})\}(\text{Na}^+)^{\Delta n_{Na^+}} (\text{H}^+)^{\Delta n_{H^+}}. \qquad (8.1)$$

Using a logarithmic function for simplicity, we obtain $\ln K = \ln K\{T, P, \ln(\text{Na}^+), \ln(\text{H}^+)\}$; its total differential is $d \ln K = (\partial \ln K/\partial T) \, dT + (\partial \ln K/\partial P) \, dP + (\partial \ln K/\partial \ln(\text{Na}^+)) d\ln(\text{Na}^+) + (\partial \ln K/\partial \ln(\text{H}^+)) d\ln(\text{H}^+)$, the last two partial differentials correspond to the desired quantities or linking numbers at constant T and P: $\Delta n_{Na^+} = (\partial \ln K / \partial \ln(\text{Na}^+))_{(\text{H}^+)}$ and $\Delta n_{H^+} = (\partial \ln K / \partial \ln(\text{H}^+))_{(\text{Na}^+)}$.

These two linking numbers are measured experimentally with the assumption that counterion or proton binding to the helical and coil states of the oligonucleotide takes place with a similar type of binding. Applying the chain rule to each partial differential and converting ionic activities to concentrations and natural logarithms for decimal logarithms, we obtain [41]

$$\Delta n_{Na^+} = (\partial \ln K / \partial T_M)[\partial T_M / \partial \ln(Na^+)] = 0.478[\Delta H_{cal}/RT_M^2](\partial T_M / \partial \log[Na^+])$$

(8.2)

and

$$\Delta n_{H^+} = (\partial \ln K / T_M)[\partial T_M / \partial \ln(H^+)] = -0.434[\Delta H_{cal}/RT_M^2](\partial T_M / \partial pH).$$ (8.3)

The first term in brackets of Eqs. (8.2) and (8.3), $[\Delta H_{cal}/RT_M^2]$, is a constant determined directly in differential scanning calorimetric experiments, and R is the gas constant; the second term in parentheses is also determined experimentally from UV melting curves or DSC curves, from the T_M-dependences on the concentration of counterions and protons, respectively. In the determination of Δn_{Na^+}, UV melting curves were measured in the salt range of 10 mM to 0.2 M NaCl, while in the determination of Δn_{H^+} for the unfolding of the i-motif structures, DSC melting experiments were carried out in the pH range of 4.2 to 6.0 and 100 mM NaCl.

8.3 MELTING BEHAVIOR OF DNA DUPLEXES CONTAINING LOOPS

Our current understanding of the structures and stability of DNA and RNA has been enhanced by thermodynamic investigations of the helix–coil transitions of model compounds, such as oligonucleotides of known sequence [34–37]. Our laboratory is currently interested in understanding the unfolding of single-stranded DNA oligomers containing a variety of secondary structures. The helix–coil transition of their melting domains takes place at relatively high temperatures, which are higher than their intermolecular counterparts. The reason for this is that the formation of intramolecular duplexes takes place with a lower entropy cost, allowing investigations of the physical properties of their 100% helical conformations over a wider temperature range. In this section, we present a full thermodynamic description of the melting behavior of three DNA oligonucleotides forming single-stranded DNA secondary structures. Specifically, we used a combination of UV and circular dichroism spectroscopies and calorimetric techniques to investigate the unfolding thermodynamics of a single-stranded hairpin (*Hairpin*), three-way junction (*Hammer*), and DNA pseudoknot (*Pseudoknot*) [42–44]. The sequences and designations of these molecules are shown in Figure 8.1. These sequences were designed in such a way to favor the exclusive formation of intramolecular secondary structures; for instance, all three molecules contained a common complementary stem with sequence AAATT/TTTAA that helps to identify their proper folding. The results clearly demonstrate that all three oligonucleotides form stable intramolecular structures with the expected melting domains.

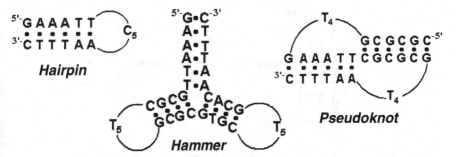

Figure 8.1. Sequences, designation, and cartoon of intramolecular complexes.

8.3.1 Temperature-Dependent UV Spectroscopy (UV Melting Curves)

The helix-coil transition of each oligonucleotide (Figure 8.1) was characterized initially by UV melting curves (Figure 8.2a). All three UV melting curves show the characteristic sigmoidal melting behavior of oligonucleotide duplexes, which indicates the cooperative unfolding of base pairs and base-pair stacks; however, *Hairpin* shows one transition, *Hammer* has two to three transitions, while *Pseudoknot* shows two distinctive transitions. The T_Ms for the transitions of each oligonucleotide, listed in Table 8.2, remain constant despite the ~50-fold increase in strand concentration (Figure 8.2b). This is consistent with the intramolecular unfolding of each oligonucleotide. The magnitude of the T_Ms of the transitions of each hairpin molecule corresponds roughly to the percentage of GC base pairs of their helical stems or melting domains. Furthermore, the T_M of the first transition of each hairpin corresponds to the unfolding of their GA_3T_2/CT_3A_2 stem. We obtained T_Ms of 35.8°C (*Hairpin*), 31.3°C (*Hammer*) and 45°C (*Pseudoknot*). The analysis of the shape of this transition for each hairpin, using standard procedures [40], yielded the following model-dependent or van't Hoff enthalpies: -37 kcal mol^{-1} (*Hairpin*); -26 kcal mol^{-1} (*Hammer*), and -12 kcal mol^{-1} (*Pseudoknot*). These values clearly indicate the effects of the adjacent secondary structure of this stem, C_5 end loops, three-way crossover junction, and T_4 loops, respectively, in reducing the enthalpy contributions. The CD spectra at 15°C are shown in Figure 8.2c, and each hairpin adopts the B-like conformation as seen by the similar magnitudes between the longer wavelength positive band and the negative band centered at 250 nm with a crossover at ~262 nm.

8.3.2 Differential Scanning Calorimetry

Typical DSC profiles are shown in Figure 8.3a. These curves clearly show that their unfolding is monophasic (*Hairpin*), triphasic (*Hammer*), and biphasic (*Pseudoknot*), consistent with the observations of the UV melting curves. The resulting thermodynamic profiles for all transitions observed in the folding of each oligonucleotide are listed in Table 8.2. The thermodynamic parameters of each transition show that the favorable free energy contribution results from the characteristic compensation of a favorable enthalpy contribution with an unfavorable entropy contribution. Favorable heat contributions

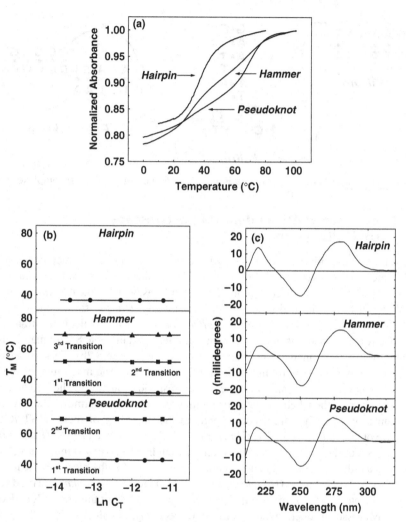

Figure 8.2. (a) Typical UV-melting curves of hairpin molecules. (b) T_M-dependence on strand concentration. (c) CD spectra of hairpin molecules at 10°C.

invoke the formation of base pairs and base pair stacks, while unfavorable entropy contributions involve the ordering of the strands, ion, and water binding. The enthalpy contribution of each transition are in good agreement with the expected enthalpy values from nearest-neighbor calculations [34, 36]. For instance, in 0.1 M NaCl we obtained enthalpy contributions of 36.7 kcal mol^{-1} (*Hairpin*) and 36.4 kcal/mol for the first transition of *Hammer*, which are in excellent agreement with the value of 39.9 kcal mol^{-1} estimated from nearest-neighbor parameters [34, 36] in 1 M NaCl; however, the enthalpy of the first transition of *Pseudoknot* is much lower, by 9.3 kcal mol^{-1}, in spite of having similar helical sequence. The overall effects are explained in terms of the actual

Table 8.2. Thermodynamic Profiles for the Formation of Hairpins at 15°C[a]

Transition	T_M (°C)	ΔH_{cal} (kcal mol^{-1})	$T\Delta S_{cal}$ (kcal mol^{-1})	$\Delta G°_{(15)}$ (kcal mol^{-1})	Δn_{Na+} (mol Na$^+$/mol)
		Hairpin			
	35.8	−36.7	−34.8	−2.5	0.063
		Hammer			
First	31.3	−36.4	−34.5	−1.9	0.027
Second	51.3	−23.6	−21.0	−2.6	—
Third	69.6	−36.2	−30.4	−5.8	0.130
		Pseudoknot			
First	45.0	−27.2	−24.9	−2.3	0.031
Second	70.6	−48.6	−40.8	−7.8	0.100

[a]All experiments were measured in 10 mM sodium phosphate buffer at pH 7. Experimental uncertainties are as follows: T_M (±0.5°C), ΔH_{cal} (±3%), $T\Delta S_{cal}$ (±3%), $\Delta G°_{(15)}$ (±5%), and Δn_{Na+} (±6%).

contribution of the adjacent secondary structure, in the case of *Hairpin* and *Hammer*, the loop and crossover junction has no effect, while the T_4 loop of *Pseudoknot* lies on the ceiling of the sugar–phosphate backbone that removes water, and it is this dehydration effect that reduces its enthalpy contribution.

In terms of the overall free energy contribution of folding each hairpin at 15°C, we obtained $\Delta G°$ contributions of −2.5 kcal mol^{-1} (*Hairpin*), −10.3 kcal mol^{-1} (*Hammer*), and −10.1 kcal mol^{-1} (*Pseudoknot*). This indicates that each molecule folds with their particular secondary structure at 15°C, and the overall thermodynamic profiles for each transition of these molecules are consistent with the sequences of their melting domains. One experimental observation is that the helical and coil states of *Hammer* and *Pseudoknot* takes place at temperatures of up to 18°C and past 95°C, respectively, in spite of having different melting behavior in the temperature range of 18–95°C. The entropy contributions, in energy terms ($T\Delta S_{cal}$), for the formation of each melting domain are shown in the third column of Table 8.2. The magnitude of these unfavorable entropy terms follow the same order as those of the enthalpies and correspond to contributions of the ordering of the oligonucleotide, uptake of counterions, and the immobilization of water molecules by the helical states of each hairpin. Of these three contributions, we have only estimated the uptake of counterions, which will be discussed next.

We measured the uptake of counterions [using Eq. (8.2)], from the T_M-dependence on salt concentration (Figure 8.3b) and from the $\Delta H_{cal}/RT_M^2$ terms obtained in the calorimetric melting curves. For each transition, the increase in the concentration of Na$^+$ from 16 to 116 mM results in the shift of melting curves to higher temperatures (data not shown). The increase in salt concentration shifts the hairpin–random coil equilibrium toward the conformation with higher charge density parameter. The T_M-dependence on salt concentration for each hairpin is shown in Figure 8.3b (with the exception of the second transition of *Hammer*, which was difficult to determine), and linear dependences are obtained with slope values ranging from 2.3°C to 2.9°C. The resulting Δn_{Na}^+ values

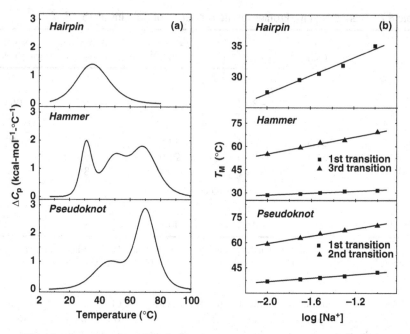

Figure 8.3. (a) Heat capacity profiles for the hairpin molecules in 10 mM sodium phosphate buffer at pH 7. (b) T_M-dependence on salt concentration for all transitions of each hairpin (with the exception of the second transition of *Hammer*, which was not determined).

are shown in the last column of Table 8.2. The magnitude of these values indicates the strength of ion binding by the helical state of each molecule, assuming that the coil states have similar ion binding. Relative to the value of 0.17 mol Na^+/mol phosphate for longer duplex DNA molecules, these values are low but consistent with the formation of five base pairs in the hairpin stems [45]. One significant observation is the higher uptake of counterions by helical stems with exclusively G·C base pairs, which is consistent with the X-ray crystallography of Loren William's laboratory indicating that the major groove of G·C base pairs are preferred ion binding sites [46].

8.4 UNFOLDING THERMODYNAMICS OF INTRAMOLECULAR DNA TRIPLEXES

The antisense and antigene strategies use oligonucleotides to modulate the expression of genes; their interaction with their targets takes place with tight affinity and specificity [47]. In the antisense approach, an oligonucleotide with a proper length binds to a messenger RNA, forming a DNA/RNA hybrid duplex. The formation of this hybrid inhibits translation of a given gene by sterically blocking the correct assembly of the translation machinery or by inducing an RNAse H mediated cleavage of their mRNA target [47]. In the antigene approach, an oligonucleotide binds into the major groove of a DNA duplex, forming a local triple helix [48, 49]. The third strand competes with the proteins that activate the transcription machinery by forming a triplex complex, thus inhibiting the expression of a specific gene [50, 51]. Since the antigene strategy targets the gene directly, one advantage of this approach over its antisense counterpart is the fewer number of copies of the gene of interest. In addition, blocking the transcription of a gene itself prevents the turnover of the mRNA pool, allowing a more efficient and lasting inhibition of gene expression [52]. However, disadvantages for this strategy are that the oligonucleotide needs to also cross the nuclear membrane and access the targeted gene within the densely packed chromatin structure [53, 54]. Another problem of both strategies is that oligonucleotides are subject to degradation by nucleases. To prevent this nuclease activity, our laboratory is investigating intramolecular triplexes. The design of these intramolecular triplexes involves the inclusion of two hairpin loops that will render them slightly more hydrophobic and compact, thus improving their interaction with hydrophobic micelles. The micelles in turn protect the oligonucleotide from nuclease degradation and improve their cellular delivery. These double-hairpin triplexes are used to target duplex DNA, via triplex formation.

Furthermore, the interaction of a third strand with a DNA duplex is sequence-specific, allowing the recognition of different DNA targets by Hoogsteen base pairing [48, 49]. These triplexes are stabilized by base-triplet stacking interactions and hydrogen bonding between the bases of the third strand and the purine bases of the duplex [55]. Triplexes are classified based on the composition and orientation of their third strand, which is placed asymmetrically in the major groove of the target duplex and hydrogen-bonded to the purine strand. The third strand of "Pyrimidine" triplexes is pyrimidine-rich and binds parallel to the purine-strand of the duplex. On the other hand, the third strand of "Purine" triplexes is purine-rich and binds the duplex purine strand in the antiparallel orientation [49, 56, 57]. In addition, it is required that a duplex have a homopurine/homopyrimidine sequence in order for the third strand to bind and thereby form a stable triplex. Hence, in order to develop oligonucleotide-forming triplexes as therapeutic agents, it is necessary to have a clear understanding on how the sequence, base composition, and solution conditions affect triplex stability. To further our understanding on the physical chemical properties of DNA triplexes, we have studied a set of intramolecular triplexes (Figure 8.4) with exclusively TAT base triplets. Furthermore, these intramolecular triplexes resemble that of H-DNA without the dangling strand, which can potentially form due to the high frequency of homopurine/homopyrimidine sequences in the human genome [58]. Sequences capable of forming H-DNA are present in regions that function in transfection, replication, recombination, and regulation of some eukaryotic genes [49, 52]. In the following sections, we report on the unfolding of intramolecular triplexes of the "Pyrimidine" motif with sequence $(A_nC_5T_nC_5T_n)$, where n is 4–9 and 11, as shown in Figure 8.4.

8.4.1 Circular Dichroism Spectroscopy

The CD spectra (Figure 8.5) indicates that all triplexes and the control duplex adopt the "B" conformation; the positive band at $\sim$277 nm has a magnitude comparable to that of the negative band at $\sim$246 nm. This indicates that binding of the third strand does not

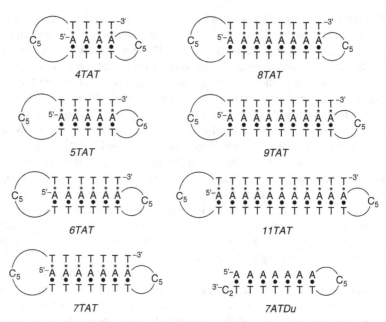

Figure 8.4. Cartoon of the structures of triplexes of the "Pyrimidine" motif and their designations.

impose major distortions in the duplex geometry. The small split of the positive band at 260–265 nm is characteristic of homopurine/homopyrimidine sequences (A_n/T_n). All triplexes exhibit increased negative bands at 210 nm and 247 nm, when compared to the control duplex (*7TADp*). The 210-nm band is considered a physical signature of triplex

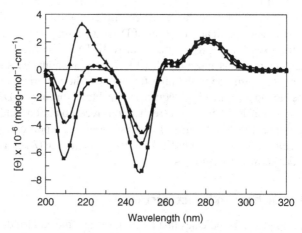

Figure 8.5. Typical CD spectra of triplexes in 10 mM sodium phosphate buffer, 1.0 M NaCl, at pH 7; *7TAT* (circles), *9TAT* (squares), and *7ATDu* (triangles).

formation [59–61], while the magnitude of the band at 247 nm is proportional to the extent of base-pair stacking contributions.

8.4.2 DSC Melting Curves

The helix–coil transition of each triplex and control duplex is characterized initially by UV melting techniques as a function of strand concentration (data not shown). All molecules yielded similar T_Ms over a 100-fold increase in strand concentration, confirming their intramolecular formation [62].

DSC experiments were conducted at two different NaCl concentrations, 0.1 M and 1.0 M, at pH 7, as shown in Figures 8.6a and 8.6b. The DSC melts indicate that these molecules unfold in monophasic (triplex or duplex→random coil) or biphasic (triplex→duplex→random coil) transitions, depending on the number of TAT base triplets on the stem. In general, the incremental incorporation of TAT base triplets yielded triplexes with larger unfolding heats and higher T_Ms (Figure 8.6c). For all triplexes, the increase in the salt concentration yielded DSC melting curves with higher T_Ms and

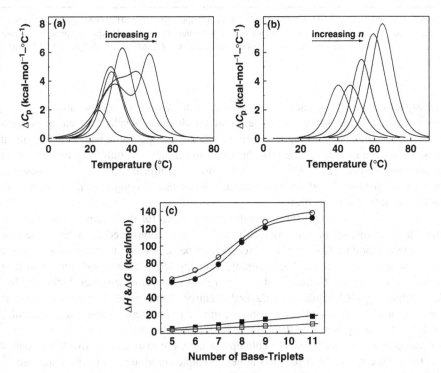

Figure 8.6. DSC melting curves of triplexes in 10 mM sodium phosphate buffer at pH 7.0, and 0.1 M NaCl (a) or 1 M NaCl (b). (c) The dependence of the total enthalpy (circles) and free energy at 20°C (squares) on the number of base triplets in 0.1 M NaCl (open symbols) and 1.0 M NaCl (solid symbols).

BUILDING BLOCKS OF NUCLEIC ACID NANOSTRUCTURES

Table 8.3. Thermodynamic Profiles for the Folding of Triplexes at 20°C[a]

Triplex	T_M (°C)	ΔH_{cal} (kcal mol^{-1})	$T\Delta S_{cal}$ (kcal mol^{-1})	$\Delta G^{\circ}_{(20)}$ (kcal mol^{-1})
4TAT	24.3	−21.1	−20.8	−0.3
5TAT	30.6	−61.0	−58.9	−2.1
	(40.6)	(−57.7)	(−53.9)	(−3.8)
6TAT	30.2	−71.6	−69.2	−2.4
	(47.0)	(−61.1)	(−55.9)	(−5.2)
7TAT	35.8	−86.7	−82.3	−4.4
	(53.2)	(−78.2)	(−70.2)	(−8.0)
8TAT	35.2	−106.0	−100.8	−5.2
	(56.6)	(−104.0)	(−92.5)	(−11.5)
9TAT	33.5	−97.2	−92.9	−4.3
	44.3	−30.4	−28.1	−2.3
	(59.8)	(−121.0)	(−106.5)	(−14.5)
11TAT	32.0	−74.7	−71.8	−2.9
	49.2	−63.0	−57.3	−5.7
	(64.7)	(−132.0)	(−114.5)	(−17.5)

[a]All DSC experiments were measured in 10 mM sodium phosphate buffer at pH 7 with 0.1 M or 1.0 M NaCl (parentheses). The parameters of the two transitions of 9TAT and 11TAT in 0.1 M NaCl are presented in two separate rows. Experimental uncertainties are as follows: T_M (±0.5°C), ΔH_{cal} (±3%), $T\Delta S_{cal}$ (±3%), and $\Delta G^{\circ}_{(20)}$ (± 5%).

slightly lower enthalpies. The higher T_Ms with the increased in salt concentration is consistent with the higher screening of the negatively charged phosphates, yielding perhaps a net improvement of stacking interactions which is compensated with a displacement of water molecules [62]. On the other hand, the lowering of the unfolding enthalpies, by an average of ∼6.1 kcal mol^{-1}, suggests the presence of heat capacity effects, possibly because the increase in sodium lowers the activity of water, enough to change the overall hydration state of these triplexes.

Table 8.3 shows complete thermodynamic profiles for the folding of all oligonucleotides; the folding of each molecule at 20°C is accompanied by a favorable free energy term resulting from the characteristic compensation of a favorable enthalpy and unfavorable entropy terms. The favorable enthalpy terms correspond mainly to heat contributions from the formation of base-pair stacks, and/or base-triplet stacks, and hydrogen bonding [60], while the unfavorable entropy terms arise from contributions of the unfavorable ordering of strands and the uptake of counterions and water molecules [60]. Further analysis of the data (Table 8.3) shows that the driving force for triplex formation is a large and favorable enthalpy change. For instance, relative to its control 7ATDu duplex, the 7TAT triplex has larger enthalpic contributions, by 39.3 kcal mol^{-1} (0.1 M NaCl) and 35.4 kcal mol^{-1} (1 M NaCl), which is due to the association of the third strand. This enthalpic contribution can be explained by the formation of Hoogsteen hydrogen bonds and base stacking interactions among the pyrimidines of the third strand, consistent with the formation of a triple helix. In this salt range, an average enthalpy of 6.2 kcal mol^{-1} is estimated for a single T/T base stack of the thymine strand in the triple-helical state.

The incremental incorporation of TAT base-triplets upon triplex formation (Figure 8.6) yielded increases in the T_M and total ΔH, which is consistent with the formation of additional base-triplet stacks, Table 8.3. Specifically, in 0.1 M NaCl (Figure 8.6a) the melting behavior of 5TAT, 6TAT, 7TAT, and 8TAT is monophasic, while 9TAT and 11TAT showed biphasic transitions. However, upon increase of salt concentration to 1.0 M NaCl (Figure 8.6b), 9TAT and 11TAT triplexes change their melting behavior from biphasic to monophasic, indicating that salt induces the formation of stronger triplexes. The T_M and total enthalpy values increased with increasing the number of base-triplets; however, this stepwise increment is not linear and follows a sigmoidal curve as shown in Figure 8.6c. The increment in enthalpy from triplex 5TAT to 6TAT is small ($\sim$10.6 kcal mol^{-1}), becomes larger 7TAT through 9TAT ($\sim$20 kcal mol^{-1}), and levels off between 9TAT and 11TAT ($\sim$10 kcal mol^{-1}). These observations are explained in terms of the folding free energy terms, as follows: Triplexes with less than four TAT base triplets do not form because this short stem is not long enough to overcome the entropy-driven nucleation step. The intercept of the two lines with the free energy axis of Figure 8.6c takes place at four TAT base triplets. The DSC melting curve of the potential triplex with four TAT base triplets indicates that this triplex does not form; the magnitude of the enthalpy (21 kcal mol^{-1}) corresponds to the unfolding of a monomolecular duplex with four AT base pairs; thus, confirming our observation of Figure 8.6c. However, 5TAT overcomes this nucleation step with a compensating enthalpy contribution; thus, the observed enthalpy is less favorable. As the formation of TAT base triplets increases, the enthalpy contribution for a single step becomes larger than its counterpart entropic contribution. The additional incorporation of base triplets to 9, and then 11, may cause the formation of triplexes with a lower number of TAT/TAT base triplet stacks, presumably by slipping the third strand toward the formation of larger loops; that is, the loops will also include thymine residues.

To calculate the enthalpic contribution of a single base triplet stack, we use the stepwise incremental of the linear portion of Figure 8.6c; by averaging the heats of 7TAT, 8TAT, and 9TAT, we obtained an enthalpic contribution of $\sim$20.5 kcal mol^{-1} for the incorporation of an additional base triplet stack, in agreement with previous reports [63]. Furthermore, the increase in salt concentration yielded lower values of the unfolding enthalpies with higher T_Ms (Table 8.3) and indirectly shows an average heat capacity effect of -505 cal $^{\circ}$C^{-1} mol^{-1}. A negative heat capacity effect in the unfolding of a macromolecule usually indicates a burial of hydrophobic surfaces. However, the unfolded state of a nucleic acid triplex is a random coil where the polar/nonpolar bases are more exposed to the solvent, yielding a more hydrophobic strand than their triplex state, so we invoke changes on the hydration state of the folded triplexes. In this range of salt concentration, the increase in salt concentration lowers the activity of water which in turn reduces the overall hydration of both triple-helical and random coil states; however, salt has a larger effect on the helical state because of its higher charge density parameter; thus the net effect is most likely a displacement of water molecules that are immobilized around charges (electrostricted water) from the triple-helical state and since energy is needed to remove this type of water [64–67], a lower energy will be expected for the unfolding of these triplexes at 1 M NaCl. This displacement of water molecules by the additional association of sodium counterions is estimated from the average enthalpic difference of 6.1 kcal mol^{-1} (at 0.1 and 1 M NaCl) of all six triplexes and the energy (300 cal mol^{-1} of water) needed to remove electrostricted water [60]; we obtain a total average displacement of 20 water molecules/mol triplex. An alternative explanation is that the increase in salt stabilizes the W-C duplex of these molecules, destabilizing the association of the third strand; the net effect is an overall reduction of base–base stacking contributions in the third strand, which is consistent with a general rule on triplex formation that a strong duplex yields a weak triplex while a weak duplex yields a strong triplex [68].

Figure 8.6c also shows that the incorporation of additional TAT base triplets yields a linear increase in the $\Delta G°_{(20)}$ values and that the Y-intercept at $\Delta G°_{(20)} = 0$ occurs at about four TAT base triplets; this means that the nucleation step for these triplexes is a minimum of four base triplets or three TAT/TAT base triplet stacks. From the slopes of the two free energy lines (0.1 M and 1 M NaCl) of Figure 8.3c, we obtained a $\Delta G°_{(20)}$ of 2.5 kcal mol^{-1} for TAT/TAT base triplet stack, then we estimate a nucleation free energy, $\Delta G°_{nucleation}$, at 20°C of 7.5 kcal mol^{-1} compared to the nucleation free energy (2–3 kcal mol^{-1}) for the formation of DNA duplexes [36].

In conclusion, we have investigated the melting behavior of double hairpin triplexes of the pyrimidine motif containing exclusively TAT base triplets as a function of salt concentration. Their melting behavior is monophasic or biphasic depending on the number of TAT base triplets and salt concentration. The results are consistent with an enthalpy–entropy coupling between duplex and triplex formation. A minimum of five TAT base triplets is needed to form stable intramolecular triplexes in the B-like conformation; however, the increase of base triplets to 9 or 11 causes strand slipping generating triplexes with a lower number of TAT base triplets in the stem and larger cytosine loops.

8.5 DNA COMPLEXES CONTAINING TRIPLEX AND DUPLEX MOTIFS

In this section we will describe the calorimetric melting behavior of higher-order complexes, namely DNA complexes containing triplex and duplex motifs. Our starting molecule is the *6TAT* triplex that was described in the previous section. What we have done to obtain *Complex-1* (see Figure 8.7) is to substitute the C_5 loop of the duplex domain of *6TAT* with the hairpin loop, GCGCT$_5$GCGC (*GCGChp*). *Complex 1* (Figure 8.7) now has a duplex stem of 10 base pairs, A$_6$GCGC/GCGCT$_6$, closed with a T_5 end loop. *Complex-2* is similar to *Complex-1* but has an additional cytosine at the 3′ end forming a C$^+$GC base triplet (see Figure 8.7). The other two complexes, *Complex-3* and *Complex-4*, are similar to *Complex-1* with one and two TAT → C$^+$GC substitutions in their helical stems (Figure 8.7), respectively. Therefore, in the following sections the melting behavior of four triplex–duplex complexes is described.

We used a combination of spectroscopy and calorimetric techniques to investigate their unfolding thermodynamics. The helix–coil transition of each complex was

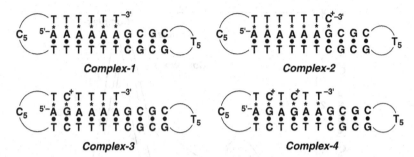

Figure 8.7. Cartoon of the structures of DNA complexes and their designations.

characterized initially by UV melting curves as a function of strand concentration (data not shown). The T_{MS} for each of the transitions of the complexes remain constant despite the ~20-fold increase in strand concentration, consistent with their intramolecular formation. Typical DSC profiles are shown in Figures 8.8 and 8.9. Clearly, the unfolding of *Complex-1* and *Complex-2* has three transitions (Figure 8.8), while the unfolding of *Complex-3* and *Complex-4* has two transitions (Figure 8.9). Thermodynamic profiles for the transitions of each complex are listed in Table 8.4. The thermodynamic parameters of each transition show that the favorable free energy contribution results from the characteristic enthalpy–entropy compensation. The T_{MS} of the first transition of each complex range from 31.4°C to 32.8°C, which are in good agreement with the T_M of 33.7 for the single transition observed in the unfolding of *6TAT*, which is also shown in Figure 8.8; however, the average enthalpies of 29.6 kcal mol^{-1} (*Complex-1* and *Complex-2*) and 22 kcal mol^{-1} (*Complex-3* and *Complex-4*) are lower than the enthalpy of 70.9 kcal

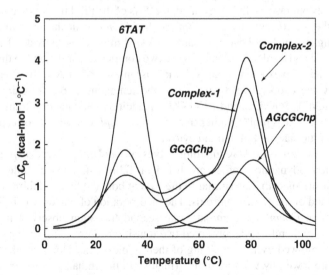

Figure 8.8. DSC melting curves of DNA complexes and of the control hairpin molecules in 10 mM sodium phosphate buffer at pH 7.0 and 0.2 M NaCl.

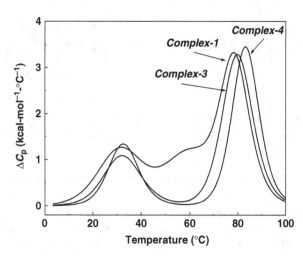

Figure 8.9. DSC curves of DNA complexes in 10 mM sodium phosphate buffer at pH 7.0 and 0.2 M NaCl.

mol^{-1} of *6TAT*. These effects can be seen clearly by inspecting the low temperature transitions of *Complex-1*, *Complex-2*, and *6TAT* shown in Figure 8.8, and the results clearly demonstrate that the stabilization of the duplex domain of each complex, by the incorporation of additional base pairs and/or incorporation of AT → GC base pairs, induce the dissociation of the third strand of the complexes.

Further inspection of Table 8.4 and Figure 8.8 shows that the T_Ms of the second transition of *Complex-1* and *Complex-2* are much higher than the T_M of *6TAT* with low enthalpy contributions; on the other hand, the T_Ms of the third transition are in between the T_Ms of the control hairpins, *GCGChp* and *AGCGChp*, while the unfolding enthalpies are much higher, by 13.8 kcal mol^{-1} and 25.3 kcal mol^{-1}, respectively. This indicates that the second and third transitions of these two complexes correspond to the sequential melting of their duplex stem—that is, initial disruption of 3–4 AA/TT base-pair stacks followed by the cooperative disruption of the remaining 5–6 base-pair stacks (1–2 AA/TT, 1 AG/CT, 2GC/GC, and 1 CG/CG). In addition, the total unfolding enthalpy of *Complex 2* is 8.7 kcal mol^{-1} higher than that of *Complex-1*, which corresponds to the disruption of the stacked third strand cytosine.

Figure 8.9 shows the DSC melting curves of *Complex-3* and *Complex-4* and the resulting thermodynamic profiles are listed in Table 8.4. Each complex unfolds in biphasic transitions with T_Ms similar to the T_Ms of both the first and third transitions of *Complex-1* and of the control hairpins. The incorporation of one or two TAT → C$^+$GC substitutions in *Complex-1* eliminates the second transition observed in *Complex-1* and *Complex-2;* therefore, the observed two transitions correspond to disruption of the third strand followed by the unfolding of their duplex stem. However, their unfolding enthalpies are lower, by 9.5 kcal mol^{-1} (relative to the enthalpy of removing the third strand from *Complex-1*) and by an average of 16.6 kcal mol^{-1}, relative to the enthalpy of the duplex domain that is calculated from nearest-neighbor data [34,36]. We can only

Table 8.4. Thermodynamic Profiles for the Folding of DNA Complexes and Hairpin Control Molecules at 5°C

Transition	T_M (°C)	ΔH_{cal} (kcal mol^{-1})	$T\Delta S_{cal}$ (kcal mol^{-1})	$\Delta G°_{(5)}$ (kcal mol^{-1})
		Complex-1		
First	32.0	−27.7	−25.2	−2.5
Second	60.2	−31.6	−26.4	−5.2
Third	78.6	−50.9	−40.2	−10.7
Total	—	−110.2	−91.8	−18.4
		Complex-2		
First	31.4	−31.5	−28.8	−2.7
Second	60.5	−25.0	−20.8	−4.2
Third	78.8	−62.4	−49.3	−13.1
Total	—	−118.9	−98.9	−20.0
		Complex-3		
First	32.5	−21.6	−19.7	−1.9
Second	80.0	−55.9	−44.0	−11.9
Total	—	−77.5	−63.7	−13.8
		Complex-4		
First	32.8	−22.4	−20.4	−2.0
Second	83.3	−51.7	−40.3	−11.4
Total	—	−74.1	−60.7	−13.4
		Control Molecules		
6TAT	33.7	−70.9	−64.3	−6.6
GCGChp	73.9	−36.3	−29.1	−7.2
AGCGChp	81.4	−37.9	−29.7	−8.2

All experiments were performed in 10 mM sodium phosphate buffer at pH 7 and 200 mM NaCl. Experimental errors are shown in parentheses: T_M (±0.5°C), ΔH_{cal} (±5%), $T\Delta S_{cal}$ (±5%), $\Delta G°_{(5)}$ (±7%).

explain the first discrepancy in terms of the solution conditions used in these experiments, pH 7 and 0.2 M NaCl; both conditions do not favor the stable formation of C$^+$GC base triplets; that is, lower pHs protonate the cytosines [41] better while counterions compete with protonated cytosines [60].

In terms of the overall free energy contribution for the folding of each complex at 5°C, we obtained $\Delta G°$s of −18.4 kcal mol^{-1} (*Complex-1*), −20.0 kcal mol^{-1} (*Complex-2*), and −13.6 kcal mol^{-1} (*Complex-3 and Complex-4*). These values are robust and indicate that each molecule folds with their particular triplex and duplex domains at 5°C. Furthermore, the overall thermodynamic profiles for each transition are consistent with their sequential unfolding of the triplex and duplex domains.

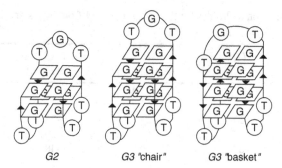

G2   G3 "chair"   G3 "basket"

Figure 8.10. Sequences and cartoons of the structure of G-quadruplexes.

8.6 MELTING BEHAVIOR OF G-QUADRUPLEXES

A tremendous amount of interest has been generated recently for the role of telomeres in maintaining the integrity of chromosomes and the correct replication of eukaryotic chromosomes [69–74]. Human telomeres are 100–150 nucleotide overhangs at the 3′ position that comprise G and T repeats of 12–16 bases [75–77]; the typical human sequence is the (TTAGGG) repeat [70,78]. The guanines in these overhangs tend to form G-quartets, which contain four coplanar guanines stabilized by Hoogsteen hydrogen bonding and the tight coordination of specific metal ions, such as K^+ [79], to O6 of guanine. The stacking of an array of G-quartets results in very stable structures called G-quadruplexes [80, 81]. Some other possible biological roles of G-quadruplexes include chromosomal alignment during recombination, DNA replication, and self-recognition, as well as regulation of gene transcription [82]. DNA aptamers also form G-quadruplexes; some examples include the thrombin-binding aptamer that inhibits thrombin-catalyzed fibrin clot formation [83, 84] and the HIV-binding aptamer, T30695 [85].

In this section we report the CD spectral characteristics and the melting behavior of a pair of intramolecular G-quadruplexes, G2 and G3, containing two and three G-quartets, respectively (Figure 8.10). The analysis of the experimental data allowed us to determine the molecular forces responsible for the high stability of these structures, the thermodynamic contributions of a single G-quartet stack, and the specific thermodynamic contributions of the loops in these structures.

8.6.1 Comformational Analysis of G-Quadruplexes by CD

CD spectroscopy with its low structural resolution provides the average solution conformation or ensemble conformation of model telomere sequences. Specifically, it yields the simple arrangement of guanines, parallel or antiparallel, in these quadruplexes. CD spectra are measured at several temperatures to determine the preferential conformation of the helical complex at lower temperatures and of the random coil state at higher temperatures. To prevent the aggregation of guanines and to obtain reversible and reproducible CD spectra, all solutions were heated to 90°C, cooled to 60°C slowly, kept at 60°C for 15 min, cooled to 2°C slowly, and maintained at this temperature for

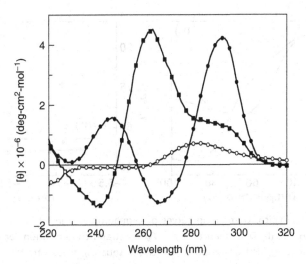

Figure 8.11. CD spectra of *G2* (circles) and *G3* (squares) in 10 mM Cs-HEPES buffer, 0.1 M K$^+$ at pH 7.5 and 20°C (using a ~6 μM in strands), *G2* at 90°C (open circles).

40 min, as reported earlier [86]. The CD spectra of *G2* and *G3* at low temperatures and of their random coils at high temperatures are shown in Figure 8.11. To assign the particular oligomer conformations and to classify what type of quadruplex is being formed, the various spectra were compared to the CD spectra of two control sequences (data not shown): d(G$_4$T$_4$G$_4$) and d(T$_4$G$_4$); the former forms a bimolecular quadruplex with guanines in the antiparallel orientation [87, 88], while the latter forms a tetramolecular quadruplex with guanines in the parallel orientation [77, 88]. Figure 8.11 indicates that the spectral characteristics of *G2* are those of the [G$_4$T$_4$G$_4$]$_2$ quadruplex; that is, *G2* has an antiparallel arrangement of guanines [89–91], which means that *G2* would need to adopt the "chair" conformation [80]. On the other hand, *G3* has two positive bands with spectral characteristics of both the antiparallel [G$_4$T$_4$G$_4$]$_2$ and the parallel [T$_4$G$_4$]$_4$ quadruplexes, indicating that *G3* is a mixture of the "chair" and "basket" conformers (see Figure 8.10).

8.6.2 UV and CD Unfolding of Complexes

Typical UV melting curves of *G2* and *G3* are shown in Figure 8.12a. All curves are sigmoidal and show a hypochromic effect at 297 nm. The helix–coil transition of *G2* takes place in monophasic transitions, while *G3* unfolded in a biphasic transition with hypochromicities of 18% and 28%, respectively. The T_Ms are in the order *G2* (52.6°C) ~ *G3* first < *G3* second (77.5°C), indicating the high stability of G-quadruplex structures in the presence of K$^+$. To check the molecularity of each complex, melting curves were obtained as a function of strand concentration; these T_M-dependences are shown in Figure 8.12b. Over a 10-fold range in strand concentration, the T_Ms remained constant, indicating that both quadruplexes formed intramolecularly.

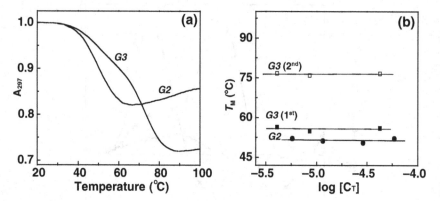

Figure 8.12. (a) Typical UV-melting curves of G-quadruplexes in 10 mM Cs-HEPES buffer, 0.1 M K$^+$ at pH 7.5. (b) Dependence of the T_M on strand concentration for each oligomer transition: *G2* (circles), first transition of *G3* (closed squares), second transition of *G3* (open squares).

8.6.3 DSC Unfolding

Typical DSC melting curves of *G2* and *G3* are shown in Figure 8.13a, *G2* shows a monophasic transition, and *G3* has a biphasic transition; the initial and final states have similar heat capacities values, indicating negligible heat capacity effects. Furthermore, the first transition of *G3* occurs 5°C above the single transition of *G2*, consistent with *G3* having one additional G-quartet. Since *G2* adopts the "chair" conformation, we assigned the first transition of *G3* to the unfolding of its "chair" conformer. This assignment is confirmed with the CD spectra of *G3* at several temperatures (Figure 8.13b), the ellipticity at 292 nm decreases at lower temperatures, and the ellipticity at 262 nm decreases at

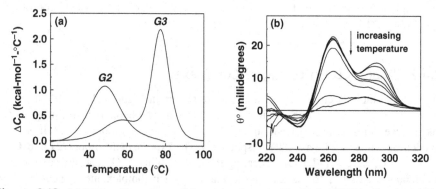

Figure 8.13. (a) DSC melting curves of G-quadruplexes in 10 mM Cs-HEPES buffer, 0.1 M K$^+$, at pH 7.5 and total strand concentration in the range of 40–100 μM. (b) CD spectra of *G3* as a function of temperature to determine the sequential melting of conformers.

Table 8.5. Thermodynamic Folding Profiles for G-Quadruplexes at 20°C

Conformer	T_M (°C)	ΔH_{cal} (kcal mol^{-1})	$T\Delta S_{cal}$ (kcal mol^{-1})	$\Delta G^{\circ}_{(20)}$ (kcal mol^{-1})	Δn_{K+} (per mol)
			G2		
"Chair"	52.6	−23.7	−21.6	−2.1	−0.4
			G3		
"Chair"	57.3	−38.3	−34.0	−4.3	−0.7
"Basket"	77.5	−36.5	−30.5	−6.0	−1.2

All experiments were performed in 10 mM Cs-Hepes buffer at pH 7.5 and 100 mM KCl. Experimental errors are shown in parentheses: T_M (±0.5°C), ΔH_{cal} (±5%), $T\Delta S_{cal}$ (±5%), $\Delta G^{\circ}_{(20)}$ (±7%).

higher temperatures. This shows that the "chair" conformer of *G3* has a lower thermal stability than the "basket" conformer.

The complete thermodynamic profiles for the folding of each quadruplex at 20 °C are shown in Table 8.5. The favorable ΔG° terms results from the characteristic compensation of a favorable enthalpy with an unfavorable entropy contributions. The favorable enthalpy terms are due primarily to the stacking of two or more G-quartets, along with additional contributions such as hydrogen bonding, base–base and/or base–quadruplex stacking of the loops. The immobilization of electrostricted water and the removal of structural water may also contribute with exothermic heats. The unfavorable entropy contribution is due to the ordering of a random coil into an intramolecular quadruplex, along with uptake of ions and water molecules.

8.6.4 K$^+$ Binding to G-Quadruplexes

UV melting curves as a function of the concentration of K$^+$ were performed in the 10–500 mM KCl range (data not shown) and the differential binding of K$^+$ was calculated using Eq. (8.2) and is summarized in Table 8.5. We obtained negative $\Delta n_K{}^+$ values for each quadruplex, indicating that quadruplex formation is accompanied by an uptake of K$^+$. The $\Delta n_K{}^+$ values of 0.4 mol K$^+$/mol of quadruplex and 0.7 mol K$^+$/mol of quadruplex were obtained for the "chair" conformer of *G2* and *G3*, respectively.

8.6.5 Thermodynamic Contributions of the Stacking of Two G-Quartets and of the Loops

The *G2* and *G3* quadruplexes have similar loop structures and differ only by one G-quartet. We determine the energetic contributions of a G-quartet stack from the thermodynamic profiles of the "chair" conformation of *G2* and *G3* [86]. This is done by simply subtracting each thermodynamic parameter of *G2* from those of *G3*, yielding the following thermodynamic profile at 20°C: $\Delta G^{\circ} = -2.2$ kcal/mol, $\Delta H_{cal} = -14.6$ kcal/mol, $T\Delta S_{cal} = -12.4$ kcal/mol, and $\Delta n_{K+} = -0.3$ mol K$^+$/mol. The favorable formation of a G-quartet stack is enthalpy-driven and is compensated with an unfavorable entropy contribution, due mainly to the immobilization of K$^+$ ions.

To determine the thermodynamic contributions of the combined two T_2 loops and of the single TGT loop of these quadruplexes [86], we subtracted each thermodynamic parameter of one or two G-quartet stacks from the experimental profile of each quadruplex. We obtained an average differential $\Delta G°$ term of 0.2 kcal/mol, which indicates that the incorporation of loops stabilizes the quadruplex structure. Using the differential enthalpy terms for the chair conformation, we calculated that all three loops contribute with an average favorable enthalpy term of -9.1 kcal/mol. This shows that the bases of these loops are stacked or constrained and/or these loops may well stack onto the surface of the G-quartets. If instead we use the enthalpy of the basket conformer of $G3$, we obtain a less favorable enthalpy difference of 1.8 kcal/mol, suggesting an extended conformation of the TGT loop. The entropy differences between $G2$ and each conformers of $G3$ range from 6 kcal mol^{-1} to 9 kcal mol^{-1}, which is due to the additional penalty in the ordering of the loops with the G-quadruplexes.

We would like to add that the *in vivo* control of the formation and dissociation of G-quadruplexes is an important strategy for developing novel therapeutics for the treatment of cancer and other diseases. Knowledge concerning the stabilities of these quadruplexes is also critical for understanding the formation, longevity, and resolution of these structures *in vivo*. The present study quantifies the magnitude of the thermodynamic forces (both enthalpic and entropic) that favor quadruplex formation. This information is useful in predicting the stabilities of putative quadruplex structures that have been identified based on the DNA sequence, but have not yet been evaluated experimentally [92]. Information concerning the magnitude of the forces stabilizing quadruplex structures will also be of particular value in predicting whether chemical approaches to manipulate quadruplex structures are likely to be effective at disrupting or stabilizing quadruplex structures *in vivo*.

8.7 UNFOLDING THERMODYNAMICS OF C-QUADRUPLEXES OR i-MOTIFS

Cytosine-rich sequences are frequently found in eukaryotic genomes located near or within regions of functional and/or regulatory importance, and they play important biological roles in replication, recombination, transcription, and chromatin organization [93]. Forty years ago, it was proposed that poly(dC) could form parallel duplexes [94] stabilized by C·C$^+$ base pairs at acidic pH, Figure 8.14. This particular base-pairing scheme has been confirmed with the crystal structure of acetyl cytosine [95], poly(dC), and poly(rC) [96, 97]. Cytosine-rich oligonucleotides can potentially adopt complex pH-dependent conformations; for instance, the crystal structure of d-TC$_5$ [98, 99], indicated that two parallel-stranded C·C$^+$ duplexes with an antiparallel orientation form a four-stranded structure. The base pairs of one C·C$^+$ duplex intercalate into those of the other duplex; hence, the structure is called the "i-motif" or intercalated motif (Figure 8.14); the base planes of the stacked C·C$^+$ base pairs are perpendicular to each other with all the bases in the anti-conformation. This type of alternating stacking of C·C$^+$ base pairs further stabilizes the i-motif [100]. The formation of i-motifs using different DNA oligonucleotides (ODNs) have been studied extensively by a variety of experimental

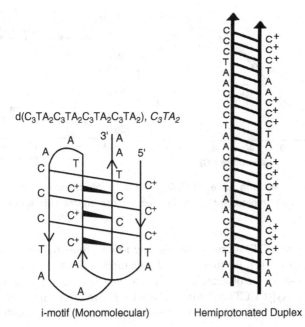

d($C_3TA_2C_3TA_2C_3TA_2C_3TA_2$), C_3TA_2

i-motif (Monomolecular)   Hemiprotonated Duplex

Figure 8.14. Sequence and cartoon of two of the possible conformations adopted by C_3TA_2 at acid pH.

techniques [98,99], [101–110]; however, calorimetric melting techniques have not been used to determinate model-independent unfolding thermodynamics.

In this section we report on the unfolding of the complementary sequence of the human telomere: [d(CCCTAA)]$_4$ (Figure 8.14). UV and DSC melting profiles show the following sequential melting: bimolecular complex(s) → intramolecular complex → random coil. From the thermodynamic profiles we were able to estimate the energetic contributions for the formation of C·C$^+$/C·C$^+$ base-pair stacks in each of these complexes at the optimum pH of 5.2.

8.7.1 UV Melting of C_3TA_2

The unfolding of C_3TA_2 was initially investigated by UV melting techniques, Figure 8.15a shows these UV melts as a function of strand concentration at the pH of 4.8. The resulting melting curves are biphasic and their T_M-dependences on strand concentration are shown in Figure 8.15b. Over a 10-fold concentration range, the T_M of the first transition increases with the increase in strand concentration, while the T_M of the second transition remains constant (Figure 8.15b). This indicates the formation of bimolecular and monomolecular complexes, respectively. Therefore, the UV melting behavior of C_3TA_2 is through the unfolding of a bimolecular complex at lower temperatures, followed by an intramolecular complex that yields the final random coil state at high temperatures. This observation is consistent with previous findings that cytosine-rich DNA sequences can adopt complex pH-dependent conformations. For instance, NMR solution studies

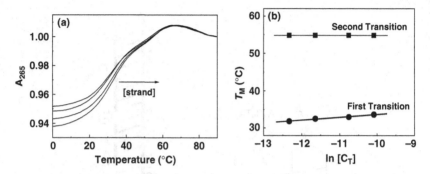

Figure 8.15. (a) UV melting curves of C_3TA_2 over a 10-fold range in strand concentration to determine transition molecularities. (b) T_M-dependence on strand concentration for each transition of C_3TA_2.

have shown that d(CCCTAA) can form multiple i-motif four-stranded complexes with different topologies [111]. Parallel-stranded structures with heteropyrimidine oligonucleotides at pH 4–5.5 were detected using IR, UV-melting, and fluorescence techniques [112]. The d([C$_3$TA$_2$]$_3$CCCT) and d([C$_4$A$_2$]$_3$CCCC) ODNs form intramolecular i-motif structures at acidic pH [101, 109]. All of these novel structures are stabilized by C·C$^+$ base pairs.

8.7.2 CD Spectroscopy

Figure 8.16 shows the CD spectra of C_3TA_2 at several pHs and at two different temperatures. At acidic pH, the CD spectra show a large positive band with a peak at 286 nm, along with a negative band centered at 254 nm. This type of spectrum has been attributed

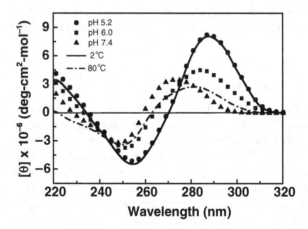

Figure 8.16. CD spectra of C_3TA_2 in 10 mM sodium cacodylate buffer as a function of pH at 2 °C (symbols), as well as in 10 mM sodium cacodylate buffer at pH 5.2 as a function of temperature (lines).

to the formation of $C \cdot C^+$ base pairs [113], characteristic of the i-motif structure [114]. As the pH is increased, a blue shift is observed followed by a sharp decrease in the magnitude of the ellipticity at 285 nm. Furthermore, the overlay of the CD spectra also shows an isodichroic point at 275 nm, suggesting the equilibrium of two species. The overall comparison of the CD spectra as a function of pH with the increase in temperature from 2°C to 80°C indicates that the spectral changes correspond to the destabilization of the complexes with the increase in pH. For instance, the spectrum of C_3TA_2 at pH 7.4 is similar to the one obtained at pH 5.2 and 80°C, which is assigned to the spectrum of the random coil state.

8.7.3 Differential Scanning Calorimetry

DSC melting curves were carried out to determine complete thermodynamic profiles for the unfolding of C_3TA_2. To ensure reproducibility of the DSC curves at the lower pHs and before a heating scan, we developed a protocol in which C_3TA_2 solutions were prepared in the DSC sample cell by first heating and keeping it at 90°C for 5 min, slowly cooled to room temperature, then kept at 4°C for several hours. All DSC scans were reproducible among samples undergoing this temperature protocol, and only these scans were used for the determination of thermodynamic profiles.

To confirm the molecularity of each complex, DSC melts of C_3TA_2 were carried out at pH 4.8 using two different total strand concentrations, 75 μM and 251 μM. The resulting DSC curves are shown in Figure 8.17a; each curve yielded biphasic transitions, and the T_M of the first transition increases with the increase of strand concentration, while the T_M of the second transition remains the same, confirming that the first transition corresponds to the unfolding of a bimolecular complex while the second transition involves the unfolding of an intramolecular complex. The curves yielded total heats of 1.25 mcal and 3.96 mcal that when normalized by the total strand concentration yielded similar

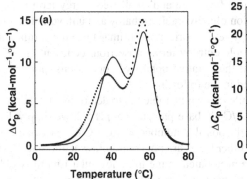
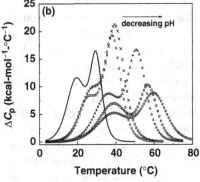

Figure 8.17. (a) DSC melting curves of C_3TA_2 in 10 mM sodium acetate buffer, 100 mM NaCl at pH 4.8, and strand concentration of 251 μM (solid line) and 75 μM (dotted line). (b) DSC melting profiles of C_3TA_2 in 10 mM sodium cacodylate buffer at pHs of 5.0, 5.2, 5.6, and 6.0, or in 10 mM sodium acetate buffer, 100 mM NaCl at pHs of 4.2 and 4.8.

Table 8.6. Thermodynamic Profiles for the Unfolding of C_3TA_2 at pH 5.2 and 10°C

T_M (°C)	ΔH_{cal} (kcal mol^{-1})	$T\Delta S_{cal}$ (kcal mol^{-1})	$\Delta G°_{(10)}$ (kcal mol^{-1})	$\Delta n_H{}^+$ (per mol)
		Bimolecular		
27.1	25.8	24.3	1.5	1.7
		i-motif		
39.9	25.0	22.6	2.4	2.3

All parameters measured in a 10 mM sodium cacodylate buffer at pH 5.2. The $\Delta n_H{}^+$ parameter was measured in the same buffer containing 100 mM NaCl. Experimental errors are shown in parentheses: T_M (±0.5°C), ΔH_{cal} (±5%), $T\Delta S_{cal}$ (±5%), $\Delta G°_{(10)}$ (±7%) and $\Delta n_H{}^+$ (±5%).

overall enthalpies, as seen in the similarity of the whole areas under these curves (Figure 8.17a). This confirms that the observed transitions correspond to the sequential transition of a bimolecular complex(s) → intramolecular → random coil state. This sequential melting behavior is similar to the one observed in the unfolding of the Dickerson–Drew dodecamer, which shows two transitions: duplex → hairpin → random coil at neutral pH [115]. This sequential melting can be explained in terms of the entropy penalty of complex formation, and consistent with conventional wisdom, that the bimolecular complex should form at lower temperatures, because of its higher unfavorable entropy contribution. We speculate that at low temperatures, only one type of bimolecular species is present, the hemiprotonated duplex or a bimolecular i-motif (Figure 8.14), but it is quite possible that both bimolecular complexes are present in equilibrium.

DSC melting curves as a function of pH are shown in Figure 8.17b. We obtained clear biphasic transitions at each pH, which show the formation of at least two protonated complexes. The T_M of each transition increases as the pH is decreased. DSC thermodynamic profiles for the formation of each complex (bimolecular and i-motif complexes) at the pH of 5.2 are shown in Table 8.6. These profiles show that the folding of each complex at 10°C is accompanied by a favorable and small free energy term resulting from the relatively large compensation of a favorable enthalpy and unfavorable entropy contributions. The favorable enthalpy terms correspond primarily to the formation of base-pair stacks, while the unfavorable entropy terms arise from contributions of the unfavorable ordering of strands and the putative uptake of counterions, protons, and water molecules, as will be discussed in the following section.

We use the thermodynamic profiles shown in Table 8.6 to determine the thermodynamic contributions for a single C·C$^+$/C·C$^+$ base-pair stack at pH 5.2; we simply divide the magnitude of each parameter by the possible number of C·C$^+$/C·C$^+$ base pair stacks of each structure, assuming that the other bases are not contributing at all. Since C_3TA_2 contains 12 cytosines, six should be protonated in the intramolecular i-motif, yielding five C·C$^+$/C·C$^+$ base-pairs stacks; the hemiprotonated duplex and bimolecular i-motif both contain eight and ten C·C$^+$/C·C$^+$ base-pair stacks, respectively. For the formation of a single C·C$^+$/C·C$^+$ base-pair stack at the pH of 5.2, we obtained free energy contributions of −0.2 kcal mol^{-1} (bimolecular complex) and −0.5 kcal mol^{-1} (i-motif), along with enthalpy contributions of −3.2 kcal mol^{-1} (hemiprotonated duplex) and

-5.0 kcal mol^{-1} (i-motif). As expected, the overall profiles are much smaller than the average thermodynamic profiles for the formation of a canonical W-C base-pair stack and consistent with the nonvertical stacking of the cytosine aromatic rings.

8.7.4 Differential Binding of Protons in the Helix–Coil Transitions of C_3TA_2

To obtain the T_M-dependence on pH for each transition, we extracted the T_Ms of the biphasic transitions shown in the DSC curves of Figure 8.17, and we obtained two sigmoidal curves (data not shown) with midpoints at pH 5.1 (bimolecular complex) and 5.0 (monomolecular i-motif) and negative slopes around the pH of these midpoints. This means, according to Eq. (8.3), that the unfolding of each complex is accompanied by a release of protons. The slopes determined around the midpoints, 40.7°C (bimolecular) and 34.5°C (i-motif), together with the $\Delta H_{cal}/RT_M^2]$ terms, measured by averaging the values of the DSC melts of each complex at pH 5 and 5.2, yielded average proton releases of 1.9 moles of proton per mole of DNA for each of these complexes. This release of protons indicates that complex formation is accompanied by the protonation of cytosines, which should be proportional to the difference in the pK_a values between folded and unfolded cytosines. In addition, marginal binding of counterions are obtained by the intramolecular i-motif complex (data not shown) [41]. The net binding of protons and marginal uptake of counterions in the formation of these complexes is in good agreement with earlier investigations of DNA triplexes containing C^+GC base triplets [63, 116], the positively charged cytosine bases excludes counterions due to the local and fixed charged density parameter [117]. At the pH range of 5 to 5.2, where optimum base-pair stacking interactions take place, we measured a similar proton uptake of 1.9 moles of protons per mole of DNA for each type of complex. If we use the Henderson–Hasselbach relationship and a pK_a of 4.6 for the N3 protonation site of cytosine in the random coil state [118], we estimate a 20% degree of cytosine protonation takes place at pH 5.2. In a similar exercise, but including the experimental values for the uptake of protons by each helical complex, we obtain 28% of cytosine protonation for each complex. This translates to a protonation of 3.3 cytosines out of the 6 needed for an optimum i-motif complex, along with 6.6 protonated cytosines out of 12 needed in the bimolecular complex.

In conclusion, we have investigated the unfolding thermodynamics of the structures formed by the cytosine-rich oligonucleotide [d(CCCTAA)]$_4$. UV and DSC melting profiles show the following sequential melting: bimolecular complex(s) $\rightarrow$ intramolecular complex $\rightarrow$ random coil. Acidic pH stabilizes the structures formed at low temperatures, by protonating the cytosines that in turn form $C \cdot C^+$ base pairs. We also determined standard thermodynamic profiles for the formation of $C \cdot C^+/C \cdot C^+$ base-pair stacks in each of these complexes at the optimum pH of 5.2. These results should help in studying the different structures formed by cytosine-rich sequences and associated equilibriums. Furthermore, it has been suggested that the DNA i-motif structure may have some biological relevance, due to the discovery of nuclear proteins that bind to cytosine-rich sequences containing repeated stretches of at least three cytosines [119, 121, 121].

8.8 CONCLUSIONS

We have reported the thermodynamic contributions on the stability and melting behavior of a variety of intramolecular DNA complexes, including hairpin loops, double hairpins, three-way junctions, pseudoknots, triplexes, G-quadruplexes, and C-quadruplexes. This information will improve our current picture of how sequence, loops, and ion binding control the stability and melting behavior of nucleic acid molecules. The resulting data will supplement existing nearest-neighbor thermodynamic parameters to help in the prediction of secondary structure from knowledge of its sequence. Furthermore, the data should help in the optimization of oligonucleotide reagents for the targeting of specific transient structures that form *in vivo*.

ACKNOWLEDGMENTS

This work was supported by Grant MCB-0315746 from the National Science Foundation.

REFERENCES

1. Johnson PF, McKnight SL. Eukaryotic transcriptional regulatory proteins. *Annu Rev Biochem* 1989;58:799–839.
2. Felsenfeld G, Davies DR, Rich A. Formation of a triple stranded polynucleotide molecule. *J Am Chem Soc* 1957;79:2023–2024.
3. Larsen A, Weintraub H. An altered DNA conformation detected by S1 Nuclease occurs at specific regions active chick globin chromatin. *Cell* 1982;29:609–622.
4. Mirkin SM, Lyamichev VI, Drushlyak KN, Dobrynin VN, Filippov SA, Frank-Kamenetski MD. DNA H form requires a homopurine–homopyrimidine mirror repeat. *Nature (London)* 1987;330:495–497.
5. Htun H, Dahlberg JE. Single strands, triple strands, and kinks in H-DNA. *Science* 1988;241:1791–1796.
6. Voloshin ON, Mirkin SM, Lyamichev VI, Belotserkovskii BP, Frank Kamenetskii MD. Chemical probing of homopurine–homopyrimidine mirror repeats in supercoiled DNA. *Nature (London)* 1988;333:475–476.
7. Shindo H, Matsumoto N, Shimizu M. Dynamics in the isomerization of intramolecular DNA triplexes in supercoiled plasmids. *Nucleic Acids Res* 1997;25:4786–4791.
8. Roberts RW, Crothers DM. Kinetic discrimination in the folding of intramolecular triple helices. *Mol Biol* 1996;260:135–146.
9. Moser HE, Dervan PB. Sequence specific cleavage of double helical DNA by triple helix formation. *Science* 1987;238:645–650.
10. Rajagopal P, Feigon J. Triple-strand formation in the homopurine: Homopyrimidine DNA oligonucleotides d(G-A)4 and d(T-C)4. *Nature (London)* 1989;339:637–640.
11. Rajagopal P, Feigon J. NMR studies of triple-strand formation from homopurine–homopyrimidine deoxiribonucleotides d(GA)4 and d(TC)4. *Biochemistry* 1989;28:7859–7870.

12. de los Santos C, Rosen M, Patel D. NMR studies of DNA (R+)n·(Y−)n· (Y+)n triple helices in solution: Imino and amino proton markers of T·A·T and C·G·C+ base-triple formation. *Biochemistry* 1989;28:7282–7289.

13. Strobel SA, Moser HE, Dervan PB. Double-strand cleavage of genomic DNA at a single site by triple helix formation. *J Am Chem Soc* 1988;110:7927–7929.

14. Sun JS, Francois JC, Montenay-Garestier T, Saison-Behmoaras T, Roig V, Thuong NT, Hélène C. Sequence specific intercalating agents: Intercalation at specific sequence on duplex DNA via major groove recognition by oligonucleotide-intercalator conjugates. *Proc Natl Acad Sci USA* 1989;86:9198–9202.

15. Hélène C, Thuong NT. Control of gene expression by oligonucleotides covalently linked to intercalating agents. *Genome* 1989;31:413–421.

16. Perrouault L, Asseline U, Rivalle Chr, Thuong NT, Bisagni E, Giovannangéli C, Le Doan T, Hélène C. Sequence-specific artificial photoinduced endonucleases based on triple helix-forming oligonucleotides. *Nature (London)* 1990;344:358–360.

17. Maher LJ, Dervan PB, Wold B. Analysis of promoter-specific repression by triple helical DNA complexes eukariotic cell-free transcription system. *Biochemistry* 1992;31:70–81.

18. Francois JC, Saison-Behmoaras T, Thuong NT, Hélène C. Inhibition of restriction endonuclease cleavage via triple helix formation by homopyrimidine oligonucleotides. *Biochemistry* 1989;28:9617–9619.

19. Cooney M, Czernuszewicz G, Postel EH, Flint SJ, Hogan ME. Site-specific oligonucleotide binding represses transcription of the human c-myc gene *in vitro*. *Science* 1988;241:456–459.

20. Hélène C, Toulme JJ. Specific regulation of gene expression by antisense and antigene nucleic acids. *Biochim Biophys Acta* 1990;1049:99–125.

21. Hélène C. The anti-gene strategy: Control of gene expression by triplex-forming oligonucleotides. *Anticancer Drug Des* 1991;6:569–584.

22. Duval-Valentin G, Thuong NT, Hélène C. Specific inhibition of transcription by triple helix-forming oligonucleotides. *Proc Nat Acad Sci USA* 1992;89:504–508.

23. Hsieh P, Camerini-Otero CS, Camerini-Otero RD. Pairing of homologous DNA sequences by proteins: Evidence for three-stranded DNA. *Genes Dev* 1990;4:1951–1963.

24. Rao BJ, Dutreix M, Radding CM. Stable three-stranded DNA made by RecA protein. *Proc Natl Acad Sci USA* 1991;88:2984–2988.

25. Yavachev LP, Georgiev OI, Braga EA, Avdonina TA, Bogomolova AE, Zhurkin VB, Nosikov VV, Hadjiolov A. Nucleotide sequence analysis of the spacer regions flanking the rat tRNA transcription unit and identification of repetitive element. *Nucleic Acids Res* 1986;14:2799–2810.

26. Behe MJ. The DNA sequence of the human beta-globin region is strongly biased in favor of long strings of contiguous purine and pyrimidine residues. *Biochemistry* 1987;26:7870–7875.

27. Behe MJ, An overabundance of long oligopurine tracts occurs in the genome of simple and complex eukaryotes. *Nucleic Acids Res* 1995;23:689–695.

28. Kato M, Kudoh J, Shimizu N. The pyrimidine/purine-biased region of the epidermal growth factor receptor gene is sensitive to S1 nuclease and may form an intramolecular triplex. *Biochem J* 1990;268:175–180.

29. Panayotatos N, Wells RD. Cruciform structures in supercoiled DNA. *Nature* 1981;289:466–470.

30. Lilley DMJ. The inverted repeat as a recognizable structural feature in supercoiled DNA molecules. *Proc Natl Acad Sci USA* 1980;77:6468–6472.

31. Pollack Y, Stein R, Razin A, Cedar H. Methylation of foreign DNA sequences in eukaryotic cells. *Proc Natl Acad Sci USA* 1980;77:6463–6467.

32. Lilley DMJ. Hairpin-loop formation by inverted repeats in supercoiled DNA is a local and transmissible property. *Nucl Acids Res* 1981;9:1271–1290.

33. Rothemund PWK. Folding DNA to create nanoscale shapes and patterns. *Nature* 2006, 440:297–302.

34. Breslauer KJ, Frank R, Blocker H, Marky, LA. Predicting DNA duplex stability from the base sequence. *Proc Natl Acad Sci USA* 1986;83:3746–3750.

35. Xia T, SantaLucia J, Jr, Burkard ME, Kierzek R, Schroeder SJ, Jiao X, Cox C, Turner DH. Thermodynamic parameters for an expanded nearest-neighbor model for formation of RNA duplexes with Watson–Crick base pairs. *Biochemistry* 1998;37:14719–14735.

36. SantaLucia J, Jr, Allawi HT, Seneviratne PA. Improved nearest-neighbor parameters for predicting DNA duplex stability. *Biochemistry* 1996;35:3555–3562.

37. Sugimoto N, Nakano S, Katoh M, Matsumura A, Nakamuta H, Ohmichi T, Yoneyama M, Sasaki M. Thermodynamics parameters to predict stability of RNA/DNA hybrid duplexes. *Biochemistry* 1995;34:11211–11216.

38. Cantor CR, Schimmel PR. *Biophysical Chemistry*. New York: W. H. Freeman and Company; 1980.

39. Marky LA, Blumenfeld KJ, Kozlowski S, Breslauer KJ. Salt-dependent conformational transitions in the self-complementary deoxydodecanucleotide d(CGCAATTCGCG): Evidence for hairpin formation. *Biopolymers* 1983;22:1247–1257.

40. Marky LA, Breslauer KJ. Calculating thermodynamic data for transitions of any molecularity from equilibrium melting curves. *Biopolymers* 1987;26:1601–1620.

41. Kaushik M, Suehl N, Marky LA. Calorimetric unfolding of the bimolecular and i-motif complexes of the human telomere complementary strand, d(C3TA2)4. *Biophys Chem* 2007; 126:(154–169).

42. Guo Q, Lu M, Chruchill MEA, Tullius TD, Kallenbach NR. Asymmetric structure of a three-arm DNA junction. *Biochemistry* 1990;29:10927–10934.

43. Lu M, Guo Q, Kallenbach NR. Effect of sequence on the structure of three-arm DNA junctions. *Biochemistry* 1991;30:5815–5820.

44. Ulyanov NB, Ivanov VI, Minyat EE, Khomyakova EB, Petrova MV, Krystyna L, James TL. A pseudosquare knot structure of DNA in solution. *Biochemistry* 1998;37:12715–12726.

45. Rentzeperis D. *Thermodynamics and ligand interactions of DNA hairpins. Doctoral Dissertation*, New York: New York University; 1995; p. 246.

46. Howerton SB, Sines CC, VanDerveer D, Williams LD. Locating monovalent cations in the grooves of B-DNA. *Biochemistry* 2001;40:10023–10031.

47. Hélène C. Rational design of sequence-specific oncogene inhibitors based on antisense and antigene oligonucleotides. *Eur J Cancer* 1991;27:1466–1471.

48. Thuong NT, Helene C. Stereospecific detection and modification of double helix DNA by oligonucleotides. *Angew Chem Int Ed Engl* 1993;32:666–690.

49. Soyfer VN, Potaman VN. *Triple-Helical Nucleic Acids*. New York: Springer-Verlag; 1996.

50. Hélène C. Control of oncogene expression by antisense nucleic acids. *Eur J Cancer* 1994;30A:1721–1726.

51. Maher LJ 3rd, Wold B, Dervan PB. Inhibition of DNA binding proteins by oligonucleotide-directed triple helix formation. *Science* 1989;245:725–730.

52. Praseuth D, Guieysse AL, Hélène C. Triple helix formation and the antigene strategy for sequence-specific control of gene expression. *Biochim Biophys Acta* 1999;1489:181–206.

53. Neidle S. Recent developments in triple-helix regulation of gene expression. *Anticancer Drug Des* 1997;12:433–442.

54. Brown PM, Madden CA, Fox KR. Triple-helix formation at different positions on nucleosomal DNA. *Biochemistry* 1998;37:16139–16151.

55. Cheng YK, Pettitt BM. Stabilities of double- and triple-strand helical nucleic acids. *Prog Biophys Mol Biol* 1992;58:225–257.

56. Keppler MD, Fox KR. Relative stability of triplexes containing different numbers of T.AT and C+.GC triplets. *Nucleic Acids Res* 1997;25:4644–4649.

57. Beal PA, Dervan PB. Second structural motif for recognition of DNA by oligonucleotide-directed triple-helix formation. *Science* 1991;251:1360–1363.

58. Schroth GP, Ho PS. Occurrence of potential cruciform and H-DNA forming sequences in genomic DNA. *Nucleic Acids Res* 1995;23:1977–1983.

59. Plum GE, Breslauer KJ. Thermodynamics of an intramolecular DNA triple helix: A calorimetric and spectroscopic study of the pH and salt dependence of thermally induced structural transitions. *J Mol Biol* 1995;248:679–695.

60. Soto AM. Thermodynamics for the unfolding of non-canonical nucleic acids: Bent duplexes, okazaki fragments, and DNA triplexes. Doctoral Dissertation, University of Nebraska Medical Center, Omaha, NE, 2002.

61. Soto AM, Marky LA. Thermodynamic contributions for the incorporation of GTA triplets within canonical TAT/TAT and C+GC/C+GC base-triplet stacks of DNA triplexes. *Biochemistry* 2002;41:12475–12482.

62. Shikiya R, Marky LA. Calorimetric unfolding of intramolecular triplexes: Length dependence and incorporation of single AT → TA substitutions in the duplex domain. *J Phys Chem B* 2005;109:18177–18183.

63. Soto AM, Loo J, Marky LA. Energetic contributions for the formation of TAT/TAT, TAT/CGC+, and CGC+/CGC +base triplet stacks. *J Am Chem Soc* 2002;124:14355–14363.

64. Rentzeperis D, Shikiya R, Maiti S, Ho J, Marky LA. Folding of intramolecular DNA hairpin loops: Enthalpy–entropy compensations and hydration contributions. *J Phys Chem B* 2002;106:9945–9950.

65. Conway BE. *Ionic Hydration in Chemistry and Biophysics*; New York; Elsevier; 1981.

66. Tunis MJ, Hearst JE. Hydration of DNA. I. Preferential hydration and stability of DNA in concentrated trifluoroacetate solution. *Biopolymers* 1968;6:1325–1344.

67. Tunis MJ, Hearst JE. Hydration of DNA. II. Base composition dependence of the net hydration of DNA. *Biopolymers* 1968;6:1345–1353.

68. Roberts RW, Crothers DM. Stability and properties of double and triple helices: Dramatic effects of RNA or DNA backbone composition. *Science* 1992;258:1463–1466.

69. Folini M, Pennati M, Zaffaroni N. Targeting human telomerase by antisense oligonucleotides and ribozymes. *Curr Med Chem Anticancer Agents* 2002;2:605–612.

70. Neidle S, Read MA. G-quadruplexes as therapeutic targets. *Biopolymers* 2000–01;56:195–208.

71. Huard S, Autexier C. Targeting human telomerase in cancer therapy. *Curr Med Chemi Anticancer Agents* 2002;2:577–587.

72. Mills M, Lacroix L, Arimondo PB, Leroy JL, Francois JC, Klump H, Mergny JL. Unusual DNA conformations: Implications for telomeres. *Curr Med Chemi Anticancer Agents* 2002;2:627–644.

73. Kyo S, Inoue M. How to inhibit telomerase activity for cancer therapy. *Curr Med Chem Anticancer Agents* 2002;2:613–626.

74. DePamphilis ML. Eukaryotic DNA replication: Anatomy of an origin. *Annu Rev Biochem* 1993;62:29–63.

75. Blackburn EH. Structure and function of telomeres. *Nature* 1991;350:569–573.

76. Sundquist WI, Klug A. Telomeric DNA dimerizes by formation of guanine tetrads between hairpin loops. *Nature* 1989;342:825–829.

77. Williamson JR, Raghuraman MK, Cech TR. Monovalent cation-induced structure of telomeric DNA: The G-quartet model. *Cell* 1989;59:871–880.

78. Moyzis RK, Buckingham JM, Cram LS, Dani M, Deaven LL, Jone MD, Meyne J, Ratliff RL, Wu JR. A highly conserved repetitive DNA sequence, (TTAGGG)n, present at the telomeres of human chromosomes. *Proc Natl Acad Sci USA* 1988;85:6622–6626.

79. Williamson JR, Raghuraman MK, Cech TR. Monovalent cation-induced structure of telomeric DNA: The G-quartet model. *Cell* 1989;59:871–880.

80. Macaya RF, Schultze P, Smith FW, Roe JA, Feigon J. Thrombin-binding DNA aptamer forms a unimolecular quadruplex structure in solution. *Proc Natl Acad Sci USA* 1993;90:3745–3749.

81. Schultze P, Macaya RF, Feigon J. Three-dimensional solution structure of the thrombin-binding DNA aptamer d(GGTTGGTGTGGTTGG). *J Mol Biol* 1994;235:1532–1547.

82. Han H, Hurley LH. G-quadruplex DNA: A potential target for anti-cancer drug design. *TIPS* 2000;21:136–142.

83. Bock LC, Griffin LC, Latham JA, Vermaas EH, Toole JJ. Selection of single-stranded DNA molecules that bind and inhibit human thrombin. *Nature* 1992;355:564–566.

84. Wang KY, Krawczyk SH, Bischofberger N, Swaminathan S, Bolton PH. The tertiary structure of a DNA aptamer which binds to and inhibits thrombin determines activity. *Biochemistry* 1993;32:11285–11292.

85. Rando RF, Ojwang J, Elbaggari A, Reyes GR, Tinder R, McGrath MS, Hogan ME. Suppression of human immunodeficiency virus type 1 activity *in vitro* by oligonucleotides which form intramolecular tetrads. *J Biol Chem* 1995;270:1754–1760.

86. Olsen CM, Gmeiner WH, Marky LA. Unfolding of G-quadruplexes: Energetic, ion and water contributions of g-quartet stacking. *J Phys Chem B* 2006;110:6962–6969.

87. Miyoshi D, Nakao A, Sugimoto N. Structural transition from antiparallel to parallel G-quadruplex of d(G4T4G4) induced by Ca2+. *Nucleic Acids Res* 2003;31:1156–1163.

88. Guo Q, Lu M, Marky LA, Kallenbach NR. Interaction of the dye ethidium bromide with DNA containing guanine repeats. *Biochemistry* 1992;31:2451–2455.

89. Kankia BI, Marky LA. Folding of the thrombin aptamer into a G-quadruplex with Sr2+: Stability, heat, and hydration. *J Am Chem Soc* 2001;123:10799–10804.

90. Lu M, Guo Q, Kallenbach NR. Thermodynamics of G-tetraplex formation by telomeric DNAs. *Biochemistry* 1993;32:598–601.

91. Dapic V, Abodmerovic V, Marrington R, Peberdy J, Rodger A, Trent JO, Bates PJ. Biophysical and biological properties of quadruplex oligodeoxyribonucleotides. *Nucleic Acids Res* 2003;31:2097–2107.

92. Todd AK, Johnston M, Neidle S. Highly prevalent putative quadruplex sequence motifs in human DNA. *Nucleic Acids Res* 2005;33:2901–2907.

93. Wells RD, Collier DA, Hanvey JC, Shimizu M, Wohlrab F. The chemistry and biology of unusual DNA structures adopted by oligopurine–oligopyrimidine sequences. *FASEB J* 1988;2:2939–2949.

94. Langridge R, Rich A. Molecular structure of helical polycytidylic acid. *Nature* 1963;198:725–728.

95. Marsh RE, Bierstedt R, Eichhorn L. The crystal structure of cytosine-5-acetic acid. *Acta Crystallogr* 1962;15:310–316.

96. Akinrimisi EO, Sander C, Ts'o O. Properties of helical polycytidylic acid. *Biochemistry* 1963;2:340–344.

97. Inman RB. Multistranded DNA homopolymer interactions. *J Mol Biol* 1964;10:137–146.

98. Gehring K, Leroy JL, Fueron A. Tetrameric DNA structure with protonated cytosine·cytosine base pairs. *Nature* 1993;363:561–565.

99. Leroy JL, Gehring K, Kettani A, Gueron M. Acid multimers of oligodeoxycytidine strands: Stoichiometry, base-pair characterization, and proton exchange properties. *Biochemistry* 1993;32:6019–6031.

100. Gueron M, Leroy JL. The i-motif in nucleic acids. *Curr Opin Struct Biol* 2000;10:326–331.

101. Leroy JL, Gueron M, Mergny JL, Helene C. Intramolecular folding of a fragment of the cytosine-rich strand of telomeric DNA into an i-motif. *Nucleic Acids Res* 1994;22:1600–1606.

102. Leroy JL, Gueron M. Solution structures of the i-motif tetramers of d(TCC), d(5methylCCT) and d(T5methylCC): Novel NOE connections between amino protons and sugar protons. *Structure* 1995;3:101–120.

103. Kang CH, Berger I, Lockshrin C, Ratliff R, Moyzis R, Rich A. Crystal structure of intercalated four-stranded d(C3T) at 1.4 Å resolution. *Proc Natl Acad Sci USA* 1994;91:11636–11640.

104. Kang C, Berger I, Lockshin C, Ratliff R, Moyzis R. Stable loop in the crystal structure of the intercalated four-stranded cytosine-rich metazoan telomere. *Proc Natl Acad Sci USA* 1995;92:3874–3878.

105. Berger I, Kang C, Fredian A, Ratliff R, Moyzis R, Rich A. Extension of the four-stranded intercalated cytosine motif by adenine·adenine base pairing in the crystal structure of d(CCCAAT). *Nat Struct Biol* 1995;2:416–425.

106. Chen L, Cai L, Zhang X, Rich A. Crystal structure of a four-stranded intercalated DNA: d(C4). *Biochemistry* 1994;22:13540–13546.

107. Nonin S, Leroy JL. Structure and conversion kinetics of a bi-stable DNA i-motif: Broken symmetry in the [d(5mCCTCC)]4 tetramer. *J Mol Biol* 1996;261:399–414.

108. Mergny JL. Following G-quartet formation by UV-spectroscopy. *FEBS Lett* 1998;435:74–78.

109. Manzini G, Yathindra N, Xodo LF. Evidence for intramolecularly folded i-DNA structures in biologically relevant CCC-repeat sequences. *Nucleic Acids Res* 1994;22:4634–4640.

110. Benevides JM, Wang AHJ, Rich A, Kyogoku Y, Vander Marel GA, Van Boom JH, Thomas GJ, Jr. Raman spectra of single crystals of r(GCG)d(CGC) and d(CCCCGGGG) as models

for A DNA, their structure transitions in aqueous solution, and comparison with double-helical poly(dG)·poly(dC). *Biochemistry* 1986;25:41–50.

111. Kananori K, Shibayama N, Gohda K, Tajima K, Makino K, Multiple four-stranded conformations of human telomere sequence d(CCCTAA) in solution. *Nucleic Acids Res* 2001;29:831–840.

112. Geinguenaud F, Liquier J, Brevnov MG, Petrauskene OV, Alexeev YI, Gromova ES, Taillandier E. Parallel self-associated structures formed by T,C-rich sequences at acidic pH. *Biochemistry* 2000;39:12650–12658.

113. Edwards EL, Patrick MH, Ratliff RL, Gray DM. A·T and C·C+ base pairs can form simultaneously in a novel multistranded DNA complex. *Biochemistry* 1990;29:828–836.

114. Kanehara H, Mizuguchi M, Tajima K, Kanaori K, Makino K. Spectroscopic evidence for the formation of four-stranded solution structure of oligodeoxycytidine phosphorothioate. *Biochemistry* 1997;36:1790–1797.

115. Marky LA, Blumenfeld KS, Kozlowski S, Breslauer KJ. Salt-dependent conformational transitions in the self-complementary deoxydodecanucleotide d(CGCGAATTCGCG): Evidence for hairpin formation. *Biopolymers* 1983;22:1247–1257.

116. Soto AM, Marky LA. Thermodynamic Contributions for the incorporation of GTA triplets within canonical TAT/TAT and C+GC/C+GC base-triplet stacks of DNA triplexes. *Biochemistry* 2002;41:12475–12482.

117. Manning GS. Electrostatic free energy of the DNA double helix in counterion condensation theory. *Biophys Chem* 2002;101–102:461–473.

118. Zimmer Ch, Venner H. Protonations in cytosine in DNA. *Biopolymers* 1966;4:1073–1079.

119. Marsich E, Piccini A, Xodo LE, Manzini G. Evidence for a HeLa nuclear protein that binds specifically to the single-stranded d(CCCTAA)n telomeric motif. *Nucleic Acids Res* 1996;24:4029–4033.

120. Marsich E, Xodo LE, Manzini G. Widespread presence in mammals and high binding specificity of a nuclear protein that recognizes the single-stranded telomeric motif (CCCTAA)n. *Eur J Biochem* 1998;258:93–99.

121. Lacroix L, Lienard H, Labourier E, Djavaheri-Mergny M, Lacoste L, Leffcos H, Tazi J, Helene C, Mergny JL. Identification of two human nuclear proteins that recognize the cytosine-rich strand of human telomeres *in vitro*. *Nucleic Acids Res* 2000;28:1564–1575.

9

NANOTOXICOLOGY

Diandra L. Leslie-Pelecky

9.1 INTRODUCTION

In 1895, Wilhelm Roentgen discovered that a strange beam coming from the cathode-ray tube he was studying allowed him to see the bones in his hand against a fluorescent screen. Roentgen's discovery of X-rays sparked such great public interest that photographers were offering X-ray "portraits" by early 1896. In February 1896, a physics professor at Vanderbilt University persuaded the dean of the medical school to be the subject of a skull radiograph. Three weeks later the dean's hair fell out.

X-rays were one of the research community's first experiences with public reaction to scientific innovation, but it certainly was not the last. Controversies over genetically modified organisms and stem-cell research already have had significant impacts on research and development. While researchers tout the promises of nanomaterials to produce revolutionary advancements in drug and gene delivery, medical imaging, and cancer treatment, media coverage has focused on questions about how well we understand the safety aspects of this emerging technology.

Nanomaterials can have optical, magnetic, electronic, and structural properties that are unrealizable in corresponding bulk materials. These unusual properties, however, complicate determining the safety of nanomaterials because nanomaterials often do not

Biomedical Applications of Nanotechnology. Edited by Vinod Labhasetwar and Diandra L. Leslie-Pelecky
Copyright © 2007 John Wiley & Sons, Inc.

behave like their bulk counterparts. Our understanding of the bulk toxicity of particular elements and alloys may not be directly translatable to the nano versions of these materials. A gold nanoparticle, for example, may not have the same chemical reactivity as bulk gold. Most nanomaterials used in biological applications contain multiple components for targeting, increasing circulation time, or other functionalities, which further complicate matters. Ironically, the very feature that makes nanomaterials so promising is the same feature that raises very important questions about potentially toxic effect.

An additional, but equally important, issue is the public perception of nanotechnology. In 1986, Eric Drexler expressed fears that self-replicating nanotechnology could spread at exponential rates. His use of the term "grey goo" to describe the eventual takeover of society by nanotechnology became one of the most prevalent public images of nanotechnology. This dire vision of the future was the theme of Michael Crichton's book *Prey*, which featured swarming nanobots intent on destroying their creators.

The casting of nanotechnology as villain surprised even Drexler [1]. In 2004, Drexler noted surprise at how warnings had dominated the public perception of nanotechnology:

I expected the contemplation of the broad societal impacts of nanotechnology to cause some discomfort, but did not expect that depictions of swarms of self-replicating nanobugs would dominate popular perceptions of advanced nanotechnology, nor did I envision that the term "nanotechnology" would come to describe a wide range of almost unrelated research fields, and that efforts to disassociate those fields from concerns about "grey goo" would spur false scientific denials of the original concepts.

Despite Drexler's recent statement that "I wish I had never used the term 'grey goo' " [2], the idea that nanotechnology represents a potential threat remains in the public awareness due to the popular press and governmental studies [3–8]. Governmental reports generally have been supportive of nanotechnology, but express concerns about rushing forward without reliable, consistent data on which to base decisions and regulatory activities. The continuing public interest emphasizes the need to accurately assess and communicate information about potentially negative effects of nanotechnology to the public. This brief summary of the literature regarding questions of toxicity is meant to provide nanomaterials researchers—especially those coming from the nonmedical sciences—an introduction to the critical issues surrounding this important topic.

9.2 NANOTOXICOLOGY

9.2.1 Parameters

The fact that nanomaterials behave in new—and sometimes unexpected—ways has important implications for the field of toxicology [9–18]. While nanomaterials enter the body in the usual ways (orally, through the respiratory system, through the blood circulation, and through the skin), they can interact with biological materials very differently than larger versions of the same material. Although size is important, it is far from being the only relevant factor. Chemical composition, surface area, surface charge, crystal structure, chemical reactivity, solubility, shape, and degree of agglomeration are also important parameters for understanding toxicity. This produces a much more extensive list of parameters than those addressed in typical toxicity studies.

Nanotechnology has changed the research paradigm. Addressing complex biomedical issues requires truly interdisciplinary teams; however, the confluence of researchers from different disciplines underscores a need for common definitions and usage. For example, does the stated size of a nanoparticle include surfactant, coating, or functionalizing molecules, or does it specify only the core? Coating molecules, or even residual impurities from synthesis procedures, can change the chemical behavior of a nanoparticle, so that comparing two gold nanoparticles is not as straightforward as one might expect. For example, some evidence of toxicity in carbon nanotubes may be due to residual metal catalysts used in their fabrication [19]. The nature of nanomaterials is that we cannot say a particular material is "biocompatible": We must study specific combinations of materials and evaluate these on an individual basis.

9.2.2 Toxicity Mechanisms

There are numerous reasons why nanomaterials may have greater toxicity than their bulk counterparts. In general, nanomaterials can penetrate into smaller structures and move deeper into passageways and tissues than larger particles. Nanoparticles may have increased mobility within human cells [20–23]. Once inside cells, they can produce oxidative stress, impair phagocytosis, inhibit cell proliferation, and decrease cell viability. The mechanisms of transport, cohesion, and adhesion are highly dependent on nanoparticle size [24]. Smaller particles may evade the reticuloendothelial system more effectively, and thus the nanoparticles remain in the body for longer times. Nanomaterials may generate mobile complexes that can enter tissue sites normally inaccessible to the proteins. Unexpected chemical reactions due to enhanced reactivity can impair functional and structural cell properties. All of these possibilities contribute to the potential toxicity of nanomaterials.

9.2.2.1 Production of Reactive Oxygen Species.

One of the most important toxicity mechanisms is the production of reactive oxygen species (ROS). For example, electron capture of O_2 can lead to the formation of the superoxide radical O_2^-, which then generates additional ROS. Normally, glutathione and other antioxidant enzymes maintain equilibrium in the body. An increasing ROS generation rate causes oxidative stress and triggers inflammatory responses. High levels of oxidative stress perturb the mitochondrial PT pore and disrupt electronic transfer, which can produce cellular apoptosis or necrosis [25]. Reactive oxygen species can damage cellular proteins, lipids, membranes, and DNA.

The mechanism by which oxidative species are produced is still being studied, but it appears that pro-oxidative organic hydrocarbons (such as polycyclic aromatic hydrocarbons and quinones) and transition metals such as copper, vanadium, chromium, nickel, cobalt, and iron are likely to participate in these reactions [9,25–34]. Many carbon nanotube production techniques require a metal catalyst, and significant concentrations of metals such as nickel, yttrium, and iron can be found in nanotubes, which may contribute to higher toxicity. Some semiconductors, when photoexcited, can create electron–hole pairs that can produce ROSs [20].

9.2.2.2 Effect of Physical Parameters. Ascribing toxicity solely to a nano-material's size, mass, or surface area is difficult because of the many variables that characterize a nanomaterial. A coated nanoparticle and a bare nanoparticle of the same material may have the same size, but distinctly different surface characteristics. Studies in which these variables can be isolated are difficult, but critical to understanding toxicity mechanisms [24, 35–37].

One can find in the literature examples in which smaller nanoparticles have greater toxicity than larger nanoparticles, especially in terms of inflammatory responses, and in which toxicity can be scaled to surface area or mass uptake [27, 38–45]. Other studies, however, show no or minimal surface area or size effect, and some show that smaller nanoparticles are less toxic than larger materials [18, 36, 46]. One possible contributor to these results is that aggregation and agglomeration significantly change the surface area of a material. If smaller nanoparticles aggregate to a greater degree than larger particles, the aggregates may not be taken up by cells as readily and thus show lower toxicity [24].

The size and shape of the nanomaterials may interfere with the efficiency of macrophage clearance. Fibrous shapes in which the length is larger than the diame-ter of the alveolar macrophage cannot be taken up and removed by macrophages [47]. In vitro studies focus primarily on the effect of nanomaterials in cell cultures, which is measured by cell viability and by the release of chemical mediators such as cytokines that are related to immunological and inflammatory reactions.

Structural properties also play a role in toxicity because of the role of structure in determining surface electronic states. Crystalline silica, for example, has a more deleterious effect on lung epithelial cells than similarly sized amorphous silica particles [48]. Materials such as TiO_2 that have multiple phases for the same chemical composition may exhibit very different toxicity in the different phases [49].

9.2.2.3 Effect of Surface Chemistry. Other forms of toxicity may be produced by unanticipated chemical reactions, including protein denaturation, membrane damage, DNA damage, and immune reactivity. The requirement of putting nanomaterials in an aqueous suspension adds another level of complexity because toxicity may be due to functionalizing molecules [36] or unexpected reactions between the nanoparticles and the functionalizing molecules [50].

Surface treatment is one route to mitigating surface reactions. Direct toxicity of materials is a problem in semiconductor quantum dots, which are possible alternatives to organic dyes. Most cadmium-based materials, such as cadmium selenide (CdSe) and cadmium telluride (CdTe), are known to be toxic, due to the release of Cd and/or Se ions [51–54]. A zinc selenide (ZnS) coating was used on CdSe to prevent Cd ion release; however, DNA damage was produced by oxidation of the ZnS cap [55]. Exposure to UV light appears to increase the amount of damage, likely due to exciton-induced chemical reactions from the semiconductors. Although surface modifications may confer biocompatibility, many coatings can be removed by chemical reactions within the body (such as oxidation), physical abrasion, or exposure to UV radiation [51,56].

9.2.2.4 Effect of Surface Charge. Polycationic macromolecules interact strongly with cell membranes, and the multiple positive charges appear to be an important factor in the cytotoxicity for cells of all kinds [50,57]. The addition of polyanionic compounds appears to protect against polycationic cytotoxic effects [14]. The polycation effect also may explain why branched and flexible polymers, which can have more attachments on a cell surface than rigid or globular polymers, have higher cytotoxic effects. The configuration of the polymer on the cell surface produces a high charge density [58,59]. Globular polymers appear to be less cytotoxic than linear polymers. Even 2-nm nanoparticles made of gold—a material known for its inertness—were toxic when made with a cationic surface charge [60]. The mechanism was proposed to be concentration-dependent lysis mediated by initial electrostatic binding. Surface charge also affects the degree of agglomeration, which can affect particle uptake [24].

9.2.2.5 Carbon Nanostructures. Carbon nanostructures, such as fullerenes and nanotubes, deserve special attention because these materials receive the most popular press. Carbon nanostructures have many potential applications because they have a unique free radical chemistry and, due to their strong attraction for electrons, antioxidant properties. Some carbon-60 (C_{60}) fullerenes have been found to bind strongly to nucleotides and may hamper self-repair in double-strand DNA [61]. Carbon nanomaterials also have high electrical and thermal conductivity, high strength, and rigidity. The many potential medical and nonmedical applications suggest that these materials may be produced on an industrial scale, increasing concern about occupational or accidental exposure.

There are a wide variety of reports regarding toxicity of carbon fullerenes (closed cage structures), single-walled nanotubes (SWNT), and multi-walled nanotubes (MWNT). Some studies suggest not only that fullerenes are nontoxic, but that they may provide protection against pathologies such as acute or chronic neurodegenerative diseases and liver disease [62–71]. Other reports show that carbon nanostructures produce O_2^- anions, lipid peroxidation, and cytotoxicity in plants and animals [72–78]. For example, uncoated fullerenes in largemouth bass show lipid peroxidation in brain tissue and glutathione depletion in gills [75]. Studies comparing carbon nanomaterials with oxide and metal nanomaterials suggest that, in inhalation studies at least, the toxicity mechanism may be quite different for carbon nanomaterials [29,79,80].

The disparity of results suggests that the problem is more complex than originally realized. For example, one of the first reports of toxicity in C_{60} was samples in which the surface was modified with PVP [81]; however, later studies showed that C_{60} can react with PVP to produce highly stable charge-transfer complexes [82]. A similar concern has been expressed that some fullerene samples are suspended in solution by dissolving C_{60} in THF and that not all of the solvent was removed prior to toxicity testing [68]. Although C_{60} may not cross the blood–brain barrier, the THF may. As mentioned, metal catalysts used in nanotube fabrication may be toxic, and sample preparation may exacerbate the toxicity of these metals [18, 77, 80]. Accurate and detailed characterization of samples used for these studies is critical.

More derivatized C_{60} species appear to be less toxic, in part due to a lower efficiency in generating ROS [72, 83]. Derivatization originally was required because carbon nanomaterials are not water-soluble. Toxicity effects ranging over seven orders of magnitude for different functionalizing molecules have been observed [72]. Aggregation becomes a factor, because interior members of the aggregate may not be appropriately derivatized. The most recent studies suggest that sidewall functionalization is significantly better at reducing toxicity than surfactant coating [84].

9.2.3 In Vivo Studies

9.2.3.1 Inhalation. The largest number of studies of *in vivo* toxicity deal with the respiratory route, and most focus on carbon black, TiO_2, and diesel particulates. There remains significant disagreement in the literature, much of which will likely be resolved by more detailed physical and chemical characterization of the materials being studied. The primary problems with respirated nanoparticles is that they can travel further into the lung, have a higher probability for chemical reactions that cause oxidative stress, and are more likely to persist in the body. Persistent nanoparticles may increase the risk of tumor formation.

Inhaled particulate matter can induce pulmonary and airway inflammations, interfere with the clearance and inactivation of bacteria in the lungs, and affect other organs by translocation. Oxidative stress and inflammation may lead to atherosclerotic plaques, failure to properly regulate heart rate, and decreased blood clotting ability.

Particles smaller than 2.5 μm can reach the alveoli, where clearance requires macrophage phagocytosis. Macrophage phagocytosis can produce sustained inflammation and some types of nanomaterials interfere with the efficiency of the clearance [47]. Smaller nanoparticles can take a significantly longer time to clear from the body and may promote translocation to interstitial sites and regional lymph nodes [85]. Dose-dependent formation of epithelioid granulomas (combinations of living and dead tissue that are signs of toxicity) observed with carbon nanostructures appear to originate in aggregates of nanotube-containing macrophages [79, 86].

Although the primary impact is in the pulmonary system, translocation of nanoparticles leads to circulatory access and allows distribution throughout the body [85, 87]. Translocation is very dependent on the specific properties of the nanoparticles. Many nanomaterials translocate to the liver, with secondary uptake by the spleen, bone marrow, heart, kidney, bladder, and brain [12]. Nanoparticles that migrate to the heart can affect arrhythmia and coagulation [88]. There is evidence that inhaled micron-sized particles affect the autonomic nervous system, so it is possible that similar or larger effects might be seen with nanoparticles [89–91]. Recent data suggest that some nanoparticles can move from olfactory nerve endings in the nose to the brain. Manganese appears to be translocated in this manner, but iron does not [92, 93].

The lung epithelial barrier is an example of how size can impact translocation. The epithelial barrier in a dog could be modeled by a three-pore system with pore radii of 1 nm, 40 nm, and 400 nm comprising 68%, 30%, and 2% of the pores, respectively. This suggests that smaller nanoparticles could more easily pass through the lung epithelial barrier [94].

At high doses, toxicity has been observed due to physical blockage produced by agglomerates. Carbon nanotubes, for example, have a strong electrostatic attraction and easily form aggregates [79]. A potential advantage of this tendency to agglomerate is that they may form unrespirable masses and thus decrease exposure [95].

9.2.3.2 Dermal Toxicology. Nanoparticle use in cosmetics (especially sunscreen) is widespread, with the primary materials being nanoscale TiO_2 and ZnO. The primary barrier to absorption through the skin is physical: the outer layer of the epidermis, which contains mostly dead skin cells. There is less information about dermal toxicity nanoparticles than about other types, and specific skin conditions may affect greatly the penetration of nanoparticles; however, the primary toxic mechanism appears to be generation of ROS. Whether the nanoparticles penetrate the epidermis depends on a number of factors. Movement of the skin or damaged skin can allow the penetration of microscale beads to the dermis and may allow movement to regional lymph nodes, where chemical reactions with proteins can affect the autoimmune system [96, 97]. There is evidence that submicron particles penetrate the skin shallowly and may penetrate more deeply via hair follicles [98, 99]. Smaller particles may penetrate far enough into the skin to interact with the immune system [100], or they may not penetrate deeply enough to be removed to the lymph nodes by macrophages and remain in the skin [101]. Even when nanoparticles remain in the dermis, molecules associated with the nanoparticle may detach and be transported through the skin.

An addition factor for dermal toxicity is the possibility of light-stimulated changes to the nanomaterials. Semiconducting anatase TiO_2 can create an electron–hole pair by absorbing a photon of light in the UVA or UVB region; interaction of this pair with water can produce ROS, including hydroxyl radicals, singlet oxygen, and superoxide [102]. Some researchers found that coating with inert oxides (silica, alumina, or zirconia) may reduce ROS generation [18]; however, photostimulated ROS production has been seen in commercial sunscreens [103].

9.2.3.3 Other Pathways. Nanoparticles can enter the gastrointestinal tract by direct methods (eating and drinking), or as a result of clearance via the mucociliary escalator in the respiratory tract. Most nanoparticles are rapidly eliminated through the bowel [104, 105]. Particle size, surface charge, and attachment of ligands or surfactants affect particle uptake [106–109]. In general, smaller particles are taken up more than larger particles and can be transported to the liver, spleen, blood, and bone marrow [110–112].

Ingestion of water-soluble radiolabeled fullerenes in rats show a 98% clearance in feces, while 90% of radio-labeled fullerenes administered intravenously were retained for at least a week with more than 70% residing in the liver [113, 114]. Intravenously injected water-soluble-single-walled carbon nanotubes functionalized with chelating molecule diethyletriaminepentaacetic and a radiotracer were found to be rapidly cleared from systemic blood circulation in mice via renal excretion with a half-life of about three and a half hours. Both functionalized SWNT and MWNT were excreted as intact nanotubes [115]. As in inhalation, aggregation plays a role in toxicity. Intestinal obstruction due to agglomerates of nanoscale zinc (Zn) powders produced mortality in some mice, while no impairments were observed with microscale Zn [116].

9.3 CONCLUSIONS

The field of nanotoxicity is in the very early stages of development, but it is clear from the available data that our knowledge of bulk material toxicity is not directly transferable to nanoscale counterparts. The toxicology community is developing methods and protocols to collect relevant and specific data on nanomaterial toxicity. Federal funding of studies specifically aimed at determining toxicity is increasing. Improved physical and chemical characterization of nanomaterials, combined with detailed reporting of these parameters, will help in comparing toxicity studies. In some respects, the question of whether carbon nanotubes are "safe" is academic because most biomedical nanosystems contain multiple materials. A definitive determination as to whether C_{60} "is toxic" is not as important as determining the toxicity of specific formulations.

Despite the potential for greater toxicity, nanomaterials will be an important tool in toxicology because of the ability to tightly control size dispersion, surface characteristics, and purity compared to naturally occurring materials. Toxicologists have an opportunity to investigate in greater detail the mechanisms responsible for toxicity, especially in the interactions of nanomaterials with cells. Once toxicity mechanisms are determined, nanoparticles that do exhibit toxic effects may find use as antibiotics or anticancer agents.

Finally, nanomaterials researchers need to be aware of the high visibility of anything "nano" in the eyes of the public. Previous experiences with stem cell research, dioxins, and genetically modified organisms should serve as lessons. Reports of potentially toxic effects of fullerenes have received far more attention in the popular press than have reports of nontoxic formulations. Special interest groups already have called for a moratorium on nanomaterials research and development [117, 118]. One highly publicized incident, even if minor, could prove a serious setback to progress. It is in the best interests of nanomaterials researchers to educate the public so that they can make informed decisions about the role that nanomaterials will play in their lives.

REFERENCES

1. Phoenix C, Drexler E. Safe exponential manufacturing. *Nanotechnology* 2004;15:869–872.
2. Giles J. Nanotech takes small step towards burying 'grey goo'. *Nature* 2004;429:591–591.
3. Helm B. The worries over nano no-nos. *Business Week Online*, February 23, 2005.
4. Royal Society/Royal Academy of Engineering. *Nanotechnologies: Opportunities and Uncertainties*, 2004. http://www.nanotec.org.uk/finalReport.htm
5. President's Council of Advisors on Science and Technology. *The National Nanotechnology Initiative at Five Years*, Washington, DC, 2005. http://www.ostp.gov/pcast/PCASTreportFINAL.pdf

6. Weiss R. Nanotech's frightening unknowns: It's like something from science fiction, yet it's being applied now—before Its effects Are properly understood. Most admit it's unstoppable, but will it prove benign or deadly? *Washington Post*, May 29, 2004.

7. Ross R. Big questions about tiny particles. *Toronto Star*, October 11, 2004.

8. Davidson K. The promise and perils of the nanotech revolution; possibilities range from disaster to advances in medicine, space. *San Francisco Chronicle* July 26, 2004.

9. Oberdörster G, Maynard A, Donaldson K, Castranova V, Fitzpatrick J, Ausman K, Carter J, Karn B, Kreyling W, Lai D, Olin S, Monteiro-Riviere N, Warheit D, Yang H. Principles for characterizing the potential human health effects from exposure to nanomaterials: Elements of a screening strategy, *Particle Fiber Toxicol* 2005;2:8.

10. Donaldson K, Stone V, Tran CL, Kreyling W, Borm PJ. *Nanotoxicology Occup Environ Med* 2004;61:727–728.

11. Hood E. Nanotechnology: Looking as we leap. *Environ Health Perspect* 2004;112: A740–A749.

12. Oberdörster G, Oberdörster E, Oberdörster J. Nanotoxicology: An emerging discipline evolving from studies of ultrafine particles. *Environ Health Persp* 2005;113:823–839.

13. Maynard A, Kuempel E. Airborne nanostructured particles and occupational health. *J Nanoparticle Res* 2005;7:587.

14. Hoet PH, Bruske-Hohlfeld I, Salata OV. Nanoparticles—Known and unknown health risks. *J Nanobiotechnol* 2004;2:12.

15. Nel A, Xia T, Madler L, Li N. Toxic potential of materials at the nanolevel. *Science* 2006;311:622–627.

16. Thomas K, Sayre P. Research strategies for safety evaluation of nanomaterials, part I: Evaluating the human health implications of exposure to nanoscale materials. *Toxicol Sci* 2005;87:316–321.

17. Holsapple MP, Farland WH, Landry TD, Monteiro-Riviere NA, Carter JM, Walker NJ, Thomas KV. Research strategies for safety evaluation of nanomaterials, Part II: Toxicological and safety evaluation of nanomaterials, current challenges and data needs. *Toxicol Sci* 2005;88:12–17.

18. Tsuji JS, Maynard AD, Howard PC, James JT, Lam C-w, Warheit DB, Santamaria AB. Research strategies for safety evaluation of nanomaterials, Part IV: Risk assessment of nanoparticles. *Toxicol Sci* 2006;89:42–50.

19. Kagan VE, Tyurina YY, Tyurin VA, Konduru NV, Potapovich AI, Osipov AN, Kisin ER, Schwegler-Berry D, Mercer R, Castranova V, Shvedova AA. Direct and indirect effects of single walled carbon nanotubes on raw 264.7 macrophages: Role of iron. *Toxicol Lett* 2006;165:88–100.

20. Michalet X, Pinaud FF, Bentolila LA, Tsay JM, Doose S, Li JJ, Sundaresan G, Wu AM, Gambhir SS, Weiss S. Quantum dots for live cells, in vivo imaging, and diagnostics. *Science* 2005;307:538–544.

21. Allen TM, Cullis PR. Drug delivery systems: Entering the mainstream. *Science* 2004;303:1818–1822.

22. Koch AM, Reynolds F, Merkle HP, Weissleder R, Josephson L. Transport of surface-modified nanoparticles through cell monolayers. *ChemBioChem* 2005;6:337–345.

23. Win KY, Feng SS. Effects of Particle Size and Surface Coating on Cellular Uptake of Polymeric Nanoparticles for Oral Delivery of Anticancer Drugs. *Biomaterials* 2005;26:2713–2722.

24. Limbach LK, Li Y, Grass RN, Brunner TJ, Hintermann MA, Muller M, Gunther D, Stark WJ. Oxide nanoparticle uptake in human lung fibroblasts: Effects of particle size, agglomeration, and diffusion at low concentrations. *Environ Sci Technol* 2005;39:9370–9376.

25. Li N, Sioutas C, Cho A, Schmitz D, Misra C, Sempf J, Wang M, Oberley T, Froines J, Nel A. Ultrafine particulate pollutants induce oxidative stress and mitochondrial damage. *Environ Health Perspect* 2003;111:455–460.

26. Fernandez-Urrusuno R, Fattal E, Feger J, Couvreur P, Therond P. Evaluation of hepatic antioxidant systems after intravenous administration of polymeric nanoparticles. *Biomaterials* 1997;18:511–517.

27. Zhang Q, Kusaka Y, Sato K, Nakakuki K, Kohyama N, Donaldson K. Differences in the extent of inflammation caused by intratracheal exposure to three ultrafine metals: Role of free radicals. *J Toxicol Environ Health A* 1998;53:423–438.

28. Dick CA, Brown DM, Donaldson K, Stone V. The role of free radicals in the toxic and inflammatory effects of four different ultrafine particle types. *Inhal Toxicol* 2003;15:39–52.

29. Warheit DB, Brock WJ, Lee KP, Webb TR, Reed KL. Comparative pulmonary toxicity inhalation and instillation studies with different TiO2 particle formulations: Impact of surface treatments on particle toxicity. *Toxicol Sci* 2005;88:514–524.

30. Warheit DB, Webb TR, Reed KL. Pulmonary toxicity studies with TiO$_2$ particles containing various commercial coatings. *Toxicologist* 2003;72:298A.

31. Li N, Alam J, Venkatesan MI, Eiguren-Fernandez A, Schmitz D, Di Stefano E, Slaughter N, Killeen E, Wang X, Huang A, Wang M, Miguel AH, Cho A, Sioutas C, Nel AE. Nrf2 is a key transcription factor that regulates antioxidant defense in macrophages and epithelial cells: Protecting against the proinflammatory and oxidizing effects of diesel exhaust chemicals. *J Immunol* 2004;173:3467–3481.

32. Nel A. Air pollution-related illness: Effects of particles. *Science* 2005;308:804–806.

33. Silbajoris R, Ghio AJ, Samet JM, Jaskot R, Dreher KL, Brighton LE. In vivo and in vitro correlation of pulmonary map kinase activation following metallic exposure. *Inhal Toxicol* 2000;12:453–468.

34. Nel A. Air pollution-related illness: Effects of particles. *Science* 2005;309:1326–1326.

35. Oberdörster G, Gelein RM, Ferin J, Weiss B. Association of particulate air pollution and acute mortality: Involvement of ultrafine particles? *Inhal Toxicol* 1995;7:111–124.

36. Yin H, Too HP, Chow GM. The effects of particle size and surface coating on the cytotoxicity of nickel ferrite. *Biomaterials* 2005;26:5818–5826.

37. Soto KF, Carrasco A, Powell TG, Garza KM, Murr LE. Comparative in vitro cytotoxicity assessment of some manufactured nanoparticulate materials characterized by transmission electron microscopy. *J Nanoparticle Res* 2005;7:145–169.

38. Brown DM, Wilson MR, MacNee W, Stone V, Donaldson K. Size-dependent proinflammatory effects of ultrafine polystyrene particles: A role for surface area and oxidative stress in the enhanced activity of ultrafines. *Toxicol Appl Pharmacol* 2001;175:191–199.

39. Oberdörster G, Finkelstein JN, Johnston C, Gelein R, Cox C, Baggs R, Elder AC. Acute pulmonary effects of ultrafine particles in rats and mice. *Res Rep Health Eff Inst* 2000;96:5–74:dicussion pp. 75–86.

40. Driscoll KE, Deyo LC, Carter JM, Howard BW, Hassenbein DG, Bertram TA. Effects of particle exposure and particle-elicited inflammatory cells on mutation in rat alveolar epithelial cells. *Carcinogenesis* 1997;18:423–430.

41. Oberdörster G, Yu CP. Lung dosimetry—considerations for noninhalation studies. *Exp Lung Res* 1999;25:1–6.

42. Oberdörster G, Ferin J, Gelein R, Soderholm SC, Finkelstein J. Role of the alveolar macrophage in lung injury: Studies with ultrafine particles. *Environ Health Perspect* 1992;97:193–199.

43. Webb DR, Wilson SE, Carter DE. Comparative pulmonary toxicity of gallium arsenide, gallium(III) oxide, or arsenic(III) oxide intratracheally instilled into rats. *Toxicol Appl Pharmacol* 1986;82:405–416.

44. Zhang Q, Kusaka Y, Zhu X, Sato K, Mo Y, Kluz T, Donaldson K. Comparative toxicity of standard nickel and ultrafine nickel in lung after intratracheal instillation. *J Occup Health* 2003;45:23–30.

45. Heinrich U, Muhle H, Hoymann HG, Mermelstein R. Pulmonary function changes in rats after chronic and subchronic inhalation exposure to various particulate matter. *Exp Pathol* 1989;37:248–252.

46. Warheit D, Webb T, Sayes C, Colvin V, Reed K. Pulmonary instillation studies with nanoscale TiO₂ rods and dots in rats: Toxicity is not dependent upon particle size and surface area. *Toxicol Sci* 2006;9:227–336.

47. Schins RP. Mechanisms of genotoxicity of particles and fibers. *Inhal Toxicol* 2002;14:57–78.

48. Monarca S, Crebelli R, Feretti D, Zanardini A, Fuselli S, Filini L, Resola S, Bonardelli PG, Nardi G. Mutagens and carcinogens in size-classified air particulates of a northern Italian town. *Sci Total Environ* 1997;205:137–144.

49. Yamamoto A, Honma R, Sumita M, Hanawa T. Cytotoxicity evaluation of ceramic particles of different sizes and shapes. *J Biomed Mater Res A* 2004;68:244–256.

50. Hoet PH, Gilissen L, Nemery B. Polyanions protect against the in vitro pulmonary toxicity of polycationic paint components associated with the ardystil syndrome. *Toxicol Appl Pharmacol* 2001;175:184–190.

51. Derfus AM, Chan WCW, Bhatia SN. Probing the cytotoxicity of semiconductor quantum dots. *Nano Lett* 2004;4:11–18.

52. Kirchner C, Liedl T, Kudera S, Pellegrino T, Munoz Javier A, Gaub HE, Stolzle S, Fertig N, Parak WJ. Cytotoxicity of colloidal cdse and CdSe/ZnS nanoparticles. *Nano Lett* 2005;5:331–338.

53. Lovric J, Bazzi HS, Cuie Y, Fortin GR, Winnik FM, Maysinger D. Differences in subcellular distribution and toxicity of green and red emitting CdTe quantum dots. *J Mol Med* 2005;83:377–385.

54. Shiohara A, Hoshino A, Hanaki K, Suzuki K, Yamamoto K. On the cytotoxicity caused by quantum dots. *Microbiol Immunol* 2004;48:669–675.

55. Green M, Howman E. Semiconductor quantum dots and free radical induced DNA nicking. *Chem Commun (Camb)* 2005; 121–123.

56. Rancan F, Rosan S, Boehm F, Cantrell A, Brellreich M, Schoenberger H, Hirsch A, Moussa F. Cytotoxicity and photocytotoxicity of a dendritic C(60) mono-adduct and a malonic acid C(60) tris-adduct on jurkat cells. *J Photochem Photobiol B* 2002;67:157–162.

57. Hoet PH, Gilissen LP, Leyva M, Nemery B. In vitro cytotoxicity of textile paint components linked to the "Ardystil syndrome". *Toxicol Sci* 1999;52:209–216.

58. Ryser HJ. A membrane effect of basic polymers dependent on molecular size. *Nature* 1967;215:934–936.

59. Dekie L, Toncheva V, Dubruel P, Schacht EH, Barrett L, Seymour LW. poly-L-glutamic acid Derivatives as vectors for gene therapy. *J Control Release* 2000;65:187–202.

60. Goodman CM, McCusker CD, Yilmaz T, Rotello VM. Toxicity of gold nanoparticles functionalized with cationic and anionic side chains. *Bioconjug Chem* 2004;15:897–900.

61. Zhao X, Striolo A, Cummings PT. C60 binds to and deforms nucleotides. *Biophys J* 2005; 89:3856–3862.

62. Chiron JP, Lamande J, Moussa F, Trivin F, Ceolin R. [Effect of "micronized" C60 fullerene on the microbial growth in vitro]. *Ann Pharm Fr* 2000;58:170–175.

63. Scrivens WA, Tour JM, Creek KE, Pirisi L. Synthesis of C-14-labeled C-60, its suspension in water, its uptake by human keratinocytes. *Am J Chem Soc* 1994;116:4517–4518.

64. Satoh M, Matsuo K, Takanashi Y, Takayanagi I. Effects of acute and short-term repeated application of fullerene C60 on agonist-induced responses in various tissues of guinea pig and rat. *Gen Pharmacol* 1995;26:1533–1538.

65. Jia G, Wang H, Yan L, Wang X, Pei R, Yan T, Zhao Y, Guo X. Cytotoxicity of carbon nanomaterials: Single-wall nanotube, multi-wall nanotube, fullerene. *Environ Sci Technol* 2005;39:1378–1383.

66. Zakharenko LP, Zakharov IK, Lunegov SN, Nikiforov AA. [Demonstration of the absence of genotoxicity of fullerene C60 using the somatic mosaic method]. *Dokl Akad Nauk* 1994;335:261–262.

67. Xiao L, Takada H, Maeda K, Haramoto M, Miwa N. Antioxidant effects of water-soluble fullerene derivatives against ultraviolet ray or peroxylipid through their action of scavenging the reactive oxygen species in human skin keratinocytes. *Biomed Pharmacother* 2005;59:351–358.

68. Gharbi N, Pressac M, Hadchouel M, Szwarc H, Wilson SR, Moussa F. [60]Fullerene is a powerful antioxidant in vivo with no acute or subacute toxicity. *Nano Lett* 2005;5:2578–2585.

69. Andrievsky GV, Kosevich MV, Vovk OM, Shelkovsky VS, Vashchenko LA. On the production of an aqueous colloidal solution of fullerenes. *J Chem Soc Chem Commun* 1995;12: 1281–1282.

70. Yamakoshi Y, Umezawa N, Ryu A, Arakane K, Miyata N, Goda Y, Masumizu T, Nagano T. Active oxygen species generated from photoexcited fullerene (C60) as potential medicines: O2* Versus 1O2. *J Am Chem Soc* 2003;125:12803–12809.

71. Yokoyama A, Sato Y, Nodasaka Y, Yamamoto S, Kawasaki T, Shindoh M, Kohgo T, Akasaka T, Uo M, Watari F, Tohji K. Biological behavior of hat-stacked carbon nanofibers in the subcutaneous tissue in rats. *Nano Lett* 2005;5:157–161.

72. Sayes CM, Fortner JD, Guo W, Lyon D, Boyd AM, Ausman KD, Tao YJ, Sitharaman B, Wilson LJ, Hughes JB, West JL, Colvin VL. The differential cytotoxicity of water-soluble fullerenes. *Nano Lett* 2004;4:1881–1887.

73. Colvin VL. The potential environmental impact of engineered nanomaterials. *Nat Biotechnol* 2003;21:1166–1170.

74. Oberdörster E. Toxicity of nC$_{60}$ fullerenes to two aquatic species: Daphnia and largemouth bass. *Abstr Pap Am Chem S* 2004;227:U1233–U1233.

75. Oberdörster E. Manufactured nanomaterials (fullerenes, C-60) induce oxidative stress in the brain of juvenile largemouth bass. *Environ Health Persp* 2004;112:1058–1062.

76. Shvedova AA, Kisin ER, Mercer R, Murray AR, Johnson VJ, Potapovich AI, Tyurina YY, Gorelik O, Arepalli S, Schwegler-Berry D, Hubbs AF, Antonini J, Evans DE, Ku BK,

Ramsey D, Maynard A, Kagan VE, Castranova V, Baron P. Unusual inflammatory and fibrogenic pulmonary responses to single-walled carbon nanotubes in mice. *Am J Physiol Lung Cell Mol Physiol* 2005;289:L698–L708.

77. Shvedova AA, Castranova V, Kisin ER, Schwegler-Berry D, Murray AR, Gandelsman VZ, Maynard A, Baron P. Exposure to carbon nanotube material: Assessment of nanotube cytotoxicity using human keratinocyte cells. *J Toxicol Environ Health A* 2003;66:1909–1926.

78. Monteiro-Riviere NA, Nemanich RJ, Inman AO, Wang YY, Riviere JE. Multi-walled carbon nanotube interactions with human epidermal keratinocytes. *Toxicol Lett* 2005;155:377–384.

79. Warheit DB, Laurence BR, Reed KL, Roach DH, Reynolds GAM, Webb TR. Comparative pulmonary toxicity assessment of single-wall carbon nanotubes in rats. *Toxicol Sci* 2004;77:117–125.

80. Lam C-W, James JT, McCluskey R, Hunter RL. Pulmonary toxicity of single-wall carbon nanotubes in mice 7 and 90 days after intratracheal instillation. *Toxicol Sci* 2004;77:126–134.

81. Tsuchiya T, Oguri I, Yamakoshi YN, Miyata N. Novel harmful effects of [60]fullerene on mouse embryos in vitro and in vivo. *FEBS Lett* 1996;393:139–145.

82. Ungurenasu C, Airinei A. Highly stable C(60)/poly(vinylpyrrolidone) charge-transfer complexes afford new predictions for biological applications of underivatized fullerenes. *J Med Chem* 2000;43:3186–3188.

83. Dugan LL, Turetsky DM, Du C, Lobner D, Wheeler M, Almli CR, Shen CKF, Luh T-Y, Choi DW, Lin T-S. Carboxyfullerenes as neuroprotective agents. *Proc Natl Acad Sci USA* 1997;94:9434–9439.

84. Sayes CM, Liang F, Hudson JL, Mendez J, Guo W, Beach JM, Moore VC, Doyle CD, West JL, Billups WE, Ausman KD, Colvin VL. Functionalization density dependence of single-walled carbon nanotubes cytotoxicity in vitro. *Toxicol Lett* 2006;161:135–142.

85. Oberdörster G, Ferin J, Lehnert BE. Correlation between particle size, in vivo particle persistence, lung injury. *Environ Health Perspect* 1994;102(Suppl 5): 173–179.

86. Service RF. American Chemical Society Meeting. Nanomaterials show signs of toxicity. *Science* 2003;300:243.

87. Oberdörster G. Pulmonary effects of inhaled ultrafine particles. *Int Arch Occup Environ Health* 2001;74:1–8.

88. Yeates DB, Mauderly JL. Inhaled environmental/occupational irritants and allergens: Mechanisms of cardiovascular and systemic responses. Introduction. *Environ Health Perspect* 2001;109(Suppl 4): 479–481.

89. Oberdörster G, Sharp Z, Atudorei V, Elder A, Gelein R, Lunts A, Kreyling W, Cox C. Extrapulmonary translocation of ultrafine carbon particles following whole-body inhalation exposure of rats. *J Toxicol Environ Health A* 2002;65:1531–1543.

90. Gold DR, Litonjua A, Schwartz J, Lovett E, Larson A, Nearing B, Allen G, Verrier M, Cherry R, Verrier R. Ambient pollution and heart rate variability. *Circulation* 2000;101:1267–1273.

91. Liao D, Creason J, Shy C, Williams R, Watts R, Zweidinger R. Daily variation of particulate air pollution and poor cardiac autonomic control in the elderly. *Environ Health Perspect* 1999;107:521–525.

92. Oberdörster G, Sharp Z, Atudorei V, Elder A, Gelein R, Kreyling W, Cox C. Translocation of inhaled ultrafine particles to the brain. *Inhal Toxicol* 2004;16:437–445.

93. Rao DB, Wong BA, McManus BE, McElveen AM, James AR, Dorman DC. Inhaled iron, unlike manganese, is not transported to the rat brain via the olfactory pathway. *Toxicol Appl Pharmacol* 2003;193:116–126.

94. Conhaim RL, Eaton A, Staub NC, Heath TD. Equivalent pore estimate for the alveolar–airway barrier in isolated dog lung. *J Appl Physiol* 1988;64:1134–1142.

95. Maynard AD, Baron PA, Foley M, Shvedova AA, Kisin ER, Castranova V. Exposure to carbon nanotube material: Aerosol release during the handling of unrefined single-walled carbon nanotube material. *J Toxicol Environ Health A* 2004;67:87–107.

96. Kim S, Lim YT, Soltesz EG, De Grand AM, Lee J, Nakayama A, Parker JA, Mihaljevic T, Laurence RG, Dor DM, Cohn LH, Bawendi MG, Frangioni JV. Near-infrared fluorescent type II quantum dots for sentinel lymph node mapping. *Nat Biotechnol* 2004;22:93–97.

97. Nel AE, Diaz-Sanchez D, Ng D, Hiura T, Saxon A. Enhancement of allergic inflammation by the interaction between diesel exhaust particles and the immune system. *J Allergy Clin Immunol* 1998;102:539–554.

98. Lademann J, Weigmann H, Rickmeyer C, Barthelmes H, Schaefer H, Mueller G, Sterry W. Penetration of titanium dioxide microparticles in a sunscreen formulation into the horny layer and the follicular orifice. *Skin Pharmacol Appl Skin Physiol* 1999;12:247–256.

99. Tinkle SS, Antonini JM, Rich BA, Roberts JR, Salmen R, DePree K, Adkins EJ. Skin as a route of exposure and sensitization in chronic beryllium disease. *Environ Health Perspect* 2003;111:1202–1208.

100. Kreilgaard M. Influence of microemulsions on cutaneous drug delivery. *Adv Drug Deliv Rev* 2002;54(Suppl 1): S77–S98.

101. de Jalon EG, Blanco-Prieto MJ, Ygartua P, Santoyo S. Plga microparticles: Possible vehicles for topical drug delivery. *Int J Pharm* 2001;226:181–184.

102. Konaka R, Kasahara E, Dunlap WC, Yamamoto Y, Chien KC, Inoue M. Irradiation of titanium dioxide generates both singlet oxygen and superoxide anion. *Free Radic Biol Med* 1999;27:294–300.

103. Brezova V, Gabcova S, Dvoranova D, Stasko A. Reactive oxygen species produced upon photoexcitation of sunscreens containing titanium dioxide (An EPR study). *J Photochem Photobiol B* 2005;79:121–134.

104. Florence AT, Hussain N. Transcytosis of nanoparticle and dendrimer delivery systems: Evolving vistas. *Adv Drug Deliv Rev* 2001;50(Suppl 1): S69–S89.

105. Hussain N, Jaitley V, Florence AT. Recent advances in the understanding of uptake of microparticulates across the gastrointestinal lymphatics. *Adv Drug Deliv Rev* 2001;50:107–142.

106. Florence AT, Hillery AM, Hussain N, Jani PU. Factors affecting the oral uptake and translocation of polystyrene nanoparticles: Histological and analytical evidence. *J Drug Target* 1995;3:65–70.

107. Hussain N, Jani PU, Florence AT. Enhanced oral uptake of tomato lectin-conjugated nanoparticles in the rat. *Pharm Res* 1997;14:613–618.

108. Hussain N, Florence AT. Utilizing bacterial mechanisms of epithelial cell entry: Invasin-induced oral uptake of latex nanoparticles. *Pharm Res* 1998;15:153–156.

109. Woodley JF. Lectins for gastrointestinal targeting—15 years on. *J Drug Target* 2000;7:325–333.

110. Hillyer JF, Albrecht RM. Gastrointestinal persorption and tissue distribution of differently sized colloidal gold nanoparticles. *J Pharm Sci* 2001;90:1927–1936.

111. Jani P, Halbert GW, Langridge J, Florence AT. Nanoparticle uptake by the rat gastrointestinal mucosa: Quantitation and particle size dependency. *J Pharm Pharmacol* 1990;42:821–826.
112. Jani P, Halbert GW, Langridge J, Florence AT. The uptake and translocation of latex nanospheres and microspheres after oral administration to rats. *J Pharm Pharmacol* 1989;41:809–812.
113. Chen BX, Wilson SR, Das M, Coughlin DJ, Erlanger BF. Antigenicity of fullerenes: Antibodies specific for fullerenes and their characteristics. *Proc Natl Acad Sci USA* 1998;95:10809–10813.
114. Kreyling WG, Semmler M, Erbe F, Mayer P, Takenaka S, Schulz H, Oberdörster G, Ziesenis A. Translocation of ultrafine insoluble iridium particles from lung epithelium to extrapulmonary organs is size dependent but very low. *J Toxicol Environ Health A* 2002;65:1513–1530.
115. Singh R, Pantarotto D, Lacerda L, Pastorin G, Klumpp C, Prato M, Bianco A, Kostarelos K. Tissue biodistribution and blood clearance rates of intravenously administered carbon nanotube radiotracers. *Proc Natl Acad Sci USA* 2006;103:3357–3362.
116. Wang B, Feng WY, Wang TC, Jia G, Wang M, Shi JW, Zhang F, Zhao YL, Chai ZF. Acute toxicity of nano- and micro-scale zinc powder in healthy adult mice. *Toxicol Lett* 2006;161:115–123.
117. Service RF. Nanotoxicology: Nanotechnology grows up. *Science* 2004;304:1732–1734.
118. Service RF. Is nanotechnology dangerous? *Science* 2000;290:1526–1527.

INDEX

Biomedical Applications of Nanotechnology. Edited by Vinod Labhasetwar and Diandra L. Leslie-Pelecky
Copyright © 2007 John Wiley & Sons, Inc.

ABOUT THE EDITORS

Vinod Labhasetwar, Ph.D.

Dr. Labhasetwar received his undergraduate degree and Ph.D. in Pharmaceutical Sciences from Nagpur University, India. He was a staff scientist at National Institute of Immunology, New Delhi, from 1988–1990 before joining the University of Michigan, Ann Arbor, as a postdoctoral fellow. After completing his training in 1993, Dr. Labhasetwar continued at the University of Michigan as a faculty member in the School of Medicine until 1997. He then joined the University of Nebraska Medical Center as associate professor of Pharmaceutical Sciences, and Biochemistry and Molecular Biology. In 2007, Dr. Labhasetwar joined the Cleveland Clinic as professor of Nanomedicine. His research interests are in translational nanomedicine. Dr. Labhasetwar's laboratory investigates nanosystems for drug/gene delivery in cancer therapy, stroke cardiovascular conditions, and other age-related disorders. He has developed multifunctional magnetic nanoparticles which he is investigating for imaging and targeted delivery of therapeutics in cancer treatment. His group also investigates nanoparticle-cell interactions and intracellular trafficking for developing targeted drug delivery at the cellular level. The research in his laboratory is primarily funded by the National Institutes of Health, American Heart Association, and the U.S. Department of Defense.

Diandra Leslie-Pelecky, Ph.D.

Dr. Leslie-Pelecky received her Ph.D. from Michigan State University. She joined the University of Nebraksa in 1996 and currently is an associate professor in the Department of Physics and Astronomy and the Nebraska Center for Materials and Nanoscience. Her research focuses on fundamental magnetic properties of nanostructured materials, and applications of nanomaterials to biomedical problems, including drug delivery, hyperthermia and magnetic resonance imaging techniques. Her research is funded primarily by the National Science Foundation, the National Institutes of Health and the U.S. Department of Energy.

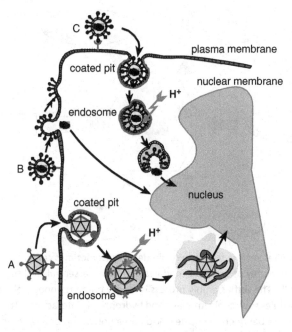

Figure 2.1. Viral entry into cells, schematic overview. Viruses bind to cell surfaces via receptor–ligand interactions. Many virus species are taken up into cells by endocytosis, like adenovirus (A) or membrane-coated viruses such as influenza virus (C). Other membrane-coated viruses such as retroviruses (B) directly fuse their membranes with the plasma membrane. Endocytosis proceeds via segregation of membrane-surrounded vesicles (endosomes) from the plasma membrane. A proton pump in the endosomal membrane mediates the acid-ification of the endosomal lumen. This pH change triggers conformational rearrangements of viral proteins, which then by interaction with the endosomal membrane can lead to the disruption of endosomes (like in the case of adenoviruses) or to the fusion of endosomal and viral membranes (like in the case of influenza virus). These membrane disruption/fusion events are essential parts of viral entry into cells, which ultimately leads to uptake/transport of viral genetic information into the cell nucleus.

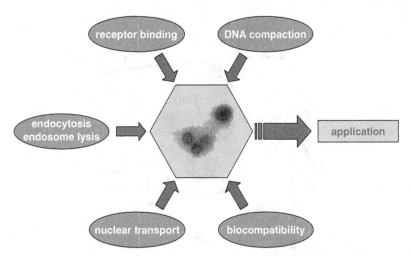

Figure 2.2. Nonviral vectors for nucleic acid delivery (sometimes called artificial viruses) are prepared by self-assembly of synthetic modules that mimic essential viral functions that allow them to infect cells. The self-assembly process is mostly based on noncovalent interactions of the individual modules such as electrostatic and hydrophobic interactions. The most important interaction is the one between the nucleic acid and a polycation or a cationic lipid, which can lead to the formation of a charged nanoparticle that is able to transfect cells. The functionalities of receptor binding, membrane destabilization (such as endosome lysis), nuclear targeting, and biocompatibility can be covalently coupled to a DNA binding/compacting moiety or can be incorporated into the complex as individual molecules by noncovalent interactions. The center of the figure shows toroidal nanoparticles that are typically formed upon mixing of plasmid DNA and polycations.

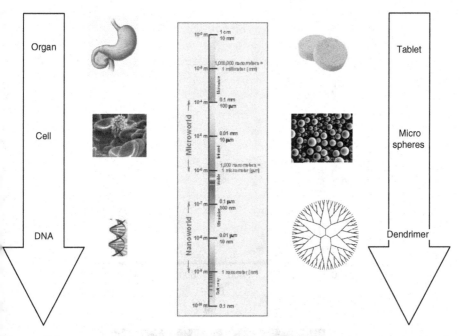

Figure 5.1. Progress of drug delivery from "macro" systems that interact at the organ level to "micro" systems that interact at the cellular level to "nano" systems that interact at the cellular level. Length scale has a significant influence on drug delivery in terms of reaching the target site, modifying the biodistribution of the drug, and enhancing the efficacy of the drug.

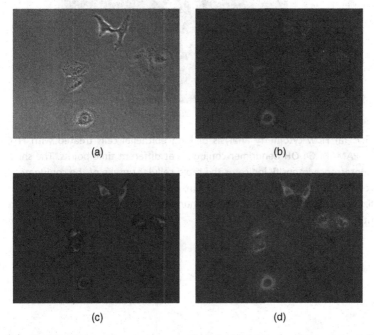

Figure 5.6. Confocal images showing lung epithelial cells. (a) Under phase contrast. (b) Localization of FITC-labeled PAMAM G4 hydroxyl terminated dendrimer in the cytoplasm. (c) Localization of lysosomal marker (lysotracker) in the lysosomes. (d) Co-localization of FITC-labeled dendrimer and lysostracker in the lysosomes. Images were captured 30 min after treatment.

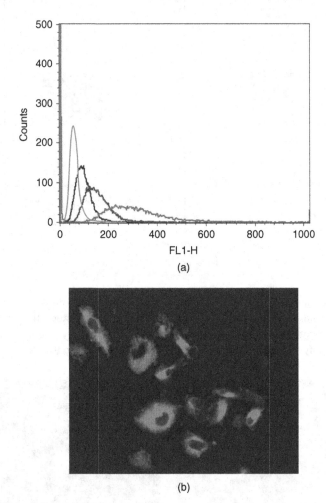

Figure 5.7. (a) Flow cytometry analysis of lung epithelial cells treated with FITC-labeled ibuprofen–PAMAM G4 OH dendrimer conjugate at different time points. The shift in intracellular fluorescence intensity indicates the rapid cellular uptake of the conjugate. Key: Red, 0 min; green, 5 min; black, 30 min; blue, 60 min; brown, 240 min (b) Confocal image showing the localization of FITC-labeled ibuprofen–dendrimer conjugate in the cytoplasm after 2 hr of treatment.

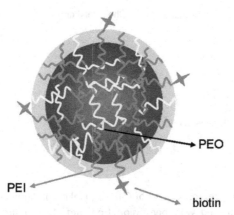

Figure 6.5. Poly(ethyleneimine) cross-linked poly(ethylene oxide) nanogel. (Reproduced with permission from Ref. 71, Figure 6.1. Copyright 2005, Elsevier Ltd.)

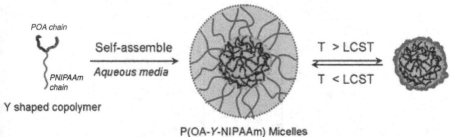

Figure 6.9. Y-shaped copolymer self-assembly to give micelle structures. (Reproduced with permission from Ref. 113, Figure 6.8. Copyright 2006, John Wiley & Sons, Inc.)

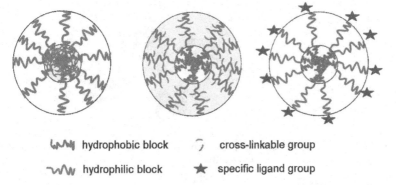

Figure 6.10. Internal structural variation in micelle gels. (Reproduced with permission from Ref. 117, Figure 6.1. Copyright 2001, Elsevier, Ltd.)

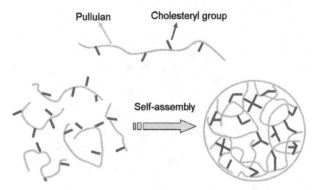

Figure 6.11. Schematic representation of CHP nanogel preparation by physical cross-linking (self-assembly). (Reproduced with permission from Ref. 128, Figure 6.1. Copyright 2004, Elsevier, Ltd.)

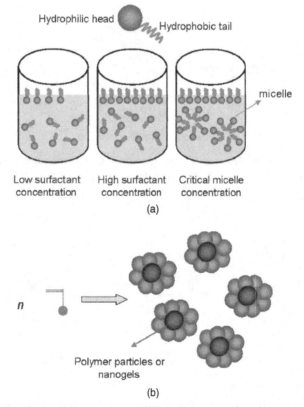

Figure 6.14. Classical emulsion polymerization technique for nanogel preparation.

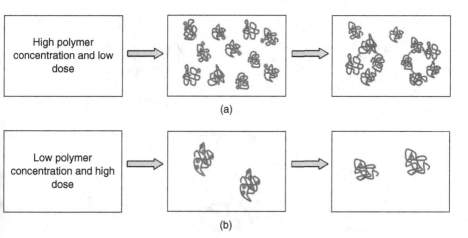

(a)

(b)

Figure 6.16. Radiation mechanism for (a) bulk/micro gel and (b) nanogel formation. (Reproduced with permission from Ref. 203, Figure 6.2. Copyright 2003, American Chemical Society.)

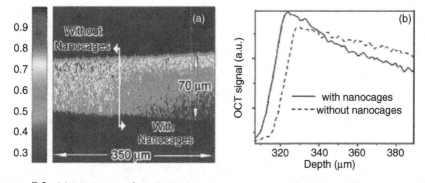

Figure 7.3. (a) OCT image of a gelatin phantom embedded with TiO2, and the concentration of TiO2 was controlled at 1 mg/mL to nimic the background scattering of soft tissues. The right portion of the phantom contained 1 nM of gold nanocages while the left portion did not contain any gold nanocages. (b) Plots of the OCT signals on a log scale as a function of depth. Note that the OCT signal recorded from the portion of phantom with gold nanocages decays faster than the portion without nanocages. (Reprinted with permission from *Nano Lett.* **5** (3), 473–477, 2005. Copyright © 2006 American Chemical Society.)

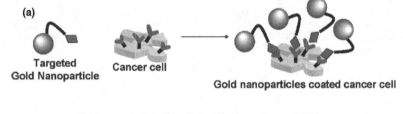

(a)

Targeted
Gold Nanoparticle

Cancer cell

Gold nanoparticles coated cancer cell

Gold nanoparticles mediated therapeutic modalities

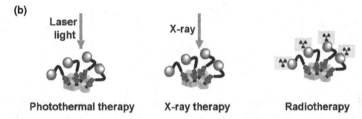

(b)

Laser
light

X-ray

Photothermal therapy

X-ray therapy

Radiotherapy

Figure 7.5. (a) Schematic illustration of tumor targeting and surface attachment of nanoparticles. (b) Gold nanoparticles mediated tumor therapeutic modalities.

Printed in the United States
By Bookmasters